ACROSS

the

BRIDGE

ACROSS *the* BRIDGE

Understanding the Origin of the Vertebrates

HENRY GEE

The University of Chicago Press • Chicago & London

The University of Chicago Press, Chicago 60637
The University of Chicago Press, Ltd., London

Published 2018
Printed in the United States of America

27 26 25 24 23 22 21 20 19 18   1 2 3 4 5

ISBN-13: 978-0-226-40286-4 (cloth)
ISBN-13: 978-0-226-40305-2 (paper)
ISBN-13: 978-0-226-40319-9 (e-book)
DOI: https://doi.org/10.7208/chicago/9780226403199.001.0001

Library of Congress Cataloging-in-Publication Data

Names: Gee, Henry, 1962– author.
Title: Across the bridge : understanding the origin of the vertebrates / Henry Gee.
Description: Chicago : The University of Chicago Press, 2018. | Includes bibliographical references and index.
Identifiers: LCCN 2017050270 | ISBN 9780226402864 (cloth : alk. paper) | ISBN 9780226403052 (pbk. : alk. paper) | ISBN 9780226403199 (e-book)
Subjects: LCSH: Vertebrates—Origin. | Vertebrates—Evolution. | Vertebrates—Physiology.
Classification: LCC QL607.5 .G44 2018 | DDC 596.138—dc23
LC record available at https://lccn.loc.gov/2017050270

♾ This paper meets the requirements of ANSI/NISO Z39.48-1992 (Permanence of Paper).

Contents

Preface ix

CHAPTER ONE: *What Is a Vertebrate?* 1

1.1 Vertebrates in Context 1

1.2 What Makes a Vertebrate? 3

1.3 Breaking Branches 9

1.4 Summary 13

CHAPTER TWO: *Shaking the Tree* 15

2.1 Embranchements and Transformation 15

2.2 Evolution and Ancestors 17

2.3 Summary 25

CHAPTER THREE: *Embryology and Phylogeny* 27

3.1 From Embryos to Desperation 27

3.2 Genes and Phylogeny 33

3.3 Summary 39

CHAPTER FOUR: *Hox and Homology* 41

4.1 A Brief History of Homeosis 41

4.2 The Geoffroy Inversion 44

4.3 The Phylotypic Stage 45

4.4 The Meaning of Homology 46
4.5 Summary 48

CHAPTER FIVE: *What Is a Deuterostome?* 51

CHAPTER SIX: *Echinoderms* 55

CHAPTER SEVEN: *Hemichordates* 65

CHAPTER EIGHT: *Amphioxus* 75

CHAPTER NINE: *Tunicates* 85

CHAPTER TEN: *Vertebrates* 101

CHAPTER ELEVEN: *Some Non-deuterostomes* 117

CHAPTER TWELVE: *Vertebrates from the Outside, In* 123
12.1 Introduction 123
12.2 The Organizer 125
12.3 The Notochord 129
12.4 Somitogenesis 133
12.5 Segmentation and the Head Problem 136
12.6 The Nervous System 144
12.7 Neural Crest and Cranial Placodes 153
12.8 The Skeleton 159
12.9 Summary 161

CHAPTER THIRTEEN: *How Many Sides Has a Chicken?* 165
13.1 Introduction 165
13.2 The Enteric Nervous System 167
13.3 The Head and the Heart 168
13.4 The Urogenital System 171
13.5 The Gut and Its Appendages 174
13.6 Immunity 175
13.7 The Pituitary Gland 177
13.8 Summary 178

CHAPTER FOURTEEN: *Some Fossil Forms* 179
14.1 Fossils in an Evolutionary Context 179
14.2 Meiofaunal Beginnings 183
14.3 Cambroernids 184
14.4 Vetulicystids 185
14.5 Vetulicolians 186
14.6 Yunnanozoans 190
14.7 *Pikaia* 192
14.8 *Cathaymyrus* 194
14.9 The Earliest Fossil Vertebrates 194
14.10 Conodonts 195
14.11 Ostracoderms and Placoderms 197
14.12 Summary 198

CHAPTER FIFTEEN: *Breaking Branches, Building Bridges* 201
15.1 Defining the Deuterostomes 201
15.2 Ambulacraria 204
15.3 Echinoderms 204
15.4 Hemichordates 205
15.5 Chordates 206
15.6 Amphioxus 210
15.7 The Common Ancestry of Tunicates and Vertebrates 211
15.8 Tunicates 212
15.9 Vertebrates 213
15.10 Cyclostomes 214
15.11 Gnathostomes 214
15.12 The Evolution of the Face 215
15.13 Crossing the Bridge 216
15.14 Conclusions 218

Notes 225
References 251
Index 297

Preface

Vertebrate origins is a topic that falls into a zoological no-man's land. If it features at all in textbooks of invertebrate zoology, it'll be half a paragraph at the end. In textbooks of vertebrate zoology, it'll be discussed for two or three paragraphs at the beginning. When I was a graduate student in the mid-1980s teaching this subject to undergraduates, I found that the primary literature on the topic was ancient, scattered, obscure, and often contradictory. Students were often told about the ideas of vertebrate origins published by Walter Garstang (1868–1949.) They were never told, however, that his early published ideas are contradicted by his later ones,[1] and that most of what he'd intended to write on the subject never materialized. I resolved that, one day, I'd sort all this out and write the textbook on vertebrate origins I wished I'd had as a teaching aid. The result was *Before the Backbone* (1996). In that book I took a largely historical look at the various ideas proposed to "explain" how vertebrates originated from other animals.

The problem is that vertebrates are so different from other animals that the gap between them has been hard to bridge. Before modern techniques of molecular phylogeny and objective ways of reconstructing evolutionary relationships, the field was clear for any number of imaginative ideas about how vertebrates could have been squeezed out of horseshoe crabs, echinoderms, crustaceans, spiders, scorpions, tunicates, lancelets, mollusks, roundworms, acorn worms, segmented worms, proboscis worms, or even protozoa.[2] In *Before the Backbone* I decided to remain above the fray in the hope that readers would be able to make up their own minds.

When *Before the Backbone* appeared, Hox genes, whose organization and function revealed deep connections between the structures of all animals, were relatively newly found. The first genomes were only then being sequenced, and these were of bacteria. Fossils were few. Extant creatures such as acorn worms and tunicates were closed books to geneticists, being very far from the model organisms labs were accustomed to handling.

Twenty years have passed and the face of science is greatly changed. Many organisms relevant to the subject have had their genomes sequenced. Molecular techniques now allow genetic inspection and even manipulation of non-model organisms. Many more fossils have been found and described. Even the shape of animal evolution is conceived differently from the way it was then. The result is that we have a much better idea of how vertebrates fit into the scheme of things, and many interesting ideas have effectively been falsified.

And yet, for all the achievements of the past two decades, the morphological chasm between vertebrates and invertebrates still remains large enough in which to conduct quite a few arguments at once, without much hope of a sure resolution to any of them. The extant relatives of vertebrates were always too few, and the more we learn about them, we find that they have evolved so far along their own courses that they are of questionable help in reconstructing what the latest invertebrate ancestor of vertebrates looked like. The fossils, while more plentiful, offer yet more problems of interpretation. There are, however, clues. They are often fragmentary, possibly misleading and subject to a degree of interpretation—but they are there.

For those who want to delve into the history of the topic, *Before the Backbone* is still in print. The aim of this book, *Across the Bridge*, is to review the state of the field as it is now; to present a synthesis of the more recent literature in a form that will be accessible to students; to see how far we have come in working out how to cross that chasm; and, finally, to venture some ideas of my own. Most of the works cited here have appeared since *Before the Backbone*, and I have generally avoided historical treatments of the subject. This book therefore complements *Before the Backbone* without supplanting it. It is an entirely new work, conceived and constructed from the ground up. This time, I hope, I'll have succeeded in writing something closer to the book I had intended to write in the mid-1980s.

In *Before the Backbone* I was wary of presenting any kind of novel synthesis, preferring to compare and contrast the ideas of others. I have been less cautious here. In this book I advance some views that will no doubt be controversial. In particular, I suggest that the peculiar Cambrian fossils known as vetulicolians and yunnanozoans, all described since *Before the Backbone* was published, represent successively closer relatives of chordates, the larger group to which vertebrates belong. In doing this, I suggest[3] that the segmented trunk region of chordates originated, with vetulicolians, entirely separately from much of the body destined to become the head and viscera, and only later became more integrated with it.

It took me quite a while for me to gather up sufficient confidence to start work on a new book on vertebrate origins. Although various people, notably Christopher J. Lowe, Michael Levine, and Nicholas D. Holland, had been encouraging me from the sidelines, the task seemed too daunting.

What nudged me forward was the opportunity to commission a special supplement of review articles for *Nature* on the origin and early evolution of the vertebrates.[4] This exercise, and the reviews that resulted, helped me get my thoughts in order. I am grateful to my colleague Ursula Weiss for allowing me to commission the supplement, Melissa Rose for handling the administration, and Jenny Rooke for copyediting.

None of it would have been possible without the willing participation of the authors: Martin D. Brazeau, Marianne E. Bronner, Lionel Christiaen, Nathaniel Clarke, Rui Diogo, Matt Friedman, John Gerhart, Stephen A. Green, Linda Z. Holland, Nicholas D. Holland, Peter W. H. Holland, Philippe Janvier, Robert G. Kelly, Michael Levine, Christopher J. Lowe, Daniel M. Medeiros, Julia L. Molnar, Drew M. Noden, Daniel S. Rokhsar, Marcos Simoes-Costa, Eldad Tzahor, and Janine M. Ziermann—and the guidance of colleagues and anonymous referees.

Per Ahlberg, Olaf Bininda-Emonds, Simon Conway Morris, Rui Diogo, Neil Gostling, Linda Z. Holland, Nicholas D. Holland, Philippe Janvier, Shigeru Kuratani, Thurston Lacalli, Mike Levine, Chris Lowe, and two anonymous referees kindly read and commented on all or part of various drafts of the book. Any errors, however, are mine.

I am grateful as always to my agent, Jill Grinberg, for her sustained confidence in me; the unfailing encouragement of Christie Henry at the University of Chicago Press, the most supportive editor one could ever hope to have, and the finest; her colleagues at Chicago, Miranda Martin and Mary Corrado; and the makers of McVitie's plain chocolate digestive biscuits, for sustenance while writing in the bitter watches of the night.[5]

As always I owe a debt of gratitude to my family for their support, as well as to the amusement and insight offered by our numerous pets, which together represent all five traditional classes of vertebrate.

Finally, I dedicate this book to the memory of Professor Robert McNeill Alexander, FRS (1934–2016), who died while I was drafting this book. Neill was Professor of Zoology at the University of Leeds when I was an undergraduate there between 1981 and 1984. Then, and since, he was a source of unfailingly wise counsel and friendship. I regard him as having been one of my mentors and keenly feel his loss.

CHAPTER 1

What Is a Vertebrate?

1.1 VERTEBRATES IN CONTEXT

Most familiar animals are vertebrates—that is, animals with backbones. We are vertebrates, as are most of our domestic animals, such as cows, horses, poultry, sheep, and pigs. The numerous animals housed at various times chez Gee—dogs, cats, chickens, rabbits, guinea pigs, hamsters, snakes, axolotls, and fish, not forgetting the frogs that crowd our garden pond each spring, are vertebrates.

Most of the animals you will meet in a zoo, from lions to lorikeets, geckos to giraffes, are also vertebrates, so much so that non-vertebrates are usually confined to a single building labeled something like "creepy crawlies." The invertebrates, though, comprise a wider and more diverse domain than that. With a proper zoological perspective, vertebrates represent one rather small branch of a riotously various and diverse array of animal life. To understand vertebrates and how they evolved, one has to have a good overview of the entirety of animal life.[1]

Perhaps the most important invertebrates, at least in terms of numbers of species, are the insects. Many of these will be familiar to the most wildlife-averse urbanite, even if they are only flies and cockroaches (see fig. 1.2). Bees, ants, wasps, butterflies, moths, beetles, dragonflies, and grasshoppers are all familiar insects. Most known animal species are, in fact, insects. And yet insects form just one branch on the much more extensive tree of arthropods, or jointed-legged animals. Besides insects, this includes spiders, scorpions, ticks, mites, crabs, lobsters, centipedes, millipedes, barnacles, and other, less familiar creatures such as pycnogonids (sea spiders) and xiphosurans (horseshoe crabs).

Other invertebrates include mollusks such as clams, squid, slugs, and snails; as well as a diverse range of worms, jellyfishes, starfishes, sponges, and so on, to name just the more familiar among a still wider array of animals. Many of these are small, rare, or obscure, and known mainly to professional zoologists, or those students who, like me, liked to explore the dusty end of the textbook in search of unpronounceable exotica.

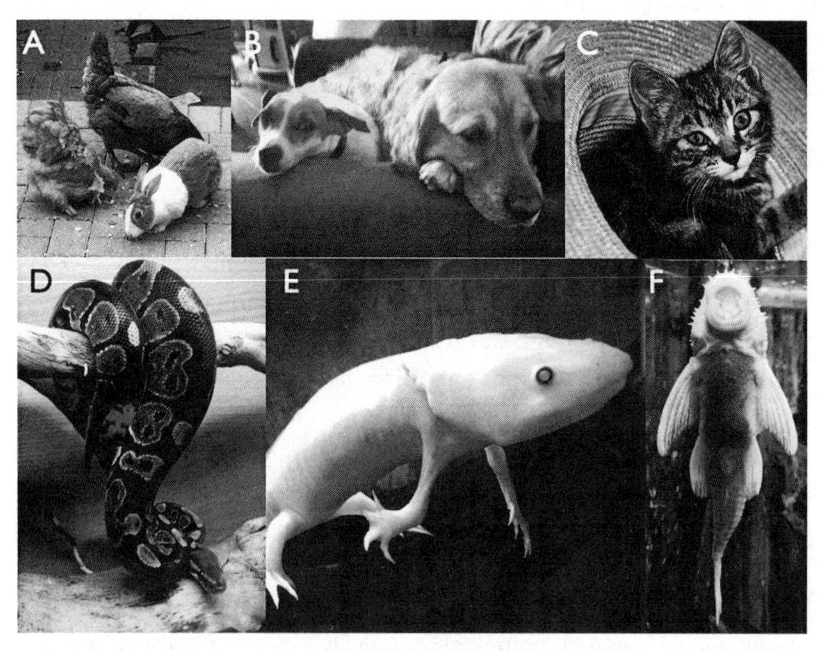

1.1 A selection of vertebrates. A: rabbit *Oryctolagus cuniculus* (mammal) and chickens *Gallus gallus* (birds); B: dogs *Canis familiaris* (mammals); C: cat *Felis domesticus* (mammal); D: royal python *Python regius* (reptile); E: axolotl *Ambystoma mexicanum* (amphibian); F: plecostoma, possibly *Pterygoplichthys* sp. (fish). Photographs by the author, from the author's menagerie.

1.2 A small selection of invertebrates. A: a house spider (arthropod); B: land snail (mollusk); C: a handful of sea gooseberries, *Pleurobrachia* (ctenophores); D: a sea cucumber (echinoderm): the rightmost object on the plate; E: jellyfish (cnidarian); F: octopus (mollusk); G: a colony of sea chervil, *Alcyonidium* (bryozoan); H: horseshoe crabs, *Limulus* (arthropod); J: bumble bee (arthropod); K: edible crab (arthropod). Photographs by the author.

Amateur microscopists will have seen the rotifers (wheel animalcules) and tardigrades (water bears) that swarm in water or crawl out of damp moss. Sharp-eyed beachcombers will have encountered sponges, tunicates, and bryozoa (moss animals). But it's a fair bet that most people will never have seen, or even heard of, priapulids, pogonophorans, placozoans, or phoronids, and those are just the ones I could immediately think of beginning with the letter *p*.[2] Yet each represents a "phylum," that is, a distinct and distinctive kind of animal life.

1.2 WHAT MAKES A VERTEBRATE?

Despite this diversity, vertebrates seem to stand apart. They are so different from other animals that recognizing a vertebrate seems almost instinctive. Could it be because we ourselves are vertebrates, and so recognize our kin, even if only from a distance? This is undoubtedly a reason, yet even when one discounts our very understandable prejudice, vertebrates do seem qualitatively different from other animals.

The presence of a distinct head is a vertebrate feature, and the characteristic vertebrate arrangement of a "face" with two eyes, set side-by-side, and a mouth beneath, might explain the almost universal feeling of kinship with all vertebrates, whereas the arrangements seen in other animals—whether a panoply of eyes, tentacles, or spiny mouthparts, or a front end that is featureless or eyeless—seem alien to us and might be greeted with horror. The emoticon of a smiley face ☺ typifies the vertebrate arrangement and has universal appeal, whereas people have to learn to love many-eyed spiders and eyeless worms. This is, in fact, proven in the breach. Tiny flatworms called planarians,[3] found in streams and ponds, are very different in their construction from vertebrates, and yet some have two large eyes at the front that make them seem curiously appealing, if not actually cuddly. You can see a couple of examples in fig. 1.3.

It's worth listing some of the many ways in which vertebrates differ from other animals. I'll go into these in much more detail later in the book, but for now it's worth rehearsing them, to get to grips with that feeling we have that there is a substantial gap between vertebrates and other animals, a chasm we need to bridge if we are to understand vertebrate origins.

I've already alluded to the presence of a head, and, in particular, a face. A head is a concentration, at one end of an animal, of entry points for air, food, and sensory information. A head, in such a broadly defined sense, is

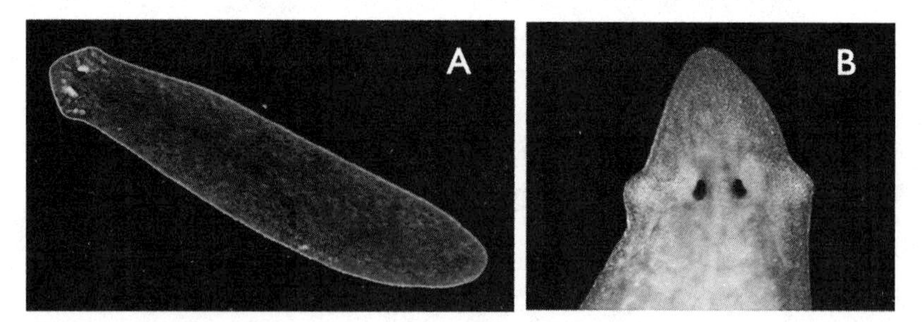

1.3 The eyes of Planaria. A: *Cura* cf. *pinguis*, from Australia. The head is on the left, and the whole animal is about 4 mm long (courtesy Miquel Villa-Farre). B: a close-up of the head of another species *Dugesia sanchezi*, just to show it's not a fluke (courtesy Alejandro Sanchez Alvarado).

only to be expected in bilaterally symmetrical animals with a preferred direction of travel. Other such animals include insects and other arthropods. These, too, have heads, but they are constructed differently from the heads of vertebrates. Insect eyes are made in a completely different way from vertebrate eyes, being constructed of many repeated units (think of pixels) rather than a single, camera-like unit with a flexible lens, as found in vertebrates. Insects' ears are found on their legs, their noses on their feet; and they breathe not through their mouths, but through many tiny pores on their bodies. This suggests that the heads of insects and vertebrates evolved entirely independently, each from headless ancestors. This is supported by what we know of the evolutionary relationships of insects and vertebrates. Insects are more closely related to various more-or-less headless worms than to vertebrates. By the same token, the closest relatives of vertebrates among the invertebrates—the sea squirts, or tunicates, and the superficially fish-like amphioxus—do not appear to have distinct heads. However, I shall explain in this book, this does not mean that tunicates and the amphioxus do not have structures comparable with what we see in the vertebrate head—it is that they are not immediately obvious. Perhaps it is truer to say that these invertebrate relatives of vertebrates do not have the smiley faces we instinctively associate with the vertebrate state.

Vertebrates are built around an internal skeleton of cartilage, which in many cases is reinforced with harder tissues such as bone, dentine, and enamel. Although cartilage of various sorts is found throughout the animal kingdom,[4] bone, dentine, and enamel are tissues unique to vertebrates. The principal mineral constituent of vertebrate hard tissues is hydroxyapatite, a form of calcium phosphate. The shells and other hard tissues of invertebrates are made of a different substance, calcium carbonate.[5] The vertebrate skeleton comprises a brain case, housing the brain and sense organs such as the eyes, ears, and nose, to which might be attached skeletal supports for jaws and gill arches, and of course the backbone made of interlocking vertebrae, from which the group gets its name.

The skeleton also includes internal supports for fins and limbs, if present. During development, the backbone replaces a longitudinal stiffening rod called the notochord, which is found at some stage in the life cycle of vertebrates as well as tunicates and the amphioxus. Because of this, the vertebrates, the tunicates, and the amphioxus are united into a larger group, the chordates.

Along with the notochord, all chordates possess, at some part of their life cycle, a system of serially repeated pouches on each side of the throat region or pharynx, which in many cases pierce the body wall and open either directly to the outside, or into a protective cavity or atrium, which communicates with the outside through a smaller number of openings. In tunicates, the amphioxus, and the larvae of lampreys alone among vertebrates, these pharyngeal pores or slits form part of a unique filter-feeding system.[6] Water is taken in through the mouth and propelled, by currents generated by cilia, outward through the pharyngeal slits. Mucus secreted by the endostyle—a region of glandular cells in a longitudinal gutter on the pharyngeal floor—is carried up the cartilage-supported bars between the slits, trapping any water-borne debris before it escapes. The food-laden mucus makes its way to the roof of the pharynx where it enters the oesophagus and the digestive system. Tunicates and the amphioxus feed like this throughout life. Filter-feeding lampreys lose this arrangement at metamorphosis. The endostyle is transformed into the thyroid gland, and in adult lampreys and all other vertebrates, the pharyngeal slits are transformed into supports for gills used to extract oxygen from water and, in fishes, to excrete excess salt. In most tetrapods (that is, land-living vertebrates) the pharyngeal slits never form at all and the elements that otherwise would have made up their bony or cartilaginous supports become incorporated into the inner ear, the jaw, or the hyoid skeleton that supports the tongue.

Pharyngeal slits are found in animals other than chordates, notably marine animals called hemichordates,[7] even though these creatures do not appear to have endostyles, notochords, or other structures found in chordates. Hemichordates come in two forms: enteropneusts (acorn worms) and pterobranchs, neither of which will be familiar to anyone but professional zoologists. Enteropneusts are blind, brainless, flaccid, and sometimes foul-smelling worms that live in marine sediment; pterobranchs are small, often colonial organisms, feeding through an arrangement of tentacles called a lophophore.[8]

Some extinct echinoderms—a group of animals that today includes starfishes, sea urchins, and sea cucumbers—appeared to have had pharyngeal slits, although no extant echinoderm does so.[9] Hemichordates and echinoderms together form a group called the Ambulacraria, and ambulacrarians and chordates together form a larger animal group called the deuterostomes.

The notochord of chordates provides support and purchase for muscles and other tissues such as nerves and blood vessels, arranged in a series of segments called somites. Although many other animals are segmented—arthropods, as well as segmented worms or annelids—these segments are constructed entirely differently. Tunicates appear to have lost their segmentation in evolution, whereas the segmentation in amphioxus differs from vertebrate segmentation in important ways.

As the notochord develops during the life of a chordate embryo, it secretes substances that induce the development, dorsal to it (that is, along the upper surface, or back), of a hollow, longitudinal nerve cord, the basis of the vertebrate central nervous system.[10] The dorsal, hollow nerve cord is a unique feature of chordates. In all invertebrates that have a central nervous system, the nerve cord, if present, is ventral (that is, along the belly) and solid. Some invertebrates have two or more nerve cords. In some animals, paired, ventral cords are joined by cross-bridges at regular intervals like the rungs of a ladder.

The formation of the dorsal nerve cord is accompanied by the migration of cells from its lateral edges, along specified routes, to various parts of the body. These cells, collectively the neural crest,[11] are responsible for many uniquely vertebrate features such as the bones of much of the head and face; parts of the organs of special sense, notably the ears; the formation of the skin, its pigmentation, and its appendages such as scales, hair, feathers, and teeth; and other parts of the anatomy such as the spinal ganglia, the adrenal glands, the nervous system that lines the intestines, and parts of some major blood vessels. In that much of the instantly recognizable vertebrate face is formed by the neural crest, one could argue that this is the single most important defining feature of vertebrates. There are, however, traces of modest neural-crest-like activity in tunicates,[12] but none at all in the amphioxus or any other invertebrate.

Vertebrates have large brains. Nothing like the vertebrate brain is seen in either tunicates or the amphioxus, although there are traces of its ground plan in the amphioxus, tunicates, and even hemichordates, if one looks hard enough.[13] Other animals have brains, notably arthropods and mollusks, and in some cases these are elaborate structures associated with complex and even intelligent behavior. One thinks of the octopus, a famously canny creature with a large and complex brain. But the brains of invertebrates are constructed differently from those of vertebrates, and are not enclosed within that other distinctively vertebrate feature—the skull.

In addition to all the features mentioned above, vertebrates have a wealth of internal features that, although less obvious, are unique to the group and serve only to widen the gap between vertebrates and other animals. These include a water management system centered on the kidneys, which has allowed vertebrates, among only a select few animal groups and uniquely among deuterostomes, to live their lives entirely away from water. The kidneys are connected to a unique system of sex organs, which are in turn connected, chemically, to a sophisticated network of internal, hormone-based signaling, complementary to that of the nervous system. Although many animals (and plants) have a degree of innate immunity to agents of disease, which can on occasion be highly discerning and sophisticated, only vertebrates have a system of acquired immunity in which the cells of the immune system can be trained to recognize and neutralize threats never before encountered. All this and lymphatic drainage, a closed blood circulation with vessels lined with a specialized tissue called endothelium, and powered by a chambered heart. Because of these internal refinements, vertebrate animals can live a life much more independently of the environments in which they are found, compared with many other animals.

At a deeper level, the genome of vertebrates seems to have been duplicated—not once, but twice—at some point in the earliest history of the group,[14] although there is some debate about whether the second duplication happened before or after the emergence of the lineage leading to the most basal extant vertebrates, that is, the jawless hagfishes and lampreys.[15] It has been thought that genome duplication allows for an increase in complexity. If two genes are produced where there was one before, each one can evolve in its own way, perhaps allowing for previously unattainable subtleties in gene regulation, morphological specification, and so on. However, what seems to happen is that many of the duplicates are lost, so the connection between gene duplication and complexity remains moot. The genomes of teleost fishes—the group of bony fishes that includes most familiar kinds, such as the cod with your fries to the guppies in your aquarium—have undergone a further duplication,[16] and although these creatures exhibit a wide range of morphology (forms as varied as sea-horses and the ocean sunfish) they are all recognizably vertebrates.

The presence in vertebrates of the head, brain, hard tissues, notochord, distinctive nervous system, neural crest, kidneys, adaptive immune system, and so on, features seen nowhere else in the animal kingdom, serves to divide vertebrates from all other animals.

1.3 BREAKING BRANCHES

At first sight, many of the characteristic features of vertebrates appear to have evolved all at once. This explains why vertebrates appear so different from anything else in the animal world. However, it is legitimate to ask whether the apparently unique features of vertebrates evolved not simultaneously, but one at a time, and, if so, in which order; and whether some of them might be found, even if in some more modest form, among invertebrates.

These are reasonable questions, because we already know that some of the features we see in vertebrates, such as the neural crest, are to some extent presaged in tunicates; that the notochord and hollow dorsal nerve cord are also found in invertebrate chordates such as tunicates and the amphioxus; and that the pharyngeal gill slits are found in hemichordates and possibly some now-extinct echinoderms.

This allows us to reconstruct an order in which these features were acquired. Pharyngeal gill slits evolved first, in the common ancestors of all deuterostomes; with the notochord and hollow nerve cord evolving later on, in the common ancestor of all chordates. The rudiments, at least, of the neural crest appeared later still, in the common ancestry of tunicates and vertebrates. Therefore it should be possible to break down all the features we see in vertebrates and try to imagine how they might have evolved sequentially.

When new species evolve, they can be recognized as different because they have traits other species do not share. As the tree of life grows, twigs thicken into branches, branches into trunks, affirming these differences. The problem is that many of the twigs lower down the branches wither and die, removing evidence of intermediate stages, so it's hard to understand how species on one evolutionary branch come to look so different from those on another. When intermediate stages are removed, branches

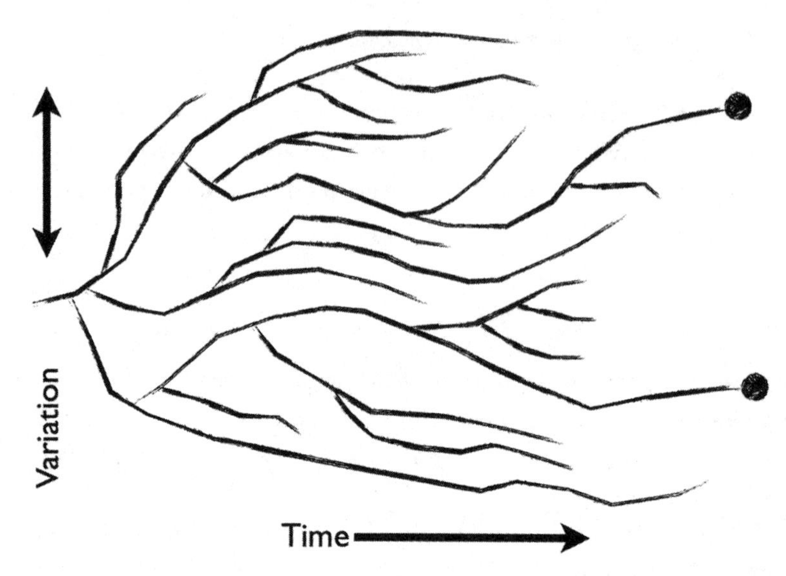

1.4 Evolution and extinction in action. Reading from left to right, species diversify and separate, but as time passes, many lineages become extinct. The effect is that the few surviving lineages (marked with the black circles), each evolving in its own way, represent only a subset, and can seem so different from one another that it's hard to imagine how they are related.

become denuded and bare, and yet seem to carry on uninterrupted. The effect will be to make the surviving species, at the ends of the branches, seem quite different from those on other branches (fig 1.4).

One could therefore recast the problem of the origin of vertebrates as what biologists call a "long branch problem." It could be that vertebrates seem so different from other animals because all the intermediate forms have disappeared. Such creatures might have had some of the traits we'd now see as quintessentially vertebrate, but not others: or showed, in one single species, a combination of traits now seen in totally separate groups. But these creatures are either undiscovered or have become extinct without trace. It would be especially interesting, for example, to find creatures that break the long branch between vertebrates and tunicates, which, despite their very different forms and lifestyles, are the closest living relatives among the invertebrates.

You might be tempted to call such a creature a "missing link," but if you are, you shouldn't.[17] The reason is that such a creature, were it ever found, has presumably not remained static, in evolutionary terms, existing for the sole purpose of our scientific enlightenment. It would have evolved from its common ancestor with vertebrates for the same length of time as vertebrates would have evolved from that same common ancestor. Furthermore, it would have accumulated unique traits all its own that might have had nothing to do with vertebrates, or any other living form. Because we cannot but interpret such a creature in the light of animals with which are already familiar, we are likely to be misled. If we came across a creature utterly unlike anything we have ever seen, how would we recognize it as an animal at all?[18]

If you need any proof of this idea in action, the tunicates themselves provide many examples. For more than a century it was thought that the amphioxus, not tunicates, was the closest relative of vertebrates. After all, it looks rather like a fish, with a clear front end and back end, and neatly arranged somites in between (fig 1.5).

The discovery that tunicates were in fact more closely related to vertebrates[19] than the amphioxus was perhaps the most significant advance in the entire field in decades. It overturned the canonical picture of the steady acquisition of vertebrate complexity, from a tunicate-like chordate ancestor, through the development of an animal with somites, to a full-fledged vertebrate.[20] The shock, however, was visceral, because tunicates do not look much like vertebrates at all. Indeed, tunicates come in all shapes and sizes. They are solitary or colonial; they live attached to one spot throughout their lives; they move about freely; or, indeed, combinations of the above. Specimens of the colonial tunicate *Botryllus* I've found on the beach near where I live defy interpretation as anything at all. That the creature is a close relative of vertebrates seems unimaginable. Although many tunicates have notochords and dorsal, tubular nerve chords

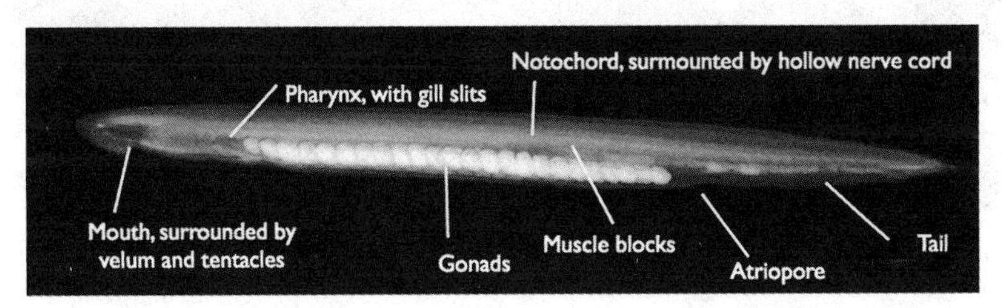

1.5 An adult amphioxus in lateral view, showing its major features. Photo courtesy of Dale Stokes.

in approved chordate fashion, most tunicates display these only fleetingly, during a brief, tadpole-like larval stage, and they are shed when the animal settles down to adult life. Although some tunicates, the larvaceans, retain this tadpole-like state into adulthood, many others have dispensed with it completely, leaving little or no sign of chordate heritage (fig. 1.6).

Clearly, tunicates have been so busily evolving away from their common ancestor with vertebrates, shedding canonically chordate features along the way, that it is now very difficult to imagine what the common ancestor of tunicates and vertebrates looked like, except that it must have looked a little like an amphioxus. However, tunicates only seem to be distant relatives because they have evolved much further from the common chordate ancestry than the amphioxus, or—in some ways—vertebrates themselves. Any and all intermediates that might have existed between tunicates and their common ancestor with vertebrates no longer exist, emphasizing the uniqueness of tunicates. Looked at from the perspective of tunicates, it is

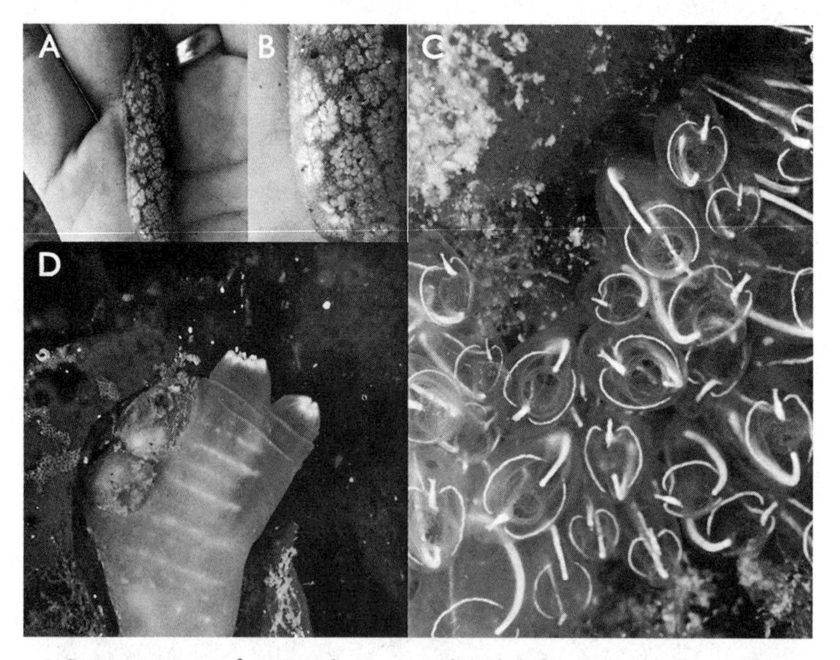

1.6 Common tunicates from British waters. A: the colonial tunicate *Botryllus* on a piece of weed; B: close-up showing individual zooids. (Photos A and B by the author.) C: a group of the gregarious but non-colonial *Clavelina*; D: a solitary adult *Ciona*. (Photos C and D courtesy of Becky Hitchin.)

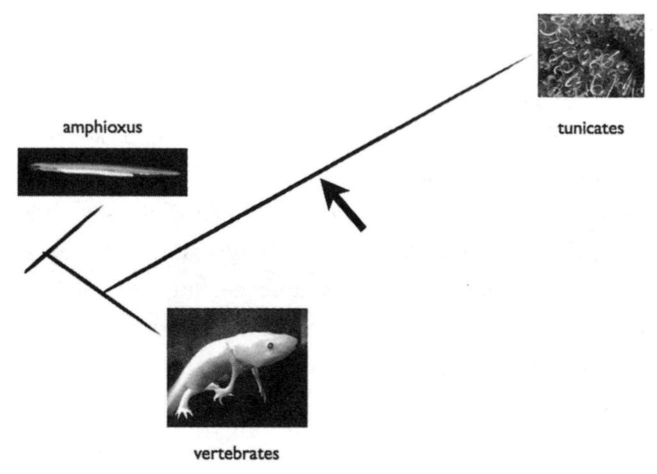

1.7 How to break a branch. In this phylogeny of chordates, the length of each branch represents a measure of relative evolutionary distance. Tunicates have evolved so far and so fast from their common ancestor of vertebrates that breaking down this evolution into steps seems hopeless. Any creatures we might find that branched off the tunicate line somewhere in the middle (bold arrow) could break that long branch, helping us to make sense of the evolution of tunicates.

they, not vertebrates, that occupy a special place in the animal kingdom (fig. 1.7).

1.4 SUMMARY

The problem with writing a book about vertebrate origins is working out where to begin. So I began at the beginning, setting the vertebrates in context as one twig in the great tree of animal life, briefly outlining the principal features that set them apart from the rest of the animal world. I showed that vertebrates join sea creatures called the amphioxus and tunicates in a more inclusive group, the chordates, united by a number of features including the notochord and the dorsal, tubular nerve cord. The chordates join hemichordates and echinoderms—collectively, the ambulacrarians—in a still more exclusive group, the deuterostomes. Perhaps the single most distinctive feature of deuterostomes is the presence of paired pharyngeal gill slits. I closed with the revolutionary discovery that tunicates, not the amphioxus, turn out to be the closest invertebrate relatives of vertebrates, even though tunicates do not look or behave very much like vertebrates. This leads us to the conclusion that the amphioxus is the most primitive chordate, and therefore that vertebrates have retained a more conservative form. The tunicates, in contrast, have explored realms of morphological space quite alien to vertebrates. Perhaps it is they, rather than vertebrates, that are special.

In the next chapter I shall introduce a little history. I shall briefly survey how scientists tried to organize animals into categories; how Darwin's ideas of evolution changed these ideas fundamentally; and how people approach classification today. This might seem like a digression, but it should give you the means necessary to understand concepts discussed further on in the book.

CHAPTER 2

Shaking the Tree

2.1 EMBRANCHEMENTS AND TRANSFORMATION

The fundamental distinction between vertebrates and other animals was recognized by the influential French savant Georges Cuvier (1769–1832), who organized the animal world into four groups or *embranchements*: the vertebrates; the articulates (arthropods and segmented worms); the radiates (radially symmetrical animals such as starfishes and jellyfishes); and mollusks (all the other soft-bodied, bilaterally symmetrical animals, in short, worms).

It is important to note that Cuvier worked in an era when most people assumed that all living things had been specially created as written in the Bible. Transformation between one kind of creature and another (what we would now call evolution) was not considered a serious possibility. Although Cuvier's recognition of the fact that fossils might represent extinct species was an immense conceptual advance, there was no sense in his scheme that members of one *embranchement* might "transform" into members of another. *Embranchements* were meant to be fixed categories discovered by the examination of nature—similar to, say, the

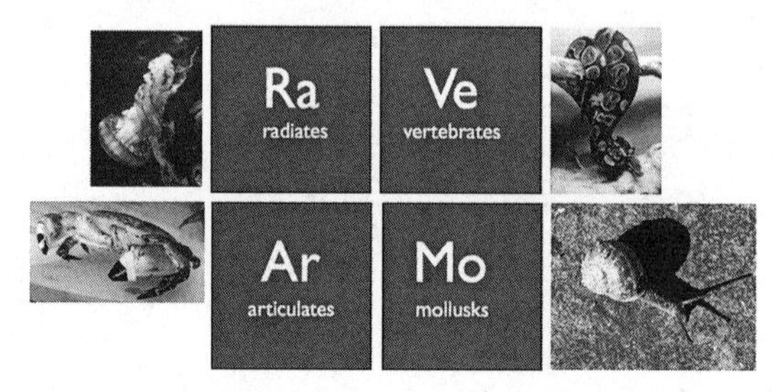

2.1 Cuvier's *embranchements*: despite their name, they are best imagined as static categories rather than branches on a tree, in the same way that we'd think of elements in the Periodic Table.

arrangements of chemical elements into the various groups in the Periodic Table (fig. 2.1).

Cuvier went to great pains to ridicule and belittle any scheme that he saw as in any way transformational. A particular victim was Jean-Baptiste Lamarck (1744–1829). His ideas of transformation (in his book *Philosophie Zoologique*, 1809) aroused Cuvier's particular scorn.

That didn't stop the reputedly straight-laced and pompous Cuvier being made fun of by his one-time room-mate and eventual intellectual adversary, the playful and mildly disreputable Etienne Geoffroy Saint-Hilaire (1772–1844).[1] It was Geoffroy who adopted wild schemes in which lobsters could be transformed into vertebrates by turning them upside down, so that their ventral nerve cord became dorsal; or suggesting that insects were in fact vertebrates that had learned to live inside their own vertebrae. It is hard to know how much Geoffroy's exuberance was meant to be taken seriously, or whether he proposed such things just to irritate Cuvier. Either way, Geoffroy's ideas presupposed that in some sense, if only purely theoretical, it was possible to transform animals from one kind into another—even across *embranchements*.

Lamarck's ideas, which involved an inherent force or *besoin* (literally, "need") that would drive organisms to change their form, were similar to other ideas current in the eighteenth and nineteenth centuries in which organisms were possessed of a life force or *vis essentialis* that would direct them toward a state of cosmic or ideal perfection.[2] Needless to say, no such life forces have been isolated so that they might be studied, as any scientist would demand. And yet, all ideas before Darwin, whether from the hard and rational Cuvier or the dreamy and disreputable Geoffroy, were based on the premise that the arrangement of animal forms was a prefigured pattern in nature waiting to be discovered. Either that pattern was a fixed point in nature, or organisms could move within that pattern, but the pattern itself was preformed and fixed, much like the Periodic Table.

Nowadays, however, we have accepted Charles Darwin's view that species can change by evolution, what he called "Descent with Modification." So much so that evolutionary theory is now the bedrock of biological thought. We tend to follow the dictum of the great evolutionary biologist Theodosius Dobzhansky that "nothing in biology makes sense except in the light of evolution."[3]

2.2 EVOLUTION AND ANCESTORS

Darwin's theory of evolution by natural selection turned the idea of a prefigured pattern completely on its head. Natural selection isn't so much a force, with some inherent if occult character, but rather the result of a set of circumstances to which all living creatures are subject—heritable variation, a superabundance of offspring, environmental change, and the passage of time.

There is nothing mysterious about any of these things. Any farmer, countryman, or indeed observant person knows that traits are passed from parents to offspring; that parents (human or otherwise) tend to have many more offspring than they can support; that they are prey to changes in the environment; and that such change accumulates with time. More importantly, the action of natural selection did not—emphatically, *not*—drive organisms along any predestined evolutionary track from simple beginnings toward increasing complexity, sophistication, perfection, or anything else.[4] Uniquely, Darwin's process did not seek to explain any prepackaged pattern. Natural selection acts only in the present, without foresight or purpose. The pattern that results is simply natural selection summed over history. The pattern will be in the form of a bifurcating diagram[5]—informally, a "tree"—but the form the tree takes in any single case cannot be prejudged or forecast. This point, about the unpredictability of trees, is absolutely crucial to a proper understanding of Darwinian evolution and yet frequently misunderstood. And yet in this point lies the key to understanding evolution in any scientific way.

I expect you have seen old-fashioned diagrams of the evolution of animal groups, done in the style of trees, in which the branches are continuous, and one species "flows" seamlessly into another to form a sequence of ancestor-descendant relationship. However, such diagrams rely on suppositions about ancestry and descent made after the fact, and which cannot be falsified.

It is, of course, possible that species X is ancestral to species Y, but it is impossible to know this for certain. On the other hand, it could be that species X and Y descend from an unknown common ancestor, Z, or even that species Y is ancestral to species X. The last possibility might seem outlandish, but given that the fossil record is often very much poorer than old-style depictions of ancestry and descent will admit, this is more likely than you might think (fig. 2.2).

Diagrams about the evolution of horses offer a good example. The fossil forms are arranged in order of increasing body size, taller molars, and diminishing numbers of toes, with the assertion that this arrangement offers evidence of evolution in action, natural selection having adapted horses for a life as open-plains grazers rather than woodland browsers. This arrangement might well be true, but blinds us to any possibility besides simple progression, as if the evolution of horses in this was somehow predetermined. Other schemes (such as those discussed above, in the case of hypothetical species X, Y, and Z) are simply not considered. Manifest destiny, on the hoof (fig. 2.3).

One might say that such diagrams are hypotheses about evolutionary history, but they are not. Hypotheses are formal statements subject to test and open to refutation: yet because they simply assert that one possible scheme is likely without examining others, such diagrams can only be appreciated subjectively, by relative plausibility, aided by the rhetorical skill of the proponent. This very far from the supposed objectivity of science, yet it was largely how the course of evolution—the *pattern*, not the *process*—was debated until late in the twentieth century.

That was until a German entomologist named Willi Hennig (1913–1976) came up with a scheme he called phylogenetic systematics, which sought to determine evolutionary pattern in a more objective way. Rather than try to weigh up the subjective importance of this adaptation or that to the success or failure of a group, Hennig simply mapped how distinctive

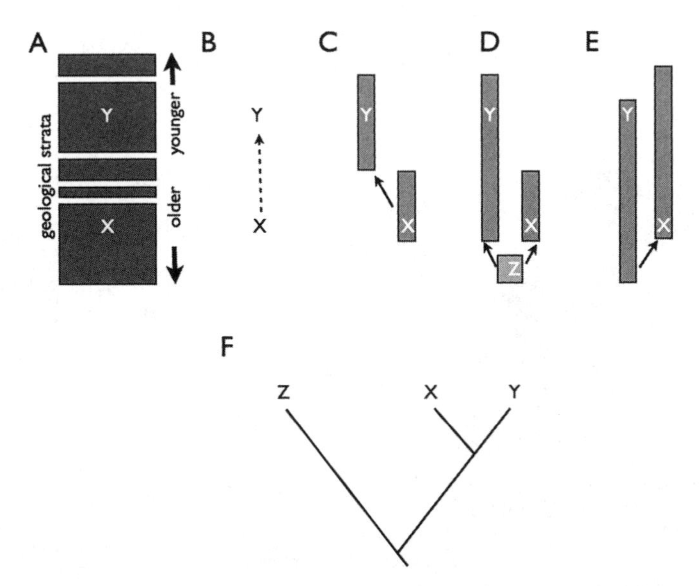

2.2 Using the fossil record to infer ancestry and descent is fraught with problems. A: Imagine you find a fossil, X, that occurs in an older stratum than another fossil, Y, which clearly belongs to a related species. Can you draw a line of ancestry and descent between them, as in panel B? It's possible that the fossil X really is an ancestor of fossil Y, but you could never know this for certain. However, the situation in panel B is really a shorthand for the following: the species of which fossil Y is a member is a descendant of the species of which fossil X is a member, as in panel C. (The inferred time ranges of the two species are represented by the gray boxes.) This seems reasonable, except that there are other plausible scenarios, such as D, in which species X and Y are descendants of an undiscovered species, Z. One might also imagine the possibility shown in E, that species X is in fact a descendant of species Y, and the fossils we've found represent an early representative of X and a late example of Y, implying that the species represented by the older fossil, X, is a *descendant* of the species represented by the younger fossil Y, turning the scheme in panel B on its head! This shows that inferring patterns of ancestry and descent from fossils is plainly open to all sorts of possibilities, some of them seemingly nonsensical. The only solution to the expression of phylogenetic relationship is to remove inference of ancestry and descent altogether. One can, however, suppose that all organisms, living and extinct, are cousins, in which case all possibilities are covered by the bifurcating diagrams known as cladograms, in which all species, irrespective of age, are drawn as terminal twigs. In panel F, the relationship between X and Y is clearly shown, together with their common relationship with Z.

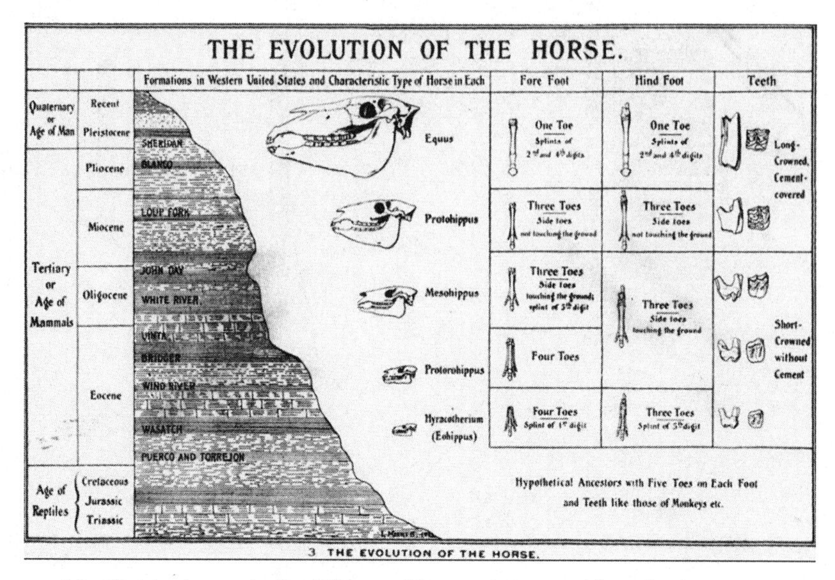

2.3 Manifest destiny on the hoof. (A very old souvenir postcard from the American Museum of Natural History: Wikimedia Commons.)

traits of an animal group were distributed across that group—how the organisms were similar, and how they differed.

In particular Hennig was interested in those similarities for which some argument might be made for homology. That is, they were similar because of shared evolutionary heritage, not because they happened to resemble one another through separate adaptations to similar circumstances without reference to common descent. Such similarities are known as convergences and reflect a phenomenon in which distantly related creatures come to look like one another by acquiring similar adaptations.

Modern horses have but one toe on each limb, and run not on the fleshy part of the toe, but the toenail—what we call the hoof. The earliest members of the horse family had more toes on their feet, and later members had reduced numbers of toes. The modern horse seems to be the ultimate expression of perfect horsiness, as if it could not possibly strive any further toward some cosmic equine ideal. If horses had any less contact with the ground they'd become airborne and float away.

However, this extreme reduction of digits is also found in an extinct animal called *Thoatherium*, a smallish herbivore that once lived on the plains of what is now South America. *Thoatherium* looked remarkably horselike. Not only were its digits reduced to a single hoof, as in the modern horse, it had also lost the splint bones—vestiges of digits appressed to the cannon bones—present in true horses. As a feat of digit reduction, *Thoatherium* had out-horsed horses long before true horses had evolved. Yet this animal was not a close relative of horses, and had evolved its horsey anatomy entirely independently. *Thoatherium*'s anatomy, therefore, was convergent on that of horses. We are confident that *Thoatherium* isn't a close relative of true horses because we know a great deal about the evolution and anatomy of horses, if rather less about the litopterns, the entirely extinct order of mammals to which *Thoatherium* belonged.[6] In practice, and without such knowledge, it's hard, even impossible, to tell the difference between convergence and homology.

The arrangement of shared homologies could be expressed as a bifurcating diagram, in other words, a tree. Hennig's approach came to be known as cladistics, and these diagrams were called cladograms (such as panel F in fig. 2.2).

It cannot be stressed enough that these diagrams were meant to reflect the *pattern* generated by evolution. That evolution had *happened* as Darwin said it had was a given—a fact frequently misunderstood by creationists, who made mischief by asserting that Hennig's adherents were anti-evolution.[7] What cladograms do *not* do is make any statement about the *process*—the particular adaptations, say, whereby a particular group had been successful.

To go back to our horse example, we cannot assert that what drove horse evolution is the progressive reduction in digits in the horse lineage, because that automatically denies the possibility that several horse lineages might have lost digits convergently. It also presupposes that digit loss is more important than some other trait, such as increased size, or change in digestive capability, or something else entirely, which might (or might not) be related to digit loss indirectly. That is to say, there could be features of an animal's anatomy that might not be subject to selection in any simple way, but change because they might be indirectly associated with some other feature that is.[8] In any case, it is essential to separate process from pattern. If we do not do this, then we cannot make any statement about how the process might have influenced that pattern without introducing circularity into our reasoning.[9]

One aim of cladistics is to produce the cladogram—the tree—in which the data are used most economically. This, the "principle of parsimony," is really an expression of the old idea of Occam's Razor,[10] but has another purpose. The most parsimonious cladogram comes with an assumption that evolution always acts in such a way that convergence is kept to a minimum. The fact is, however, that we know perfectly well that evolution is very far from neat and tidy, and that convergence is rife.

The problem with convergence is that it is impossible to estimate. Given two unicorns of different species, how would you know (without any other clues) that they each inherited their horn from a horned common ancestor, or if their common ancestor was hornless and they each evolved horns independently? You cannot. However, a cladogram with a single evolutionary event is more parsimonious than a cladogram with two separate ones. It is simplest to assume, therefore, that the horn evolved just *once* in the common ancestor of both species, rather than *twice*, separately, in the lineage of each species—*whether or not this is true*. Parsimonious cladograms are not meant to represent the absolute truth, for that is unknowable, but a lower bound on what evolution can achieve, and thus turn from unfalsifiable scenarios into properly constructed hypotheses. The principle of parsimony can be applied to assess the relative value of competing cladograms made with the same data set.

It is possible to criticize the principle of parsimony as woefully unrealistic, given the prevalence of convergence. I like to compare the business of parsimony in cladistics with an analogous situation in population genetics, in which the spread of alleles through a population, under the influence of natural selection, is measured using a toy model of a population called the Hardy-Weinberg Equilibrium.

The Hardy-Weinberg Equilibrium assumes that the proportions of alleles in a population will remain the same from generation to generation provided that two assumptions are met. First, that the population is infinite in size, and, second, that any member of that population can mate with any other. This seems quite unrealistic at first, because we know perfectly well that populations are finite, and that individuals do not mate randomly. But that's exactly the point—by making such assumptions, the Hardy-Weinberg Equilibrium sets up a situation that's ideal for the testing of ideas about what population geneticists are interested in—namely, how population size and mate choice affect the spread of alleles under selection. In the same way, by seeking to minimize convergence, cladograms allow notions of convergence to be tested explicitly, and different cladograms to be compared, one with another.

Another distinctive feature of cladograms is that ancestry and descent are never explicitly shown. All creatures in a cladogram are presented as terminal twigs, not putative ancestors along one branch or another. The reason is that ideas about ancestry and descent can never be falsified. The fossil you dig up tomorrow *might* be your ancestor, but you can never *know* this for certain. However, because you have assumed that evolution operates as Darwin said it did, and that we have very good reason to suspect that all life has a single common ancestry,[11] and so every creature, living or extinct, must be related to every other, you will be as certain as science conceivably allows that the fossil is (or was) your cousin in some degree. Indeed, all organisms, living or extinct, are cousins of every other organism, and you could say that the purpose of cladistics is not to defend ideas about ancestry and descent (because these are indefensible) but to make predictions about degrees of cousinhood.

Cladistics was originally applied to morphology—the bristles of flies and the shapes of bones—but it is now routinely applied, in one form or another, to molecular sequence data. The sequences of amino acids in a protein, or more usually the sequence of nucleotides in a strand of the genetic material DNA, may or may not be closer to the evolutionary essence of an organism, depending on your point of view. Such sequences are, however, inherently digital and atomistic in a way that morphology or anatomy frequently aren't, and are thus ideally suited for the production of cladograms, which is computationally intensive.

Cladistics has moved away from the principle of parsimony as its gold standard, as other criteria for judging trees have moved to the fore, such as maximum likelihood or the use of Bayesian statistics combined with other methods that compensate for various distortions introduced by the variable quality of data, and so on. The technical niceties do not matter, because the result is the same—a tree-like diagram in which all the organisms compared are presented as terminal twigs. This kind of presentation allows for a very graphic demonstration of the particular traits characteristic of each group, showing the order in which they are presumed to have appeared.

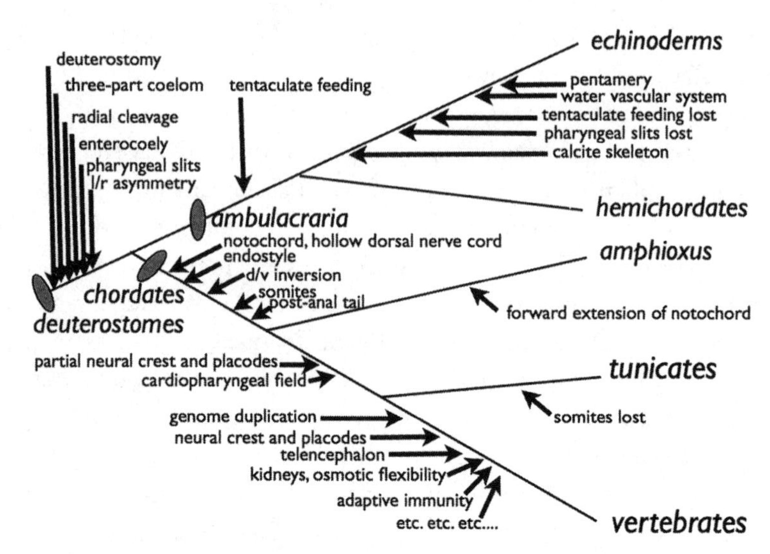

2.4 Phylogeny showing the interrelationships of extant hemichordates, echinoderms, tunicates, the amphioxus, and vertebrates—collectively, the deuterostomes—with the points at which many of the major distinguishing features of each group appeared. This phylogeny is now supported by a multiplicity of sources (see text) but the number of long branches is apparent by the clustering of major features without side-branches to break them up—showing that there were perhaps many more kinds of deuterostome, some showing intermediate forms between creatures now extant, but that these forms are now lost. It is this scarcity of intermediate forms that, in many ways, explains how hard it has been to understand the origin of vertebrates from invertebrates.

Here for example is a diagram showing the interrelationships of the various deuterostome groups, noting the segments in which various distinguishing traits appeared (fig. 2.4).

This kind of diagram makes it clear where the long branches are. These are the segments where a lot of traits are bunched together without side-branches to interrupt them—in particular the lineage segment leading to extant chordates, and the lineage segment leading to vertebrates in particular.[12] In other words, this kind of graphic is ideal for helping us to address a problem as knotty as the origin of vertebrates. What you will not see in a diagram like this is any explicit notion of ancestry and descent. Tunicates and vertebrates evolved from a common ancestor; these two groups evolved from a common ancestor with the amphioxus; and these three evolved from a common ancestor with the Ambulacraria.

Diagrams like this challenge more than a century of speculation about how animals were related to one another, based on unfalsifiable notions of ancestry and descent. This doesn't make hypotheses based on cladistics inherently any closer to the truth than those that are not, but it does make it easier to evaluate competing hypotheses.

2.3 SUMMARY

This chapter has been a history of evolutionary thought in microcosm. Before Darwin, living things were ordered in much the same way as inanimate matter, each form occupying its station in the world and incapable of leaping from one carefully constructed category to make its home in another.[13] What was important was the pattern, preconceived and preordained.

Darwin did away with all that, bringing *process* to the fore. Any pattern that resulted was simply the result of that process, acting in the continuous present, but summed over history. This idea has, however, been frequently misunderstood, and natural selection has been miscast as a director of long-term evolutionary processes. However, by discarding ideas of prefigured pattern, Darwinian selection raises the problem of how, precisely, we reconstruct phylogeny, that is, the tree-like order of life.

In this chapter I introduced the idea of cladistics and the principle of parsimony, showing how these tools can be used to create evolutionary hypotheses subject to falsification, and as ways to estimate the extent to which various parts of a group's evolutionary history are not as well known as they might be. Such spots can be identified by long branches marking several distinct evolutionary novelties. Such branches have the potential to be broken up into shorter line segments, such that the various novelties might be laid out in their order of acquisition. This, essentially, is the problem facing those who seek to understand the origin and early history of vertebrates.

CHAPTER 3

Embryology and Phylogeny

3.1 FROM EMBRYOS TO DESPERATION

A school of thought especially popular in Germany in the late eighteenth and early nineteenth century was that organisms were driven toward an ideal state of perfection. This school, known as *Naturphilosophie* (nature philosophy), is particularly associated with Johann Wolfgang von Goethe (1749–1832), who, as well as being a towering figure in literature, made many seminal contributions to natural history.[1] Although nature philosophy was, at least to start with, meant to express a kind of ideal, Darwin's idea of natural selection gave it renewed impetus. Scientists such as Ernst Haeckel (1834–1919) took Darwinian natural selection and used it as a force to drive organisms from initial simplicity to eventual complexity. Despite the fact that this is a terrible perversion of natural selection, this idea of evolution is the one that inhabits most people's heads today, fueling the idea of an order of creation in which there are hierarchies of organism with Man, naturally, as the most ideal or perfect.[2]

Haeckel and his colleagues were especially interested in embryology—how organisms develop from a single fertilized egg to achieve the adult state. They found that the progression of forms in embryology, from the simple to the complex, mirrored what they saw of the progression of adult forms in nature, from the simple to the complex, giving rise to the cliché that "ontogeny recapitulates phylogeny." Although nature-philosophical ideas of destiny, progression, transcendance, and the ideal state have largely fallen by the wayside, it was the intellectual descendants of the nature philosophers—in particular Haeckel's student Wilhelm Roux (1850–1924)—who created the study of embryology much in the form we know it today, in which the minute progress of development in the embryo is cast against the grand sweep of evolutionary history.

Here, then, is a cartoon view of the early embryonic life of animals (fig. 3.1). All organisms start as a single cell, an egg. The fertilized egg divides, into two, and then four, and is soon a ball of seemingly undifferentiated cells. The cells in this ball do not divide without limit. Liquid produced by the cells pools in the middle of the mass, producing a void called the blastocoel, bounded by a single layer of cells. This hollow ball of cells is called the blastula.

After a while a dent forms in one end of the blastula, which then pushes inward, extinguishing the blastocoel until the ball is two layers thick, with a hole at one end. The ball of cells is now called the gastrula; the process of invagination and layering that creates it is called gastrulation; and the new void, replacing the blastocoel, is called the archenteron. The outer layer of cells in a gastrula is called the ectoderm, and the inner layer, the endoderm. Cells from the endoderm move into the remnants of the bastocoel to form a third layer of cells, the mesoderm, which forms blood vessels and muscle.

The early embryologists noticed a correspondence between these steps of development, observed across the animal kingdom, and the apparent complexity of animals themselves. The initial ball of cells, seemingly undifferentiated into layers or tissues, is vaguely reminiscent of simple creatures such as sponges, or the small shapeless creeping animal called *Trichoplax*, sole member of the phylum Placozoa. The gastrula, with its two layers surrounding a void, is reminiscent of simple polyps such as the hydra, in which the body has two so-called germ layers of cells, the ectoderm on the outside and the endoderm on the inside, around the gut, with a single point of entry and exit serving as both mouth and anus. Such animals are termed diploblasts, to denote that they are formed of two cell layers.

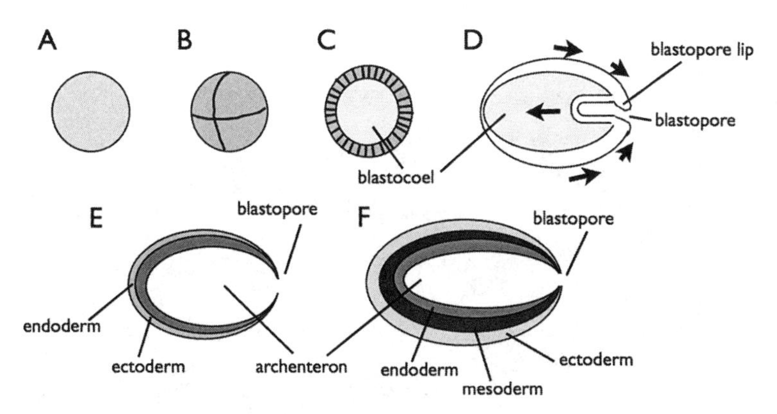

3.1 Cartoon embryology, in which the early development of an animal vaguely recapitulates the evolution of animal life. A: A single fertilized egg, reminiscent of the single-celled ancestral organism. B: The egg divides to become a mass of cells, reminiscent of very simple creatures such as sponges. C: A cross-section reveals the ball of cells to be hollow. This is the *blastula*, and the void at its center is the *blastocoel*. This stage is reminiscent of simple organisms such as the colonial alga *Volvox*. D: *Gastrulation* starts when one end of the blastula folds inwards to form two layers of cells, obliterating the blastocoel and creating a new cavity, the *archenteron*. E: Gastrulation is complete. The embryo has two layers of cells, the ectoderm and endoderm, and a single entrance into the archenteron called the *blastopore*. This two-layered or *diploblast* state is reminiscent of a simple polyp such as *Hydra* in which the blastopore does double duty as mouth and anus. F: Cells from the other layers slough into the remnants of the blastocoel to create a third, intermediate layer of cells, the mesoderm. In acoel worms, which are the simplest bilaterally symmetrical, three-layered (*triploblast*) animals, the mesoderm forms a solid block of tissue and the blastopore continues to serve as mouth and anus. In more derived triploblasts, with a through gut, the blastopore forms the mouth in protostomes, and the anus in deuterostomes.

The ectoderm forms the skin and nervous tissue; the endoderm forms the gut with its lining of glandular and digestive cells.[3] The simplest animals with mesoderm—that is, triploblasts—are tiny worm-like creatures called acoels. In these animals the mesoderm forms a mass of tissue separating ectoderm from endoderm. Like polyps, though, they have a single entry into the gut that does double duty as mouth and anus.

The correspondence between embryo and animal tends to break down in animals of more elaborate structure. In triploblasts more complex than acoels the mesoderm splits to form a cavity, the coelom, such that mesoderm lines the inside of the ectoderm and the outside of the endoderm with a cavity in between—this is the coelom—in which the internal

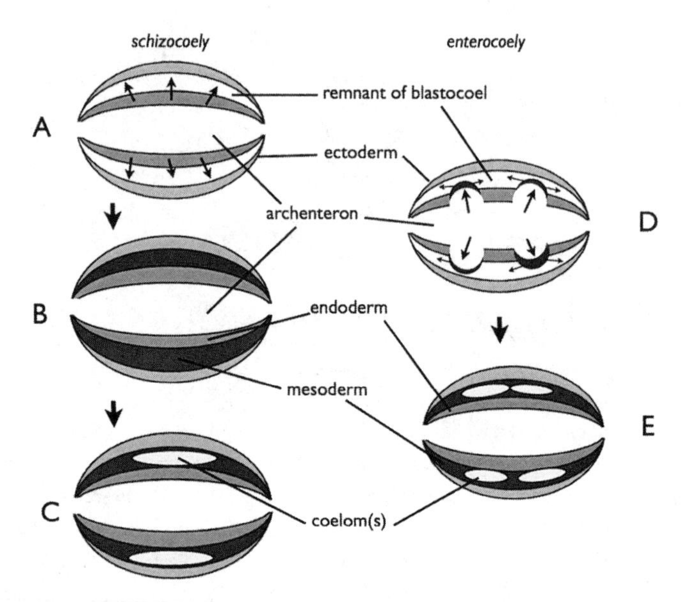

3.2 The two principal styles of coelom formation. This occurs during (or just after) gastrulation, but is shown separately here for simplicity. In *schizocoely* (A-C), endoderm cells (arrows) invade the remnants of the blastocoel to create mesoderm (B), which then splits to create separate mesodermal linings for the ectoderm and endoderm, known as the *somatopleure* and *splanchnopleure* respectively. The cavities formed by this splitting process, lined by mesoderm, are the *coeloms*. Animals have variable numbers of coeloms. Most internal organs are formed within either the somatopleure or the splanchnopleure and project into the coeloms. The other style of coelom formation is *enterocoely* (D-E) in which outpocketings of endoderm project into the blastocoel remnant (D). These sheets of cells form the mesoderm, which lines the ectoderm and endoderm as before, forming coeloms.

organs are suspended. This mode of mesoderm formation, by splitting, is called schizocoely. Mesoderm can also form by the inward invagination of endodermal pouches into the remnant of the blastocoel, a process called enterocoely (fig. 3.2).

Either way, a mesoderm is formed with cavities of its own. The cavities are called coeloms, and animals with coeloms are called coelomates.[4] This kind of body plan is seen at its simplest in annelids, or segmented worms, in which each segment is based around coelom. Animals with three layers of cells all seem to have an elongated shape, with a preferred direction of travel, and are bilaterally symmetrical. These are the Bilateria.

All bilaterians more sophisticated than acoels have a through gut, with a distinct mouth and anus, and their further classification depends on whether the mouth or anus develops from the blastopore.[5] In protostomes, the blastopore forms the mouth, and a new anus is formed. In deuterostomes, however, the reverse happens: the blastopore becomes the anus and a new mouth is formed (fig. 3.3).

This division is especially interesting in the context of this book as chordates and ambulacrarians are, of course, deuterostomes. Classically, such groups as brachiopods (lamp shells) and chaetognaths (arrow worms) were regarded as deuterostomes, partly on the basis of anus formation, and because the coeloms form by enterocoely, which is also thought to be a general characteristic of deuterostomes. Brachiopods are especially interesting as the adults have a distinctive feeding structure called the lophophore, a set of tentacles arranged on a bilaterally symmetrical loop-shaped frame. This is seen in other marine animals, the phoronids and bryozoans (moss animals), as well as colonial hemichordates called pterobranchs.

The three major groups of protostomes are the mollusks, annelids, and arthropods. The presence of segmentation in these groups is especially

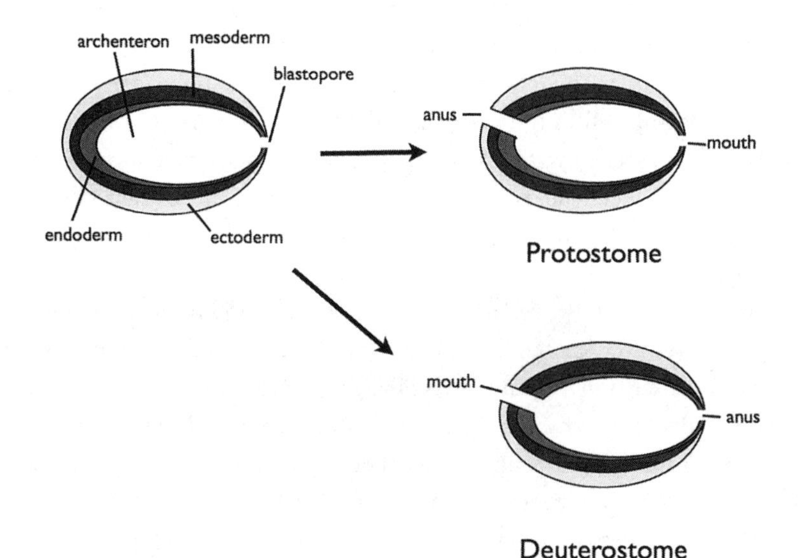

3.3 Protostomes and deuterostomes are two fundamental divisions of the animal kingdom, and membership of one or the other depends on the fate of the blastopore. In protostomes, the blastopore becomes the mouth and a new anus is formed; in deuterostomes, the reverse is true.

interesting—the whole business of segmentation captivated minds as diverse as Geoffroy and Goethe—and it was thought that arthropods and annelids shared a common ancestry. The onychophores (velvet worms), creatures of tropical forests that have segments and multiple limbs but lack the hard exoskeleton of arthropods, were seen as relics of a transition between the annelid and arthropod state. But segmentation is also a feature of chordates, in particular vertebrates and the amphioxus, leading researchers to propose, over the years, schemes in which chordates might have been derived from arthropods or annelids.[6]

Another important feature of early embryonic life used to classify animals is known as cleavage. This refers to the way that cells are organized in the very earliest stages of cell division, immediately after the egg is fertilized. In some embryos, successive cell divisions happen at right angles to one another, producing a ball of cells like a stack of blocks. This style of division is known as radial cleavage, and is seen as a characteristic of deuterostomes. In protostomes, however, the plane of each successive division is offset from the one before by a small angle, producing a spiral pattern. This is known as spiral cleavage.

In some protostomes, development is also strongly determinate. That is, the lineage of any cell in the adult can be traced to particular cells in the early blastula. The mesoderm of a mollusk, for example, can be traced to one specific cell in the early blastula; the roundworm *Caenorhabditis elegans* has precisely 959 cells as an adult, excluding sperm or egg cells.[7] In deuterostomes, cell fate is, in general, indeterminate, in the sense that any cell in the blastula might in principle give rise to any part of the developing animal.

Now, all of this is to some extent a caricature, and there are many exceptions. It's now known that some rather fundamental differences exist between closely related animals. As I've mentioned, some brachiopods develop by deuterostomy and, like deuterostomes generally, appear to form their coeloms by enterocoely,[8] but other brachiopods, on further inspection, turn out to be protostomes.[9] Yet others have a slit-like blastopore that zips up except at the ends, to produce both mouth and anus from the same, single opening, a condition known as amphistomy. Tunicates, which are deuterostomes, have highly determinate development.[10] In some hemichordates, coeloms can form by schizocoely[11]; and both forms of coelom formation can be observed in different coelomic compartments in the same acorn-worm individual.[12] And so on.

The point is that features of early development seen at the microscopic scale were used to classify the larger order of life, a clear relic of the nature-philosophical creed that ontogeny recapitulates phylogeny—even though, on reflection, evolution is under no obligation to follow any scheme based on an outmoded world view of unobservable life-forces and strivings toward some ideal cosmic perfection. More pertinently for everyday working zoologists was that as exceptions to the embryological rules accumulated, it became extremely difficult to come up with an overall phylogeny for animals that could be supported without recourse to subjective ideas about which traits were more or less important in evolution. You be the judge. Which is more important—spiral cleavage or schizocoely? Coelom formation or germ layers? How can you tell? The result was a profusion of ideas about animal evolution and very little consensus. Some zoologists even suggested that the major animal groups, or phyla, arose independently from single-celled ancestors.[13] This seems less of a working hypothesis than a counsel of despair and a sign that a radical new approach was necessary. The tree was old, rotten, and shaky—another couple of storms and it would disintegrate altogether.

Cladistics, however, was not in and of itself a panacea. The major groups, or phyla, are recognized on the basis that each has a unique body plan, different from that of any other animal, and so would have nothing in common—by definition: an inviolability reminiscent of Cuvier's *embranchements*. With nothing else to work with, cladograms of phyla tend to come up with a "star" phylogeny or "polytomy" in which everything is equally distantly related to everything else, a result that contains essentially zero phylogenetic information.

3.2 GENES AND PHYLOGENY

The solution came with the availability of information from genes—the digital information of nucleotide sequences—that could be processed by methods of phylogenetic reconstruction based on cladistics. Divorced from particular embryological traits and thus subjective notions of evolutionary process, molecular evidence was seen as having the potential to shed some light on the murky problem of the large-scale pattern of animal evolution.

The surprise is that division based on the old embryological characters has retained as much support as it has. Protostomes and deuterostomes, diploblasts and triploblasts have survived as recognizable natural groups into the molecular age, as to an extent have groups uniting animals on the basis of cleavage—although coelomates, as a group, have not. But more on this later.

The earliest influential paper in this area[14] compared the sequences of a single gene isolated from 22 animals belonging to ten phyla.[15] The resulting phylogeny suggested that multicellular animals (metazoans) had at least two separate origins, with diploblasts and bilaterians evolving separately from among single-celled organisms. The earliest bilaterians soon split into acoelomate flatworms and coelomates, which then split to give something that recalled Cuvier's *embranchements*—chordates, echinoderms, arthropods, and other coelomate protostomes.

This first essay in metazoan molecular phylogeny was overturned when more and better genetic information became available. The dual (diphyletic) origin of animals was especially contentious, and most subsequent metazoan phylogenies have produced a single (monophyletic) origin.[16]

The current, well-supported consensus is that the metazoa form a monophyletic group within a larger group of eukaryotes, the opisthokonts, whose members also include fungi and a variety of single-celled organisms, a part of what we used to call "protozoa," with other "protozoa" belonging to other, larger groups within the eukaryotes (fig. 3.4). The closest relatives of metazoans among unicellular creatures are the choanoflagellates.[17] These creatures have a number of biochemical traits seen elsewhere only in metazoa, speaking to a shared common ancestry.[18]

Within the Metazoa (fig. 3.5) there is currently some contention about the branching order of the most basal nodes.[19]

Sponges usually take the most basal branch on the grounds of perceived primitiveness, but recent evidence from the analysis of genomes suggests that ctenophores (comb jellies) might in fact be more basal. This is strange, as ctenophores are diploblasts and much more organized than sponges. Yet genetic evidence shows that they have traits quite unlike those found in any other animal. They are predators, with nervous systems, yet these are constructed differently from those found elsewhere, and have a unique

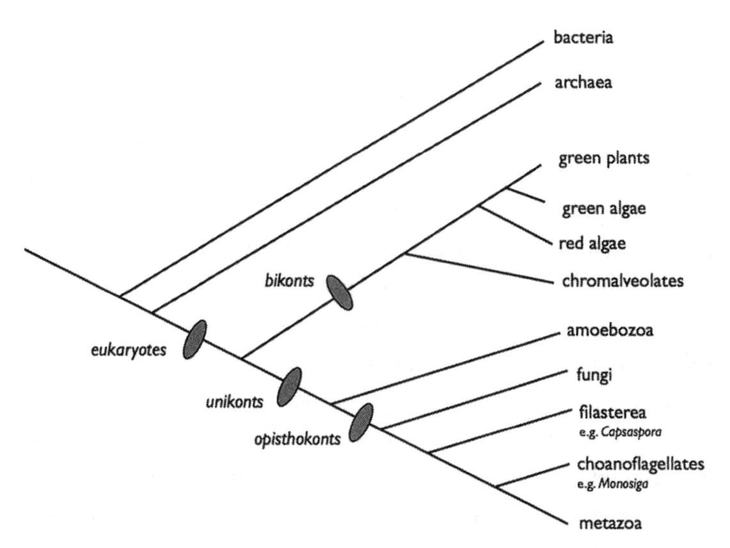

3.4 The place of metazoa (multicellular animals) in the tree of life. Although most organisms live their lives in the confines of a single cell, multicellularity has evolved many times—not just in metazoa, plants, and fungi, but in algae such as seaweeds. What should be obvious from this diagram is that single-celled organisms form a rich and complex world, greatly undeserving of the dustbin taxon "protozoa" taught in biology classes when I was a schoolchild. The consequence, however, is that the higher-order classification of eukaryotes is a subject of vigorous and ongoing debate well outside the scope of this book: this diagram is a greatly simplified reading, from various sources.

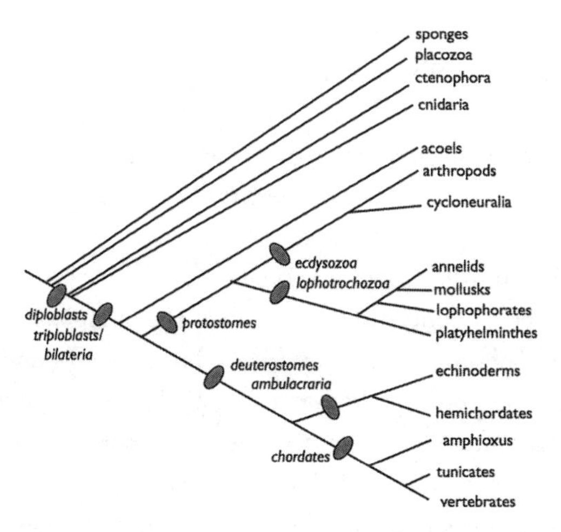

3.5 The major groups within the metazoa: a very simplified diagram based on several sources. Although there is good molecular support for much of this phylogeny, there is a degree of uncertainty about the branching order of the basal nodes (sponges and ctenophores); considerable debate about the position of acoels; and continuing doubt about the position of some other, minor phyla.

biochemistry.[20] More recent work places ctenophores in a more traditional position—that is, alongside cnidaria, with sponges as the most basal animal group.[21] The final word on the position of ctenophores has almost certainly not been said, but whatever the outcome, the tale shows that organisms exist on their own account and have traits worth exploring for their own sake.[22] They are not there to act as staging posts in a grand tale of evolving complexity created after the fact to suit our own subjective prejudices about a ladder of nature with humans at the top.

The rest of the diploblasts—cnidaria, essentially—take a traditional position, branching above sponges but below triploblasts; however, a note of caution is needed. The simplicity of many cnidaria is more apparent than real. For example, work on the genetics of the sea anemone *Nematostella vectensis* reveals a complexity quite out of tune with the simplicity of its body.[23] The *Nematostella* genome has more in common with vertebrate genomes than with those of the standard genomes of invertebrates such as that of the roundworm *Coenorhabditis elegans* and the fruit fly *Drosophila melanogaster*. This implies both that vertebrates have rather conservative, "primitive" genomes compared with many seemingly simpler animals, and also that animal evolution is not necessarily a story of steadily increasing complexity.

All remaining metazoans are triploblasts and bilaterians. It is assumed that most primitive bilaterians were small, simple, and worm-like[24]: rather, indeed, like the acoels, a group of worms formerly lumped in with the phylum Platyhelminthes (flatworms, including those cuddly planaria from chapter 1 and the altogether less cuddly parasitic flukes and tapeworms) but now seen as very much on their own.[25] All other bilaterians are either protostomes or deuterostomes.

As noted above, deuterostomes include the ambulacrarians (echinoderms and hemichordates) and chordates, but no longer include—as was once thought—the so-called lophophorate phyla (brachiopods, phoronids, and bryozoa) despite the similarity of the lophophore with that of colonial hemichordates. Indeed, there is increasing evidence that the lophophore was acquired several times convergently.[26] Neither do deuterostomes now include the planktonic arrow-worms (chaetognaths), now of uncertain affinity[27]; or beard worms (pogonophores), linked with deuterostomes in times past[28] but now firmly allied with annelids.[29] More recent work placed a simple, wormlike animal called *Xenoturbella*, and possibly acoels, among the deuterostomes[30] but the general view now is that these creatures are very simple bilaterians.[31] I discuss the curious cases of *Xenoturbella* and chaetognaths in more detail later on.

Aside from these, deuterostomes are no longer united just on the basis of deuterostomy: although the anus of all known deuterostomes forms from (or close to) the blastopore, this is known to be the case in some protostomes, as we have seen. This makes deuterostomy, as a trait, a rather rough and fuzzy guide to defining a major animal group. A more reliable criterion is the presence of pharyngeal slits.[32] Given that pharyngeal slits are found in all deuterostome phyla unequivocally grouped as such on the basis of gene sequence data, one can hypothesize that the latest common ancestor of deuterostomes was a triploblastic, bilaterian animal that had pharyngeal slits at some stage in its life cycle. At present we can say little else about how it looked, whether it was sessile or motile; whether it lived in the sunlit plankton or buried itself in the seabed; whether it developed directly from its egg, or had some intermediate larval form.[33]

The protostomes—again, grouped by gene sequence homology rather than on the basis of the embryological trait from which the group gets its name—fall into two large groups or superphyla, the lophotrochozoans[34] and ecdysozoans.[35] Although relatively new names on the block, these groups have received solid support as monophyletic groups.[36]

The lophotrochozoans include platyhelminthes (sans acoels), mollusks, annelids, brachiopods, phoronids, bryozoa, and some other, even less familiar creatures. Some of these have lophophores; others have a distinctive larval form called a trochophore, and the embryos of many undergo spiral cleavage.

The ecdysozoans include arthropods and their allies such as onychophores and tardigrades; roundworms (nematodes); and some lesser-known creatures such as mud dragons (kinorhynchs), cycliophorans, and loriciferans. The non-arthropod ecdysozoans are collectively called the cycloneuralians. The group as a whole gets its name from the habit of moulting, or "ecdysis," a literally superficial trait, in that ecdysozoans habitually shed their external skeleton or cuticle during growth. When the ecdysozoa was first proposed,[37] commentators were struck by the removal of arthropods from annelids, with which they had been grouped (velvet worms stayed with the arthropods)—and the close relationship between arthropods and the hard-to-place, apparently pseudocoelomate nematodes.

More recent work has shown that the common heritage of the Ecdysozoa might be more than skin deep, and velvet worms might be the key. You'll recall that these creatures, found beneath leaf litter on tropical forest floors, are superficially wormlike, but have many pairs of stumpy legs. They do look indeed like a cross between an earthworm and a centipede, making their status as evidence for a close relationship between annelids and arthropods almost a given. But this is the case—perhaps a rare one—where fossil evidence actually acts to inform rather than confound ideas of evolutionary relationship, and thereby hangs an exemplary tale of how fossil forms can combine, in a single organism, features now found in entirely distinct creatures.

It turns out that velvet worms are the last vestige of a large and varied group of creatures called lobopodians. These lived in the sea in the Cambrian Period more than 500 million years ago. Like modern velvet worms, they were wormlike and had many pairs of soft, apparently unjointed limbs. In addition, and quite unlike modern onychophores, they sported a wide range of claws, spines, and armor, betraying a close relationship with arthropods.[38]

Perhaps the most famous lobopod is *Hallucigenia*, from the Cambrian Burgess Shale of British Columbia. *Hallucigenia* was originally interpreted as a wormlike animal supported by seven pairs of long, unjointed stilt-like spines; on its back was a single row of tentacles, each ending in what looked like a mouth. Named *Hallucigenia* by its describer Simon Conway Morris on account of its "bizarre, dreamlike appearance,"[39] it was used as an inspiration for science-fictional aliens[40]; but it turned out that *Hallucigenia* had been described upside down. The long stilts were spines on the animal's back. The tentacles with mouths were (paired) legs ending in feet. It was still a very strange-looking animal, but became less so when other fossil lobopodians were described. Still unclear, though, was the nature of its head. Recent work has shed light on this, too.[41] It turns out that *Hallucigenia* had a circular mouth and a series of teeth in its throat—traits not typical of arthropods, but seen in cycloneuralians.

This new work on a very strange fossil, then, provides some hints of what the common ancestor of ecdysozoans looked like: a soft, wormlike creature with a series of limbs and a circular mouth. The legs persisted into arthropods, the circular mouth into cycloneuralians—but no animal currently alive has all these features together.

3.3 SUMMARY

In this chapter I showed how the early embryologists mapped the features and processes of the early development of animals and cast them against the greater history of life. They were prompted to do this by an underlying creed, nature philosophy, driven to recognize a cosmic canvas of evolution in each developing embryo.[42] However, as knowledge of the peculiarities of each animal group accumulated, more and more exceptions were found, with the result that it became impossible to forge any kind of consensus on the overall shape of the animal kingdom, at least based on shared features of morphology. Knowledge from genetic sequences came to the rescue: in the past twenty years or so, a new and surprisingly robust picture of animal evolution has emerged. To be sure, many problems remain. The surprise, perhaps, is that the modern conception retains so much of the old view. Deuterostomes, for example, survive as a coherent group, even though defined on a different set of features.

In the next chapter I look further at the genetic substructure of animals and how this affects our ideas of shared evolutionary heritage, that is, homology.

CHAPTER 4

Hox and Homology

4.1 A BRIEF HISTORY OF HOMEOSIS

Do genes really offer a less subjective, more digital kind of anatomy? The more people looked at genes, the more they became aware that they are not simply beads on a string recording information for our edification. Genes have their own architecture and natural history, and these have produced perhaps the most surprising insights into animal interrelationships.

These started with Goethe, and in plants. Goethe supposed that the organs of plants—leaves, sepals, petals, and so on—were variations on the theme of a leaf-like template. Other pre-Darwinian thinkers, especially Geoffroy, took up the baton, suggesting that other repeated structures such as vertebrae had similar properties.

From time immemorial, observers of nature had been captivated by "sports" in which creatures exhibited some gross but discrete distortion of anatomy: calves with two heads, for example, which seemed normal in all respects apart from the extra head. One of these scientists was William Bateson (1861–1926), who, dissatisfied with the notion that evolutionary change was a matter of what Darwin called "insensible gradations" of form, sought to discover and catalogue every instance of discrete variation he could find. The result was an immense bestiary, *Materials for the Study of Variation* (1894), recording hundreds of cases of sports, including one in which an insect had little legs growing out of its head, in place of antennae—the rest of the animal appearing quite normal. Bateson coined a term, *homeosis*, for cases in which one structure appeared to replace another in this neat fashion.

More than half a century later, Edward B. Lewis (1918–2004), working at the California Institute of Technology, was working with a very similar mutation in the fruit fly *Drosophila melanogaster* (the workhorse of experimental genetics, then as it is now): the homeotic mutation in which legs replaced antennae, a mutation named *antennapedia*. This was just one of Lewis's miniature bestiary. Another was a fly with the *bithorax* mutation, in which a normally wingless segment of the thorax was replaced by a duplicate of a wing-bearing neighbour, producing a fly with two pairs of wings, rather than just the normal single pair. By dint of a program of very careful breeding experiments, Lewis found that all the genes whose mutations produce homeotic sports were physically clustered together on the fly's chromosomes: *antennapedia* was found in one cluster, *bithorax* in another.[1] Work on other animals revealed that the two clusters of *Drosophila* were aberrant. In general, animals have a single cluster of homeotic genes which includes both *antennapedia* and *bithorax* genes.

When geneticists homed in on the homeotic genes they found that each had a small, signature sequence[2] that, when the gene was translated into a protein, formed a kind of prong that interacted with the DNA of other genes.[3] The sequence became known as the homeobox, and the genes as homeobox or Hox genes. Hox genes, then, produced proteins known as transcription factors that influenced the behavior of other genes. They turned out to be important in development, in particular for specifying the physical identity of particular segments or body regions. The *antennapedia* gene, when not mutated, ensured that antennae (and not legs) grow out of the appropriate part of the insect head. The *bithorax* gene, when not mutated, ensures that segments of the insect thorax bear the requisite number of wings.

Other researchers, notably Christiane Nüsslein-Volhard and Eric Wieschaus, found that Hox genes were part of a cascade of transcription factors with even more fundamental functions such as specifying which end of the animal was which; how the animal was divided into segments; and so on. Hox genes arrive fairly late to the party, establishing the character of segments already laid out by other genes. The reason that Hox genes were first noticed is that flies with mutations in these genes would occasionally survive to adulthood. Mutations in the more "upstream" genes, producing, say, fly grubs with two tail ends joined in the middle, were almost invariably lethal.[4]

When researchers turned from flies to mice (that other stalwart of genetics research labs), they found that they, too, had a cluster of Hox genes. Or, rather, mice had four clusters, reflecting a double duplication of the whole genome sometime in vertebrate evolution. Human beings, too, have four Hox clusters. Some fish, though, have seven or even eight,[5] as I discussed earlier. The amphioxus, on the other hand, has just one,[6] showing that the duplication of the genome in vertebrates happened after the divergence of the amphioxus from the vertebrate lineage.[7] All other invertebrates that have been investigated have a single Hox cluster.[8]

More startling still is that the expression of Hox genes appears to be collinear. That is, Hox genes at one end of the cluster are expressed toward the anterior end of the animal, whereas those at the other end are expressed toward the posterior end of the animal, and the genes in the middle are expressed in linear order in the body—whether in flies, mice, humans, roundworms or anything else with an identifiable anterior-posterior body axis.

Later work has complicated matters, in that genes outside the Hox cluster may also have a homeobox, and that other families of transcription-factor genes are distinguished by their own signature sequences. It now appears that the Hox cluster is part of a larger structure, including other families of transcription factors such as NK[9] and ParaHox.[10] Some elements of this wider NK-Hox-ParaHox cluster can be found in apparently primitive metazoa such as sponges,[11] and the Hox cluster in general might be the "zootype"[12]—a defining mark of metazoa[13] or at least bilateria.[14]

There are, however, further complications. Some animals, even if their ancestors might have had Hox clusters, appear to have rearranged, modified, or even lost them.[15] Of particular interest is a small group of tunicates, the appendicularians or larvaceans.[16]

Second, the expression of Hox genes depends very much on the context established by other families of transcription factors, perhaps explaining why animals, even if they all have Hox clusters, otherwise look very different. Even though Hox genes from flies can stand in for an otherwise disabled cognate Hox gene in a mouse[17] and human Hox genes can "rescue" gene expression in a fly,[18] it remains the case that flies are flies and mice are mice and humans are humans, and Jeff Goldblum will wake up tomorrow unsullied by any Kafkaesque transformational nightmare. Fly Hox genes specify the identity of segments in the fly body, whereas in mice and humans they seem to be more involved in the regional identities of parts of the mid- and hindbrain.

Nonetheless, the repeated revelations about Hox genes and their activities do have an air of Goethe and Geoffroy about them. Genes are more than just passive markers, but are intimately involved in the specification of major features of animal form. This perhaps explains why, even when the embryonic features once used to classify the animal world have sunk under the weight of caveats and exceptions, to be replaced by genetic sequence comparison, the general picture of animal evolution remains very similar in overall outline: the classical deuterostomes are still deuterostomes, even if defined by different means.

4.2 THE GEOFFROY INVERSION

Another speculation, attributed to Geoffroy, has also echoed down to the present day. As I discussed above, Geoffroy noted that the internal organs of vertebrates appeared to have been inverted with respect to those of invertebrates: the vertebrate heart is ventral, rather than dorsal, the major nerve cord is dorsal, not ventral, and so on. Could vertebrates be arthropods or annelids, turned upside down? This surprising idea has more than a grain of truth.[19] Vertebrates and arthropods share the same anterior-posterior axis. That is, the front end of an arthropod is homologous, in a general way, with the front end of a vertebrate. The dorso-ventral axis (back/belly) is also the same, but inverted. Technically, the cognate of an arthropod gene expressing a dorsalizing factor in a vertebrate will express a ventralizing factor, and so on.

For example, the protein product of a gene called *chordin*, which in the frog *Xenopus* promotes the formation of the central nervous system and dorsal mesoderm, turns out to be homologous, in terms of sequence, with a protein encoded by a gene in the fruit fly *Drosophila* called *short gastrulation* (or *sog* for short), which counters the dorsalizing effects of another gene, *decapentaplegic* (*dpp*). In other words, *sog* is a ventralizing agent, whereas its vertebrate homologue, *chordin*, is a dorsalizing agent. Detailed comparison of the modes of action of *chordin*, *sog*, and other factors with which they interact provides strong evidence that the dorsoventral axes of frogs and flies are the same, but inverted with respect to each other.

Later results from animals known, from other evidence, to be closer relatives of vertebrates (and chordates in general, given that even nonvertebrate chordates have a dorsal nerve cord) are more nuanced. Hemichordates have both dorsal and ventral nerve cords and appear to be in a permanent state of semi-inversion. Whether the situation in hemichordates is ancestral (and thus informative about chordates) or peculiar to these animals is as yet unknown. The situation is complicated by the fact that the cognates of genes expressed in the neural tube in vertebrates are expressed in the skin of hemichordates, as if they had a "skin brain."[20]

4.3 THE PHYLOTYPIC STAGE

Another recapitulation from history comes from the observations of early embryologists such as Haeckel and Karl Ernst Von Baer (1792–1876) that embryos in a particular group or phylum tend to look very similar at a particular stage, midway through the course of development.

Very early embryos of different species tend to look very different from one another. For example, embryos provisioned with a small amount of yolk tend to be ball-shaped or tubular, whereas those provisioned with a lot of yolk will develop as a disc attached to the greater ball of yolk, as a lentil might be pressed against a watermelon. Human embryos develop first as a disc, even though the embryo is nourished by a placenta rather than a yolk. This could be a nod to our reptilian, egg-laying ancestry. Very late embryos, in contrast, tend to look more and more like the adults of their particular species, losing any more generalized features.

Somewhere in between, though, embryos of all vertebrates look very similar. This is usually when the tail has started to elongate, the limbs have started to appear as buds, and the head has become distinct. At this stage it is very difficult to tell the difference between the embryo of a whale, a bird, a bat, or indeed a human.[21] This midway point in development is called the phylotypic period or phylotypic stage, and has been observed in other animals besides vertebrates,[22] and even in plants. Some researchers have treated this idea with circumspection[23] given much-debated suspicions that Haeckel might have doctored some of his own drawings to emphasize certain features at the expense of others.[24] However, like Geoffroy's inversion, there is now genetic evidence that the phylotypic stage is real. Work on a variety of animals has shown that genes reckoned to be evolutionarily ancient are expressed preferentially at the phylotypic stage, whereas "younger" genes are expressed at earlier or later stages of development.[25] Thanks to the technique of what has been termed "phylostratigraphy," the phylotypic stage of any group can now be recognized entirely genetically, without recourse to morphology.

Work on the genes of organisms, therefore, has helped clarify the evolutionary family tree of animals, but also reaffirmed some very old ideas such as inversion and the phylotypic stage. It has also required that Geoffroy's concept of homology be reassessed, given that Geoffroy coined the term in an entirely pre-Darwinian context.

4.4 THE MEANING OF HOMOLOGY

What do we mean by homology? Today we use the term to refer to the shared evolutionary heritage of a trait. Before Darwin, however, Geoffroy used more practical means to determine homology. Structures were homologues if they had similar or corresponding relationships with other structures. The two bones in our forearm—the radius and ulna—are homologs with similar bones in the forearms of other vertebrates, because they occur in the same pairing, and are always found between the single upper-arm bone (humerus) and the smaller bones of the wrist and hand. Even though the radius and ulna in any given species might differ from the corresponding bones in other vertebrates as regards their shape and function, they still "correspond." In a similar way, the radius and ulna in the forelimb are serially homologous with the tibia and fibula in the similarly constructed hindlimb, suggesting a common plan for limb construction in vertebrates. But not, however, in other animals with limbs. The limbs of—say—arthropods are constructed differently from those of vertebrates. They serve similar functions, such as walking, swimming, and flying, and can be said to be analogs—that is, solutions to similar functional problems made from different starting materials. The distinction between homology and analogy was formalized by the great Victorian naturalist Richard Owen (1804–1892) in a discourse entitled *On the Nature of Limbs*, recently reissued with useful notes and commentary[26]:

> The hard parts of the leg of a Crab or an Insect may be "analogous" to the bones of the limb of a Quadruped, but they are not homologous with them: and where there is no special homology, there can be no relations of a higher or more general homology between the parts. (*On the Nature of Limbs*, p3)

Geoffroy and Owen, not to mention Darwin, were working long before the nature of inheritance had been discovered, still less the role of various genes in development, and how the genes of different animals might be evolutionarily related, the study of which is now known as "evolutionary developmental biology" or "evo-devo" for short. The observation that—say—clusters of Hox genes are found in animals otherwise very different in morphology shows that homology is very much more than skin deep: a phenomenon known as "deep homology"[27]—but raises new and challenging questions in its turn.

At first glance, say, the eyes of insects are very different in structure from the eyes of vertebrates. As insects and vertebrates each evolved from animals without eyes, one might suppose that insect eyes and vertebrate eyes are analogs—as analogous as the wings of birds and butterflies. And yet genetic research has found that eyes are specified by homologous genes in each case. The gene *eyeless* in the fly *Drosophila* (which, when mutated, produces eyeless flies) is homologous with the gene *Pax-6* in vertebrates, mutations in which cause eye defects in mice and humans.[28] To what extent does this make eyes homologous? The answer could be that the structures of the eyes are not, in themselves, homologous, but the genetic programs specifying them might be. On the other hand, many genes do multiple duty during development: a gene expressed to perform one task early on in development might be redeployed to do something completely different later on. In vertebrates, the gene *sonic hedgehog* (*Shh*) is particularly busy—involved in the patterning of the dorsal nerve cord by the notochord, as well as the layout of the limbs, and other things besides. A common theme in evolution is that networks of interacting genes are co-opted for different tasks in development, and in evolution. From this we can begin to think of evolution in terms of interacting genetic networks, but in such a landscape the concept of homology might need yet further refinement.

As a final note in this chapter, I should add that the development of structures, even quite complicated ones, is influenced by rather simpler things than the transcription of genes, such as the balance of physical forces set up as structures in the same organism that grow at different rates. Developing embryos tend to curl up into comma shapes not through genetic instruction, but because the cells of the ectoderm tend to divide more quickly than those of the endoderm, which is sometimes hampered by the presence of larger amounts of yolk.

As I'll show later in this book, some key features of chordate anatomy rely on the mechanical movements of cells, in particular the phenomenon called convergent extension in which a mass of cells destined to form the notochord in the tunicate tadpole larva gets extruded out posteriorly to form a long line of single cells; and the process whereby cells in the amphioxus neural tube knit together during closure. Although these processes rely on gene expression—for proteins involved in cell-to-cell communication and adhesion—the process is macroscopic and relies on the disposition of physical forces.

4.5 SUMMARY

In this chapter I looked at how insights from scholars of the past, in particular Geoffroy, resonate today in studies of animal form, function, and relationships.

First, I described the startling revelation that all animals seem to have a cluster of genes, called homeobox genes or Hox genes, whose disposition is crucial in specifying how structures are arranged in the developing body, and how the position of each gene in the Hox cluster corresponds with the relative position along the body where the gene is expressed, a phenomenon known as collinearity. The discovery of Hox genes by Edward Lewis, with additional work on the deeper structure of the genetics of development by Nüsslein-Volhard and Wieschaus, earned the three of them a Nobel Prize in 1995, but the roots of their work can be traced back to Goethe, Geoffroy, and Bateson. Vertebrates seem to be unique in the animal world by their possession of at least four Hox clusters, a reflection of the wholesale duplication and reduplication of the genome in the ancestry of vertebrates.

Another insight of Geoffroy was that vertebrates appear to be "inverted" with respect to invertebrates. For example, the nerve cord of vertebrates is dorsal, but ventral in invertebrates. Although the idea appears fanciful, it is now supported by studies on gene expression. A later idea, rooted in ideas of Haeckel and Von Baer, is that of the "phylotypic stage," in which all embryos of a particular group tend to resemble one another most closely at an intermediate stage of their development. This, too, has found recent correspondence in studies of gene expression. Genes expressed in the phylotypic stage tend to be phylogenetically more conserved than those expressed at an earlier or later stage.

One of Geoffroy's trickier concepts is homology—that is, the evolutionary relationship of parts by virtue of their correspondence with other parts irrespective of function. Although set out very clearly by Owen as regards comparative anatomy, the correspondence between homologous anatomies and homologous genes is not always easy to make. Different genes seem to be involved in the development of anatomically homologous structures, and the same genes are found, again and again, involved in the specification of entirely different structures at different times in development. Nowadays, researchers are seeking to effect a rapprochement by considering how networks of genes, or modules, relate to various anatomical features.

CHAPTER 5

What Is a Deuterostome?

In this section of the book I'll look in a little more detail at the basic zoology, anatomy, development, and fossil record of the various deuterostome groups. This short chapter is a refresher on those features that unite the deuterostomes as a natural group.

By tradition, deuterostomes are defined on the basis of three features of their development.[1] The one that gives the group its name is deuterostomy—that is, the blastopore forms the presumptive anus,[2] with the mouth forming from a secondary opening, hence *deutero-* ("second") and *-stome* ("mouth"). In addition, deuterostome embryos tend to show radial (rather than spiral) cleavage, and the coelom forms by enterocoely, that is, the in-pocketing of endoderm from the archenteron to form mesoderm with coeloms at the same time, rather than by schizocoely, or the cavitation of existing mesoderm.

In general, deuterostomes have six coelomic spaces, arranged in three pairs along the anterior-posterior axis—paired protocoels at the anterior

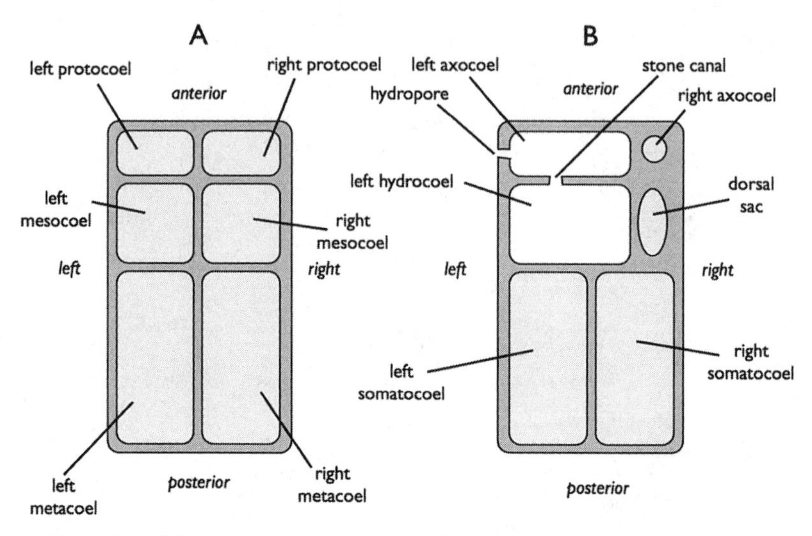

5.1 The bodies of deuterostomes are primitively bilaterally symmetrical and contain three pairs of coeloms. A: the deuterostome ground state, most obvious in hemichordate adults and echinoderm larvae. B: as modified in echinoderms. The left protocoel (axocoel) and mesocoel (hydrocoel) dominate: the right-side equivalents are either vestigial or do not form at all. The left hydrocoel is the rudiment of the pentameral structure that will come to dominate the entire body plan of the adult animal, expunging almost all traces of the original bilateral plan.

end, paired mesocoels behind these, and a pair of metacoels bringing up the rear (fig. 5.1).

The evolution of somites has heavily modified this arrangement in chordates, although the most primitive extant chordate, the amphioxus, shows this tripartite arrangement in its early larval stages.

These definitions have been subject to a great deal of qualification. It is possible, for example, that deuterostomy is the primitive condition for bilateria, in which case it cannot be used to define a more exclusive group.[3] However, genetic and genomic evidence shows that deuterostomes do indeed form a monophyletic group—that is, a natural group united by evolutionary heritage. The group consists of much the same phyla as it did before: echinoderms, hemichordates, and chordates.[4] At the same time, organisms thought at various times to be deuterostomes, or close to deuterostomes, on the basis of embryological characters are, on molecular reappraisal, found to group within protostomes instead. These include brachiopods and other invertebrates such as priapulids[5] and chaetognaths.[6]

More recent work has established pharyngeal slits as perhaps the most important and reliable defining feature of deuterostomes.[7] This has emerged from advances in anatomy and molecular biology that have revived Metchnikoff's nineteenth-century concept of ambulacraria, a union of echinoderms and hemichordates,[8] as well as reaffirming the coherence of deuterostomes as a whole.

Earlier deuterostome phylogenies tended to be constructed to emphasize the increasing sophistication of chordates and vertebrates. Echinoderms would be placed at the base. Hemichordates, having acquired pharyngeal slits, would be the next branch, forming a clade with chordates. The most basal chordates were tunicates, with the amphioxus, having acquired clear somitic mesoderm as in vertebrates, as the closest living sister taxon to vertebrates.[9]

The recent work in which tunicates are found to be more closely related to vertebrates than is the amphioxus[10] has, in its turn, showed that simple notions of progress cannot be applied to deuterostome evolution. The loss of traits is as important as their gain: the most prominent examples being the loss of pharyngeal slits in echinoderms, and the loss of somites in tunicates.[11]

I should however add a note of caution. Molecular phylogenies are no panacea, sweeping in with results of uniformly objective solidity to replace centuries of morphology-based speculation. Some molecular phylogenies have produced uncomfortable results. While affirming deuterostome monophyly, they have challenged chordate monophyly by, for example, placing the amphioxus closer to echinoderms[12]; tunicates closer to hemichordates,[13] or otherwise a sister-taxon to all other chordates.[14] The current consensus[15] is shown in fig. 2.4.

Another feature of deuterostomes is a pronounced left-right asymmetry, especially in the disposition of internal organs. Although deuterostomes (and chordates in particular) look more or less bilaterally symmetrical from the outside, the arrangement of the internal organs is pronouncedly asymmetrical. Everyone knows that the human heart is on the left side of the body, not placed exactly in the middle, and so on. Occasionally this symmetry can be reversed in patients whose internal organs are a mirror-image of the usual situation, a condition called *situs inversus viscerum* or just *situs*.

Left-right asymmetry is determined by the expression in embryonic life of the signaling molecule *Nodal*, the transcription factor *Pitx*, and others.[16] There is some debate about how the genetic signals of *Nodal* are transduced, but these arguments are really beyond the scope of this book.[17] In echinoderms, *Nodal* signaling plays a part in the determination of the dorsoventral axis as well as left-right asymmetry. In vertebrates, it is involved in the specification of the anterior-posterior axis and other processes such as the formation of endoderm and mesoderm.

Nodal is secreted by tissues on the left side of the developing chordate embryo[18]; in echinoderms, in contrast, it is secreted from the right.[19] This accords with Geoffroy's suspicions, now backed up with molecular evidence (see chapter 4), that the chordates are "upside down" with respect to other animals. Although the *Nodal* pathway for determining left-right asymmetry does appear to have been conserved in deuterostomes,[20] it seems to play a role in symmetry throughout the Bilateria[21] and indeed the animal kingdom, popping up in the coiling of snails[22] and even as a determinant of biradial (as opposed to radial) symmetry in the cnidarian *Hydra*. As Watanabe et al. (2014) suggest, axial patterning in animals in general involves a core group of genes: *Nodal*, *Pitx*, and β-catenin. This has been co-opted and elaborated for symmetry determination in protostomes and more in-depth coordination and integration in deuterostomes.

Now we're ready to look in more detail at the various groups of deuterostomes.

CHAPTER 6

Echinoderms

The name "echinoderm" means "spiny-skinned," a concept that will be appreciated by anyone who ever seen or trodden on a sea urchin. Echinoderms are, and always have been, exclusively marine. Beach-combers will be familiar with starfishes or sea stars (asteroids), sea urchins (echinoids), sea cucumbers (holothurians), brittle stars (ophiuroids), and feather stars (crinoids) though probably not sea lilies, which, like feather stars, are crinoids, but sessile and stalked residents of the deep sea (fig. 6.1).

Such are the five modern classes of echinoderm, though the rich fossil record of the group reveals a diversity of form stretching back to the first good records of the group in the Cambrian Period. Approximately seven thousand species survive today.

Echinoderms are defined and constrained by a unique set of traits. The enormous gap between vertebrates and other animals seems as large as it does perhaps because we are vertebrates. Were we echinoderms, we'd be justified in staking an equal claim to uniqueness. Echinoderms have perhaps the most distinctive body plan of any animal group—so distinctive,

6.1 A pentameral arrangement of echinoderms. A: an echinoid, *Strongylocentrotus* (David Monniaux, Wikimedia Commons); B: an asteroid, *Solaster* (courtesy Becky Hitchin); C: a crinoid, *Antedon* (courtesy Becky Hitchin); D: an ophiuroid, *Ophiocomina* (courtesy Becky Hitchin); and E: a holothurian, *Holothuria* (Wikimedia Commons).

in fact, that their abundant fossil record doesn't shed as much light as one would think on their origins.

The most distinctive feature of echinoderms is an endoskeleton made of jointed plates, each one comprising a single crystal of calcite, a form of calcium carbonate. Under the microscope the structure is seen to form a distinctive mesh-like matrix called "stereom," a characteristic of the group found in all echinoderms, living or extinct, and nowhere else,[1] unless, as in some sea cucumbers, it is secondarily reduced or lost.

Given that the skeleton is found just beneath the skin, it's tempting to think of it as superficial. It is, in fact, mesodermal in origin. The skin covering is thin, ciliated, glandular, and punctuated by various structures including the tube-feet or "podia" and other outpocketings from the underlying coelom, and, in some starfishes and sea urchins, the peculiar scissor-like pedicellariae used to clear the skin of detritus and even catch prey. Under the skin but above the skeleton is a fine meshwork of nerves, the principal nervous system in echinoderms. With the possible exception of the ganglion at the head of the stalk in sea lilies, extant echinoderms have no clear central nervous system, and the existence of a central nervous system in any extinct echinoderm is highly debatable.

In echinoderms, the right and left metacoels are known as somatocoels and form most of the adult body cavity. A kind of torsion during development in which the anterior mouth migrates to the ventral surface leaves the originally right and left somatocoels lying one atop the other like the two layers of a sponge cake: the left somatocoel in the ventral (oral) part of the adult, the right in the dorsal (aboral) half.

Similar asymmetries are observed in the protocoels and mesocoels. The left protocoel (or axocoel) connects to the outside through a small canal and opening, the hydropore—and internally to the left mesocoel (hydrocoel) by a small, internal vessel, the stone canal. The left hydrocoel is the key to the final form of the radial symmetry of the creature, as it forms the germ of another distinctive echinoderm feature, the water-vascular system. During development, the left hydrocoel develops to form a complete ring around the mouth, budding off extensions into each of the arms, each of which buds off hundreds or thousands of paired tube-feet, or podia. These blind-ended coelomic sacs penetrate the skeleton and reach the surface in "ambulacra," arranged rather like the trees on each side of a French country road. In contrast, the right protocoel (or axocoel) and right mesocoel (the dorsal sac) are neither connected to the outside nor each other, and in some echinoderms do not form at all (fig. 5.1).

All modern echinoderms, of whatever class, have the distinctive calcite skeleton; the water-vascular system; the developmental peculiarity that converts a bilaterally symmetrical larva into a pentameral adult; and a more fundamental asymmetry in which the left protocoel and particularly the hydrocoel are massively elaborated, whereas the right protocoel and hydrocoel are vestigial or even absent. Can developmental genetics shed light on these issues? How much information can fossil forms reveal?

The calcite skeleton arranged as stereom mesh is the single most distinctive feature of all echinoderms, living and extinct. The processes of biomineralization that create the skeleton are regulated by genes found only in echinoderms, and entirely different from the genes that, in vertebrates, create tissues such as bone and dentine—which are in any case reinforced by calcium phosphate, rather than calcium carbonate.[2]

For all that echinoderm and vertebrate biomineralization are radically different, they might be united by deeper homologies.[3] Some proteins from sea urchins are sufficiently similar to vertebrate dentine matrix proteins that they can cross-react with antibodies raised against them.[4] The discovery of biomineralization in two enteropneust hemichordates, *Saccoglossus bromophenolosus* and *Ptychodera flava*, might shed further light on the earliest history of biomineralization in deuterostomes in general and echinoderms in particular. The biomineralization consists of the formation of minute ossicles of calcium carbonate. This differs from echinoderm biomineralization in many ways. Echinoderm calcium carbonate is magnesium-rich calcite, whereas in the two enteropneusts so far examined it is magnesium-poor aragonite. The aragonite structure bears only a passing resemblance to stereom. Nonetheless, three genes involved in enteropneust biomineralization seem to have homologs with biomineralization-related genes in the sea urchin *Strongylocentrotus purpuratus*.[5] Livingston and colleagues (2006) suggest that the common ancestor of deuterostomes had genes which, when expressed in collagen, had the ability to sequester calcium minerals—an ability subsequently expressed in different ways in the various deuterostome groups. It is still early days, but this work could help find a way to illuminate the presently murky field of the origin of mineralization.

As noted above the internal organs in deuterostomes show a pronounced asymmetry dictated by the activities of the asymmetrically expressed gene *Nodal*. In the sea urchin the expression of *Nodal* is reversed compared with that of vertebrates. *Nodal* signals expressed on the right side of the larva suppress the right coeloms from developing. Inhibition of *Nodal* signaling after gastrulation causes the right-side rudiment to develop, producing twinned urchins; similarly, expression of the gene *Pitx*—a component of the *Nodal* signaling pathway expressed on the left side in vertebrates—is expressed on the right side in echinoderms.

All extant echinoderms have pentameral symmetry, dictated by the formation of the water-vascular system that, once it emerges from the left hydrocoel, imposes such symmetry on the rest of the animal. Pentameral symmetry is thus intimately connected with the water-vascular system, and stages in its emergence can be seen in various fossil echinoderms. It is likely that echinoderms were primitively bilaterally symmetrical, as hemichordates still are, and that the present asymmetry has evolved by

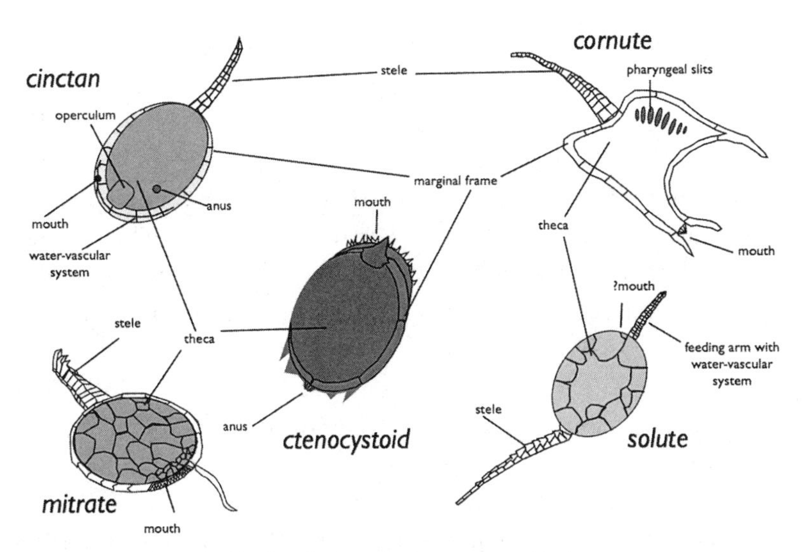

6.2 Cartoons of some of the fossil echinoderms mentioned in the text.

the co-option of the *Nodal-Pitx* gene regulatory circuit in a distinctive and idiosyncratic manner.

At the same time, fossil echinoderms (fig. 6.2) yield clues to the extinction of pharyngeal slits, otherwise ubiquitous in deuterostomes, but of which no trace exists in extant echinoderms.[6]

Ctenocystoids comprise a group of echinoderms from the Cambrian Period. They appear to have been bilaterally symmetrical, with a mouth at one end and an anus at the other, with little or no trace of either a water-vascular system or asymmetry in between. Zamora and Rahman (2014) suggest that they are the most primitive known echinoderms.

Cinctans, also known from the Cambrian, are bilaterially symmetrical creatures that look like tiny ping-pong paddles. A more-or-less circular body, framed by a number of large, marginal plates, is attached to a stiff stalk or "stele"—the handle of the paddle. A hole in a marginal plate at the opposite end to the stele is interpreted as the mouth. On either side of the mouth, engraved in the marginal plates, are grooves, possibly signs of what might be a lophophore-like filter-feeding arrangement rather than a true water-vascular system. The left-hand groove is longer than the one on the right—in some cinctans the right-hand groove is much reduced or absent—a sign of developing asymmetry. The lophophore of pterobranch hemichordates is derived from the mesocoels (hydrocoels), so to that extent the lophophore is a homologue of the echinoderm water vascular system (though the lophophore of brachiopods and so on is a convergent adaptation). The operculum, a large plate in the margin, is articulated as an outlet valve, and is best interpreted as a pharyngeal slit, or, possibly, the common outlet from an atrium as found in tunicates and the amphioxus. Indeed, fluid-dynamic simulations show that cinctans would have fared much better as active pharyngeal filter-feeders than as passive suspension feeders, so perhaps they depended much more on a deuterostome-style pharynx than on a lophophore or water-vascular system, even were these present.[7] Cinctans, then, are very primitive echinoderms—they have the characteristic calcite skeletons, but retain the pharyngeal system of deuterostomes, and the water-vascular system, if present, might have been more like that of a pterobranch and not much like that of an echinoderm.

Solutes are much more like modern echinoderms than cinctans, for all that they are not radially symmetrical. Two appendages are attached to opposite poles of a blobby, potato-shaped body. One is a stalk, and in some forms there are signs that solutes were attached to the substrate as juveniles.[8] At the other end is an arm, complete with an ambulacrum similar to that in modern echinoderms. Convention suggests that this arm would have been a left-sided structure, the water-vascular system on the right side having been suppressed. There are no signs of pharyngeal slits. Solutes, then, have a calcite skeleton and a water vascular system, but do not show signs of pentameral symmetry or of pharyngeal slits.

Somewhere between cinctans and the solutes are the stylophora, an assemblage of fossil echinoderms of such peculiar appearance that to this day there remains no agreement about the homologies of their appendages; which end is which; and even which way is up.

All stylophora have an irregularly shaped, plated "theca" or "body." Sometimes the skeleton is made of large, irregular plates, as in solutes; in other forms the body is clothed with a shagreen of smaller plates—perhaps flexible in life—supported by more robust plates at the margins, as in cinctans. Like cinctans, stylophora have a single appendage, and there is a large aperture at the other end of the animal. Even specialists on echinoderms disagree about the role of the appendage and its relationship with the aperture. Some think the appendage was a feeding arm, bearing some kind of ambulacrum and thus a water-vascular system, and the aperture at the other end was an anus. Others take—quite literally—a contrary view, that the appendage was some kind of stalk or locomotory organ; the aperture at the other end was the mouth; and that there was no water vascular system. Smith (2005) offers a guide to these interpretations and tends to favor the second view.

The cornutes, a subgroup of the stylophora, have the most irregular thecae, which are often perforated with an array of what look like excurrent openings, and which are now interpreted as pharyngeal slits. If they are, then the arrays represent only the left pharyngeal slits, demonstrating once again the suppression of right-hand structures seen in echinoderms generally.

The other subgroup of the stylophora, the mitrates, are more bilaterally symmetrical, at least superficially. They show no evidence for external pharyngeal slits, but there is compelling evidence for internal pharyngeal slits, protected by an atrium, in at least two genera of mitrate.[9]

Stylophora, then, have calcite skeletons and—at least following the interpretation of Smith (2005)—pharyngeal slits developed to some degree, and yet no convincing evidence for a water-vascular system.

Although solutes had a water-vascular system, they had just one feeding arm, so the animal was irregular, although on a bilateral plan. Some other extinct echinoderms had a triradial plan, notably the Cambrian helicoplacoids; curious, spindle-shaped creatures in which three ambulacra wound their way around the body in a screw-thread pattern.[10]

The evidence from these fossil forms, then, prompts us to cross a kind of bridge, helping us to explain the evolution of echinoderms by breaking down the acquisition and loss of features into a series of stages (fig. 6.3).

All echinoderms, living or extinct, have a calcite skeleton. The most primitive echinoderms were probably bilaterally symmetrical (ctenocystoids), although with a tendency to suppress right-hand structures to varying degrees (cinctans), a tendency completed in extant echinoderms. Some extinct echinoderms seem to have had a deuterostome-like system of pharyngeal slits, either debouching directly to the surface (cornutes), or protected by an atrium (cinctans, mitrates). The water-vascular system was likewise not universally present. Cinctans might have had a system of external tentacles instead, like a lophophore, and stylophora lacked a water vascular system altogether. Pentameral symmetry, as is found in all extant echinoderms and many extinct ones, evolved after the acquisition of the water-vascular system and the loss of the pharyngeal filter-feeding habit.

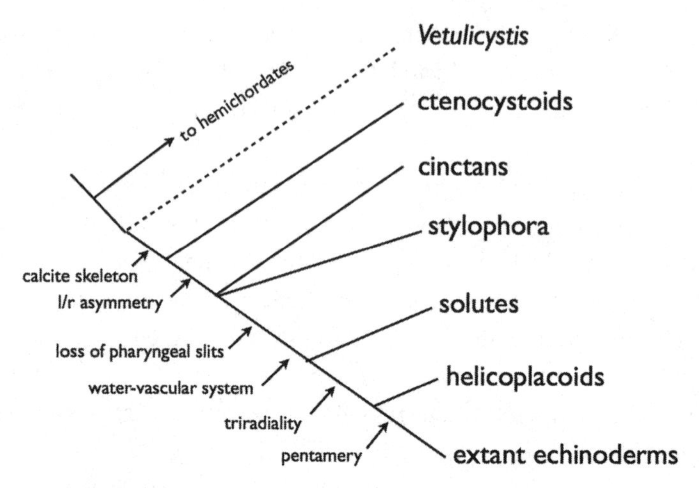

6.3 A cartoon phylogeny of echinoderms to illustrate the arguments in the text, showing how echinoderms evolved from unmineralized forms, through the loss of pharyngeal slits; the acquisition of a water-vascular system; to triradiality and finally pentamery. This diagram omits a wealth of fossil echinoderms not discussed here (cystoids, blastoids, edrioasteroids, etc.). It is not a formal cladistic analysis, and will in any case not be supported by everyone in the field.

There is, however, an elephant in the room—and that's what echinoderms were like before they acquired calcite skeletons. Of course, one must ask oneself whether such a thing is possible by definition. However, the echinoderm lineage must have acquired the trait at some time, so the question is legitimate. The essentially unmineralized Cambrian creature *Vetulicystis* might be just such a creature, though its status as a primitive echinoderm is, not surprisingly, controversial.[11] I discuss it further in chapter 14.

A related question is whether the calcite skeleton must necessarily define echinoderms exclusively. To continue with proboscidean-related metaphors—we know that all elephants have four legs, but that doesn't mean that all four-legged animals are elephants. We know that pharyngeal slits once had a wider distribution than they do now, so why could the same not be true for the calcite skeleton nowadays seen only in echinoderms? We know that hemichordates, at least, might share with echinoderms some genetic mechanisms predisposing to biomineralization[12]—in which case it is possible that the lineage leading to both echinoderms and hemichordates might have shown some degree of biomineralization.

It was this line of reasoning that led to one of the most audacious ideas of vertebrate origins—that the distinctive calcite skeleton of echinoderms might, once upon a time, have been found in chordates, which then lost them. This was Jefferies' "calcichordate theory," which for all its strangeness, resisted all serious attempts to dislodge it from its formulation in the late 1960s to the early 1990s.

Even though it is now known to be in error, the calcichordate theory contributed in no small measure to the modern interpretation of some fossil echinoderms as well as to present-day chordates. In its time it questioned the idea that the echinoderm skeleton, because it was found in modern echinoderms only, needn't have once had a wider distribution—a question for which there was no easy answer. It was also the only scheme of vertebrate origins that was based on fossils; accommodated otherwise mysterious phenomena relating to asymmetries in echinoderms and chordates; and pursued its aims with a frightening and remorseless erudition that few could match.

Several scholars had wondered whether the serial arrangement of what looked like pharyngeal openings in some cornutes, especially the Ordovician form *Cothurnocystis*, might not speak of a resemblance in at least some extinct echinoderms to chordates.[13] Where these authors had hinted toward resemblances, noting that some fossil echinoderms had chordate-like features, Jefferies reasoned that these resemblances could be taken at face value. These echinoderms resembled chordates because they really were chordates.

In a long series of papers starting in 1967 (the findings of which are summarized in a book in 1986), Jefferies placed cornutes in the ancestry of chordates in general, with mitrates being assignable to the ancestries of either the amphioxus, tunicates, or vertebrates. In the late 1980s and early 1990s Jefferies turned to solutes,[14] suggesting that they formed an assemblage standing at the root of both echinoderms and chordates, with some being on the stems of each.

The obvious objection to this idea is that the ancestry of chordates requires no fewer than three independent losses of the calcite skeleton—in the ancestries of the amphioxus, tunicates, and chordates. Whereas this might seem unlikely, evolution has no requirement to be parsimonious, least of all for our benefit. However, a more serious challenge came from Peterson, who reasoned that the particular phylogeny of deuterostomes required by the calcichordate theory was not easily generated even when the fossils were coded using Jefferies' own characters.[15] It was no surprise that Jefferies responded to state that, au contraire, his characters predicted the phylogeny he said it did.[16] And as Smith (2005) notes, Peterson's critique doesn't imply that some of Jefferies' anatomical reconstructions of cornutes and mitrates need be completely wrong. It would be hard to argue that some mitrates didn't have a pharyngeal filter-feeding system very much like that of a tunicate,[17] even though the void in the base of the stylophoran "tail" Jefferies reserved for a fish-like brain would be more plausibly reconstructed as accommodation for muscle.[18]

Like all the best theories, the calcichordate theory succumbed only to the discovery of new evidence from realms unexplored at its inception, namely, new discoveries in molecular and cell biology. To my mind, the presence in modern echinoderms of genes relating to biomineralization,[19] of which no trace is known to exist in chordates, speaks strongly against the calcichordate theory, and I shall not discuss it further here.[20]

The genomes of most animals appear to have at least one cluster of homeobox or Hox genes, one of whose roles is to specify regional identity in the anterior-posterior body axis (see chapter 4). Echinoderms are no exception, and appear to have a cluster of Hox genes similarly disposed to that found in chordates, for all that the body plan of extant echinoderms is radically (and radially) altered from the ancestral bilateral pattern.[21] The expression of these genes, however, is very different from that in chordates. Rather than being expressed in neural or epidermal tissues, the Hox genes of the sea urchin *Strongylocentrotus* at least are generally expressed, in a spatially sequential and dynamic way, in the developing somatocoels of the adult, although only one gene is expressed in the mesocoel-derived pentameral rudiment itself. The absence of Hox expression in echinoderm nerve cords is especially noteworthy, as it suggests that echinoderm nerve cords, where present, are not homologues of nerve cords in chordates.[22]

CHAPTER 7

Hemichordates

The hemichordates include the solitary enteropneusts and the small, tentacled, and mainly colonial pterobranchs.[1] Enteropneusts (fig. 7.1) are worms usually found burrowing in sand, gravel, or mud on the floors of shallow seas.

There are 110 or so known species[2] but given the retiring habit of these creatures it is likely that many remain undiscovered. After all, it is not as if they fly squawking from tree to tree, showing off their iridescent plumage. This potential for discovery is illuminated by the recent discovery of the Torquaratoridae, an entirely new family of deep-sea enteropneust, the description of which refuted the earlier idea, based on very poor photographs, that so-called "lophenteropneusts"—strange forms of tentacled enteropneust—lurked, monster-like, somewhere in the deeps.[3] There are parts of the ocean floor that we probably know less well than the surface of Mars, and still scope for us to say "here be dragons," even if only of a rather soft and squishy kind. As presently constituted, there are four families of

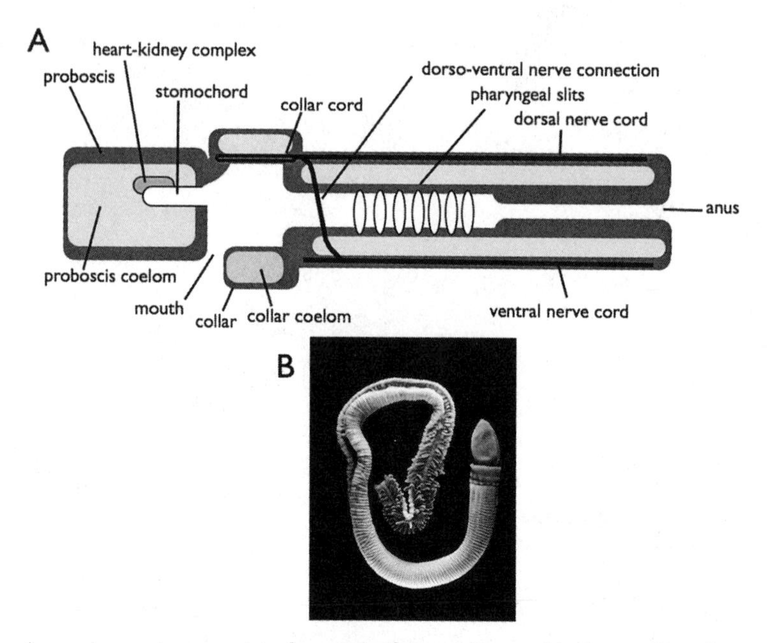

7.1 A: a cartoon enteropneust to show major features discussed in the text. B: a photograph of an adult enteropneust, *Schizocardium* (courtesy Paul Gonzalez).

enteropneust—the Harrimaniidae, Spengelidae, Ptychoderidae, and the above mentioned Torquaratoridae.

Unlike echinoderms, all enteropneusts are bilaterally symmetrical. They may range in length from less than a millimeter[4] to more than two meters. Harrimaniids develop directly, hatching from the egg as miniature worms with little left to do but grow; spengelids and ptychoderids, though, have a planktonic larva called a tornaria. The developmental mode of torquaratorids remains unknown. It was the resemblance of the tornaria to some of the planktonic larvae of echinoderms that prompted Metchnikoff, in 1881, to propose a close relationship between these two otherwise disparate phyla. On the other hand, it was the presence in the adults of large numbers of prominent pharyngeal gill slits that first suggested a link with chordates.[5]

In many ways, enteropneusts are the pivot around which the other deuterostomes revolve,[6] possibly because they appear to retain without too much alteration many of the features now regarded to be characteristic of deuterostomes in general, such as pharyngeal slits and the tripartite organization of the coelom. Such features are mirrored at the molecular level[7] such that hypotheses of vertebrate and chordate origin have now moved from once-fashionable considerations of how adult forms might have developed from pedomorphic larvae[8] to more testable ideas based on adult forms.[9]

Enteropneusts show the tripartite coelomic organization of deuterostomes at its most straightforward. At the most anterior is a single, unpaired protocoel that supports a very flexible, muscular prosome, or proboscis. The worm uses this both to dig with and to gather food. A coating of cilia transports the food backward to a mid-ventral mouth. Enteropneusts gather food in either of two ways. That which it doesn't manage to gather by deposit-feeding with the proboscis, it filters from water passing through its array of paired pharyngeal pores, arrayed on the dorsolateral surfaces of the trunk.[10]

The mouth sits at the ventral margin of the collar region, or mesosome. The collar, supported by paired mesocoels, appears to support the proboscis as an acorn is supported in its cup, leading to the vernacular name of "acorn worms" for these animals. The collars of most enteropneusts are of a diameter similar to that of the proboscis and trunk. In torquaratorids, however, the collar can be disproportionately wide, almost forming wing-like extensions like the lapels of shirts worn by 1970s movie gangsters. It was this width that suggested the presence of a lophophore-like structure when all that was known of these worms came from poor-quality photographs. The metasome or trunk—that is, the elongate hind section of the enteropneust body—is perforated by paired pharyngeal slits and supported by paired metacoels. The gut runs straight through the mesosome and metasome to a terminal anus.

This simplicity is deceptive, because enteropneusts have many peculiarities. One would expect that in a plain and unadorned deuterostome the coeloms would all originate by enterocoely, and that is indeed the case in enteropneusts. All except, that is, for the pericardial coelom, a small and, perhaps, sixth coelom within the proboscis, which is formed by schizocoely, splitting from the ectoderm.[11] The pericardial coelom contains a pulsatile vesicle called the heart, associated with a glandular complex known as the kidney. Although enteropneusts do not have a "closed" circulatory system, the heart pumps blood into the kidney—really a collection of blood vessels and filtering cells known as podocytes, embedded in an extracellular matrix. This produces a urine that leaves the worm through the protocoel pore.[12] The arrangement of heart, kidney, and excretory pore is reminiscent of the axial complex associated with the hydropore, a similar structure found in echinoderms, and presumably homologous with it.

The heart-kidney complex is supported by a short forward extension of the pharynx called the stomochord. It was the seeming resemblance of the stomochord to the chordate notochord that prompted Bateson (1886) to align enteropneusts with chordates and coin the term "hemichordate."[13] The resemblance between the two is striking, in that the cells in both structures are enlarged by internal spaces or vacuoles, the whole being bounded by a sheath, as a sausage is bound by a thin yet tight skin.

On the other hand, the stomochord of hemichordates is unlike the notochord of chordates, in that it does not seem to be necessary for the formation of the overlying neural tube. A direct comparison between neurulation (the development of the nerve cord in embryogeny) in chordates and in the indirect-developing enteropneust *Ptychodera flava* shows that, in the latter, the process happens in the absence of a notochord.[14] In addition, the chordate notochord expresses a panoply of transcription factors essential for notochord development in chordates that are not seen in stomochord development. One of these is called *brachyury*, and although it is expressed at various times in the development of *Ptychodera*, in various embryonic tissues, it never appears in the stomochord.[15]

If the stomochord is not a homologue of the chordate notochord, might it be a homologue of anything else? Several factors, including its location in the anteriormost part of the endoderm; its association with tissue of a glandular nature; and the presence in the stomochord-forming region of a transcription factor called *FoxE*, suggest that it might indeed be a homologue of the endostyle, the gutter running along the ventral edge of the pharynx in chordates. *FoxE* is likewise found in the endostyle of ascidians and the amphioxus and in the homologous thyroid gland of vertebrates.[16]

There is much less doubt about the series of slits perforating the the trunk of enteropneusts. These structures are supported by cartilage of endodermal origin. They look very similar to the pharyngeal slits in the amphioxus, a resemblance that goes down to the molecular level.[17] A cassette of genes associated with the development of pharyngeal slits has been found in the genomes of two enteropneusts.[18] It is likely that pharyngeal slits of more or less this sort were characteristic of the earliest deuterostomes.

The hemichordate nervous system resembles that of echinoderms, in that it largely consists of a net or plexus of nerves just beneath the skin. This plexus is particularly dense in the proboscis, as one would expect for the probing, mobile nature of this organ. Enteropneusts, however, have, in addition, two very prominent nerve cords. One is ventral, and runs along the ventral midline of the trunk. The other is dorsal, and runs backward along the dorsal midline from the collar. The two are linked by lateral nerve rings in typical invertebrate style.

There has, however, been a great deal of discussion about the collar cord, that is, the region of the dorsal cord in the collar, as this is hollow and forms by a process that resembles the formation of the dorsal nerve cord in chordates. To what extent might the collar cord be a homologue of the characteristic dorsal, hollow nerve cord of chordates? The formation of both structures involves homologous patterning genes, and the developing collar cord is responsive to patterning signals from underlying endoderm in a way reminiscent of the role of the notochord in chordates.[19] The collar cord does seem to be a centralized structure, to be thought of as something more than just a localized condensation of an epidermal nerve net,[20] and it is possible that some form of nervous-system centralization runs deep within deuterostomes more generally.[21]

But hold: enteropneusts have two nerve cords, not just one. Neither are they afterthoughts, as condensation of neural markers to the ventral and dorsal sides happens very early in enteropneust development.[22] Some molecular markers of what might be thought as chordate neurulation are found not in the dorsal cord of enteropneusts, but in ventral tissue. Other neural markers are expressed on the dorsal side, but not just in neural tissue.[23] This finding raises two broader issues: one, the homology of body axes between hemichordates and chordates, and, second, whether markers associated with the nervous system in chordates need be so restricted in hemichordates.

The anterior-posterior body axis of enteropneusts is homologous with that of chordates, inasmuch as the same genes that specify anterior-posterior structure in hemichordates are deployed in chordates, and in the same order along the axis. The big difference, though, is that in chordates they are generally expressed in the central nervous system, whereas in enteropneusts they are expressed in the ectoderm, in circumferential rings around the animal.[24] Ectodermal expression of genes found in the chordate central nervous system is known in the amphioxus and tunicates, as well as protostomes such as the fruit-fly *Drosophila*,[25] suggesting that the "skin brain"[26] of enteropneusts is an expression of an ancient bilaterian feature. This lends weight to the idea that much of the pattening of the chordate central nervous system is specific to chordates.

Lowe and his colleagues have also found that specific markers associated with the nervous system in chordates, including important centers in the developing brain that "organize" surrounding tissue called the Anterior Neural Ridge (ANR, at the most anterior), the Zona Limitans Intrathalamica (ZLI, in the diencephalon), and, behind that, the Midbrain-Hindbrain Boundary (MHB), are expressed throughout the epidermis in hemichordates. Perhaps surprisingly, some of these markers are not expressed in amphioxus and tunicates, suggesting, first, that some of the key organizing principles of the chordate brain are basal deuterostome in origin; and, second, that some have been secondarily lost in the amphioxus and tunicates.[27] All this suggests that whereas there might be a tendency toward central-nervous-system formation in deuterostomes, there is no easy equivalence to be drawn between either of the hemichordate nerve cords—or even the epidermal nerve net—and the central nervous system of chordates.

Lowe et al. (2006) find—among other things—that molecular markers such as *chordin*, associated with the dorsal side of the animal in chordates, are expressed in ventral tissue in hemichordates—as it is in the fly *Drosophila*. *Chordin* is an antagonist of BMPs (Bone Morphogenetic Proteins), ligands of the diverse Transforming Growth Factor Beta (TGFβ) family. BMPs are associated with the ventral side of chordates, but the dorsal side in flies—and hemichordates.[28] This means that the inversion of the dorso-ventral body axis predicted by Geoffroy (see chapter 4) must have happened in the ancestry of chordates alone, after they had diverged from hemichordates and echinoderms. This inversion would, if otherwise unmodified, have placed the chordate mouth on the dorsal surface, so a ventral migration of the mouth—or the creation of an entirely new mouth—must also have happened during chordate evolution.

Signaling with the *Nodal* pathway specifies left-right asymmetry in protostomes, but does much more in deuterostomes. In vertebrates it has been co-opted into endoderm and mesoderm formation as well as all three body axes. *Nodal* is also involved in dorso-ventral and left-right axis specification in echinoderms,[29] though the extent of its involvement in hemichordate dorso-ventral axis specification unclear.[30] Recent work on the enteropneust *Ptychodera flava* does, however, show its involvement in mesoderm formation and potentially anterior-posterior patterning.[31]

The anterior-posterior axis of enteropneusts is set down very early in embryo formation by a "posterior organizer" involving the canonical β-catenin/*Wnt* signaling pathway. In *Saccoglossus kowalevskii*, β-catenin accumulates at the vegetal pole, the future site of gastrulation.[32] This is required for specifying endoderm, and, later, mesoderm. On this specification depends, in turn, the determination as anterior or posterior of the adjacent ectoderm, and thus the entire frame of the animal in which (say) neural-specific markers are expressed. The definition of a posterior organizer and the involvement of the β-catenin/*Wnt* signaling pathway have been observed in echinoderms[33] and vertebrates,[34] suggesting conservation across deuterostomes.

Another signaling system, based on the Fibroblast Growth Factor (FGF) family, is known to be essential for mesoderm formation in chordates. Work on *S. kowalevskii* has shown involvement of various members of the FGF family in specification of mesoderm, so it could be that mesoderm specification by FGF signaling is another deuterostome-wide feature.[35]

At the time of writing, the draft genomes of two enteropneusts have been published.[36] They are of the harrimaniid *Saccoglossus kowalevskii*, which develops within a few days directly from an egg; and the ptychoderid *Ptychodera flava*, which spends some months in the plankton as a larva before metamorphosis. The essential difference between the two modes of development is that whereas the direct-developing *Saccoglossus* lays down posterior structures early in its embryonic life, the tornaria larva of *Ptychodera* is essentially composed of anterior structures—aside from a complete gut—and the posterior parts of the animal form quickly after metamorphosis.[37] In this way the larval forms have much in common with larval forms of other deuterostomes such as the dipleurula of some echinoderms and the tadpole larvae of tunicates, in that they are, in essence, animated front ends, waiting for the back ends to catch up.[38] A comparison of gene expression in the direct-developing *Saccoglossus* with the indirect developer *Schizocardium californicum* shows that whereas the larval ectoderm expressed transcription factors characteristic of the front end, expression of Hox genes that would define the trunk is absent.[39] This division between head and trunk seems to betray tendency in deuterostomes. Later in the book, I speculate that the invention of a trunk with somitic patterning was key to the evolution of chordates, and that early stages in the process can be seen in some fossil forms, notably vetulicolians and yunnanozoans (see chapter 14).

Despite their different modes of development, the genomes of *Saccoglossus* and *Ptychodera* are very similar and reveal a great deal about what the earliest deuterostomes were like, before they diverged into echinoderms, hemichordates, and chordates. The species have identical Hox clusters, each containing twelve genes. The organization of the cluster is more like that of chordates than echinoderms, though the hemichordate clusters contain three genes specific to the Ambulacraria.[40] More broadly, the genomes of the two worms share the traces of seventeen conserved linkage groups previously identified in chordates,[41] broadening this feature of the genomic landscapes from chordates to deuterostomes generally.

Perhaps the most notable finding revealed by these genomes is a tightly linked cluster of six genes, found in other deuterostomes and which is involved in the patterning of the pharyngeal slits. Four of the genes encode transcription factors, *nkx2.1*, *nkx2.2*, *Pax1/9*, and *FoxA*, and the two others contain elements that regulate the expression of *Pax1/9* and *FoxA*. These genes are expressed in the pharyngeal slits and adjacent endoderm, and are found, in the same tightly ordered cluster, even in deuterostomes in which pharyngeal slits are embryonic and transient and do not persist to adulthood (such as humans) or are not formed at all (such as the crown-of-thorns starfish *Acanthaster planci*). There is no trace of this cluster in other bilaterians, suggesting that it is a deuterostome novelty. The gene *Pax1/9* had previously been found to be connected with pharyngeal development in deuterostomes.[42]

Among several other notable findings, the two hemichordate genomes[43] reveal that stem-deuterostomes may have acquired several genes by horizontal transfer from microbes. Some of these encode enzymes involved in the manipulation of sialic acid, a molecule often used to modify the properties of glycoproteins. These sugar-rich proteins are found in mucus, which forms the basis of the cilia-based feeding found in hemichordates and also the amphioxus. This discovery underlines the suggestion that such pharyngeal-based feeding is a fundamental feature of deuterostomes.

The fossil record of enteropneusts is sparse, as one would expect for soft-bodied creatures, and those that exist show enteropneusts not that different from those that exist today, shedding little light on their evolution. *Spartobranchus* and *Oesia*, two fossil enteropneusts from the Middle Cambrian Burgess Shales of Canada, are associated with horny tubes, suggesting a link with pterobranchs.[44]

At first glance, pterobranchs seem polar opposites to their enteropneust cousins. Where enteropneusts are (mostly) large and solitary, live in shallow seas, and have a meager fossil record, pterobranchs are (mostly) tiny and colonial, live at great depths, and have a spectacular fossil record. As it turns out, enteropneusts and pterobranchs are more similar than they first appear. One might think of them as two sides of the same coin.

There are twenty-five or so species of pterobranch, grouped into three genera, *Cephalodiscus*, *Rhabdopleura*, and *Atubaria*. They are generally small (millimeters to a few centimeters) and mainly live in colonies in a system of secreted, proteinaceous tubes called a coenecium, where they filter material from seawater with paired arrays of tentacles, known as the lophophore. Like enteropneusts, they are creatures of three parts: a flexible prosome or "mouth shield" at the front is used mainly as a locomotory organ, and also secretes the coenecium. The lophophore is an extension of the mesosome. The metasome is bulbous, is covered in cilia, and contains the digestive system. The body may terminate with a long stalk and a sucker used to anchor the animal to the inside of the coenecium. In contrast with the straight gut of enteropneusts, the pterobranch gut is U-shaped, the anus emerging just behind the mesosome on the dorsal side. Pterobranchs have a stomochord and an excretory system in the prosome similar to enteropneusts, and, in *Cephalodiscus* and *Atubaria*, a single pair of pharyngeal slits. *Rhabdopleura* has no pharyngeal slits at all. The nervous system is a simple net, although it contains a brain of sorts in the mesosome at the base of the lophophore; nerve tracts along each tentacle and the ventral margin of the stalk; and, in *Cephalodiscus*, a pair of branchial nerves.[45]

As enteropneusts were once thought creatures of inshore waters, only lately found to inhabit the deeps, pterobranchs were once thought typical residents of deep, cold water, and have more recently been found in shallower and warmer seas.[46] *Rhabdopleura compacta* is found off the south coast of England, where it colonizes disarticulated mollusk shells, especially those of the dog cockle *Glycimeris glycimeris*. A relative, *R. normani*, can be found off Bermuda. They had avoided detection presumably because they are small and possibly easy to confuse with other small, tentacled or tube-dwelling creatures such as bryozoa. Nevertheless, this accessibility has meant that they are now easier to study than heretofore, although there is still much to learn.[47]

Pterobranch development appears to be uncomplicated. Cleavage is radial; the mesoderm is formed by enterocoely; and development produces small, ciliated larvae that develop directly into adult zooids.

The retiring nature of pterobranchs belies their fossil history, for they are no less than the modern representatives of the graptolites,[48] colonial organisms common in the oceans between the Cambrian and the Carboniferous and with a rich fossil record.[49] The fossils themselves are the carbonized films of coenicia, the zooids being preserved only extremely rarely.[50]

The interrelationships of hemichordates are currently a matter of some debate. In general, analyses support the classical position, that enteropneusts and pterobranchs are reciprocally monophyletic sister taxa.[51] However, some other work suggests that pterobranchs emerge from within the enteropneusts[52] and in particular might be closest to the harrimaniids.[53]

CHAPTER 8

Amphioxus

A small group of animals known variously as acraniates, cephalochordates, or lancelets constitutes the most basal known living chordates. The 25 or so known species[1] are all very similar, and I shall refer to them collectively as the amphioxus, the vernacular name[2] for historically the most familiar form, the European species *Branchiostoma lanceolatum*. "Branchiostoma" means "gill mouth." "Amphioxus," however, means "pointed at both ends," and that's possibly the first impression one will have on seeing one of these creatures.

The amphioxus can be found in tropical and temperate shallow seas throughout the world. In China one is more likely to come across *B. belcheri*; in Japan, *B. japonicum*; in Florida, *B. floridae*. *Branchiostoma* is up to about six centimeters long as an adult (*Asymmetron* is much smaller, less than 3 cm), and looks vaguely fish-like until you notice that although it has fins in the midline, it has neither a distinct head nor any visible organs of special sense such as eyes. Its translucence suggests that the skin is not covered with scales or any other thick integument. The segmented muscle blocks below the surface speak of a closeness to vertebrates, however, and it looks like a very small fish fillet.

Although it can swim rapidly if disturbed, the animal spends most of its time buried tail-first in sandy seafloor sediment, the front end poking out just above the surface as it filters particles of detritus from seawater. The mouth is guarded by a fringe of tentacles or cirri, which guard against the ingestion of overly large particles. Inside the mouth, the water is driven by ciliary action through mucus-lined pharyngeal slits. The slits are very similar in construction to those of enteropneusts and mark the amphioxus as a card-carrying deuterostome. The amphioxus pharynx, however, carries the additional refinement of the endostyle, a definitive feature of chordates. The pharynx, in the adults at least, is protected by an atrium, as in tunicates and the larval lamprey, and water leaves the animal through a median atriopore at the posterior end of the atrium.

The chordate heritage of the amphioxus is demonstrated most dramatically by a prominent notochord that is present in the animal throughout its life, but, uniquely in chordates, extends all the way from one tip of the animal to the other.[3] As in other chordates, the amphioxus notochord is formed from a stack of expanded, vacuolated cells held within a fibrous skin, although, unlike vertebrates, the notochord of the amphioxus also contains muscle.

The formation of the notochord in vertebrates is connected with the expression of the gene *brachyury* in the presumptive notochord as well as in posterior mesoderm, a pattern also seen in the amphioxus. The expression of *brachyury* in tunicates, however, is restricted to the notochord.[4] The amphioxus has two divergent copies of the *brachyury* gene, products of a recent, amphioxus-specific duplication.[5] Both are expressed in the notochord, though it is not clear whether this peculiarity is connected with the muscular nature of the notochord or its extension to the extreme anterior end. The *hedgehog* family of signaling molecules is important in vertebrates. One of them, *Sonic hedgehog* (*Shh*), is expressed in the notochord, the neural tube, and the left side of the developing body.[6] The amphioxus has a single hedgehog gene, *AmphiHh*, expressed in similar places,[7] illustrating the homology of the notochord and neural tube with those of vertebrates as well as pointing up, once again, asymmetrical tendencies in deuterostome development.

As in other chordates, the notochord is surmounted by a single, tubular nerve cord. And, as with the notochord, this extends from one end of the animal to the other. At the anteriormost end is a simple light-sensitive eyespot, set just before the cephalic vesicle, a very slight swelling of the nerve cord.

The mesoderm on either side of the notochord—that is, the paraxial mesoderm—is divided into a series of segments or somites. It is this segmentation that perhaps betokens the greatest similarity of the amphioxus with vertebrates. However, much of the elaboration of the vertebrate body, especially in the head, comes from the formation of an embryonic tissue called neural crest. Little or no sign of this exists in the amphioxus.

The amphioxus genome looks very much like the vertebrate genome in its general features and many regions are syntenic—that is, there are extensive regions in the vertebrate and amphioxus genomes in which cognate genes are more or less contiguous and arranged in the same order, suggesting that the genomic landscape of the vertebrate genome is very ancient and highly conserved. More than 90% of the human genome is encompassed within 17 ancestral chordate linkage groups revealed in the amphioxus genome. This seems all the more remarkable given that the ancestries of the amphioxus and the vertebrates diverged some 550 million years ago.[8] Even compared with vertebrates, though, the amphioxus genome is evolving very slowly: more, even, than the slowest-evolving known vertebrate genome, that of the elephant shark *Callorhinchus milii*.[9]

The main difference between the amphioxus genome and that of vertebrates is that the latter has undergone two rounds of whole-genome duplication some time in the evolution of vertebrates from the other chordates.[10] That the amphioxus genome has retained a simple, pre-duplication configuration is shown by the presence of a single *Hox* cluster,[11] found on four separate chromosomes in humans. In general, where amphioxus has a single gene, vertebrates have two, three, or four "paralogs"—that is, more or less divergent copies. Amphioxus and vertebrates retain many transcriptional enhancers, evidence of many features of development held in common. In contrast, the amphioxus genome exhibits elaborations all its own, particularly in genes connected with immunity. Whereas vertebrates have an elaborate system of acquired immunity (more on this in chapter 13), the amphioxus has a surprisingly complex system of innate immunity.[12]

The simplicity of the nervous system, combined with the extension of the notochord to the anteriormost tip, and the absence of neural crest, once gave rise to the idea that the amphioxus lacked any structure corresponding with the vertebrate head, and, conversely, that the head was the most obvious defining feature of vertebrates.[13] However, the simplicity of the front end of the amphioxus is more apparent than real. Studies on gene expression combined with highly detailed mapping of amphioxus neuroanatomy show that detailed comparisons can be made between the front end of the amphioxus neural tube and the central nervous system of vertebrates.

The vertebrate brain, like Gaul, is divided into three parts: the forebrain, further divided into anterior telencephalon and posterior diencephalon; a midbrain; and a hindbrain. The central nervous system of the amphioxus, though superficially simple, is more complex and vertebrate-like than it first appears, and work comparing the relative antiquity of genes expressed in different brain regions shows that much of the overall structure of the vertebrate brain has deep chordate roots—the main exception being the dorsal parts of the telencephalon.[14] Indeed, no structure comparable with anything found in the vertebrate telencephalon,[15] such as olfactory bulbs, is detectable in the amphioxus. Evidence for the equivalent of a diencephalon is strong. One such is the presence of markers for the infundibulum,[16] the downward extension of the hypothalamus—a part of the diencephalon that in vertebrates connects to the posterior part of the pituitary gland. Another structure, the lamellar body, may correspond with another diencephalic structure in vertebrates, the pineal gland or "third eye."[17]

Many genes in the central nervous system of vertebrates, most notably the Hox genes, have very specific anterior boundaries of expression within defined neural segments, known as rhombomeres, in the developing brain and nerve cord. Comparisons of gene expression in the amphioxus with established patterns in vertebrates has revealed a far greater degree of structure in the amphioxus cerebral vesicle and nerve cord than had at first been apparent. *AmphiHox3*, the first amphioxus Hox gene to be studied in depth, is expressed in the amphioxus nerve cord, with an anterior expression boundary corresponding to the boundary between the fourth and fifth somite. Comparison with the cognate mouse gene, *HoxB3*, showed that the cerebral vesicle and parts of the nerve cord corresponded with the vertebrate brain.[18] Later studies showed that the developing cerebral vesicle expressed genes characteristic of the vertebrate forebrain.[19]

The vertebrate brain has three organizing centers; the Anterior Neural Ridge (ANR), Zona Limitans Intrathalamica (ZLI), and Midbrain-Hindbrain Boundary (MHB). Gene expression studies show that the amphioxus central nervous system has regions that might correspond with the ANR, ZLI, and MHB, though they do not appear to act as organizers.[20]

Like the rest of the nervous system, the eyespot at the most anterior end is more vertebrate-like than it first appears. It expresses genes characteristic of the vertebrate retina[21]; its ultrastructure is similar enough to the paired eyes of vertebrates to be considered homologous; and it connects with a light-detecting center in the forebrain marked by the expression of *Pax-6*, as seen in the paired eyes of larval lampreys.[22] As mentioned above, another photoreceptor, the lamellar body, may be homologous with the pineal gland or "third eye" of vertebrates.[23]

Anatomical studies of the cerebral vesicle, dissected cell-by-cell in heroic detail by Thurston Lacalli and colleagues,[24] show that the cerebral vesicle is equivalent to the diencephalon and part of the midbrain sufficient to define a distinct midbrain-hindbrain boundary; a distinct hindbrain; and a spinal cord. Behind the amphioxus equivalent of the brain, motor neurons are apparently arranged on either side of the midline, corresponding with the somites. But rather than nerve tracts extending to the musculature, the muscles have "tails" that contact the nerve cord directly, a system apparently unique to the amphioxus.[25]

In summary, gene expression studies and detailed neuroanatomy show that the amphioxus central nervous system, far from being simple and primitive, has regions corresponding with the diencephalic part of the vertebrate forebrain, as well as a midbrain, a hindbrain, and a distinct spinal cord. It is in many ways a vertebrate central nervous system writ small. Lacalli's detailed cell-by-cell reconstruction of the amphioxus central nervous system was made possible only by the fact that the larval amphioxus nervous system comprises only a few hundred neurons (20,000 in the adult; compare with the 86 billion or so on the human brain).[26]

Early development of the amphioxus is typically deuterostome.[27] Cleavage of the fertilized egg is radial, and the blastopore is the presumptive anus. The dorsal cells of the inner layer of the gastrula become the mesoderm. The notochord forms by enterocoelous infolding of this layer of cells; somites start to pinch off from grooves on either side, starting at the front; and the overlying neural plate starts to flatten. Ectodermal cells on either side of the neural plate migrate over the top, fusing at the midline, after which the neural plate rolls up to become the nerve cord. The animal hatches from the egg at about the three-somite stage.

A key chordate character is the presence of a structure in the dorsal lip of the blastopore known as the "node" or "Spemann Organizer"—a region which, when transplanted, can direct the organization of an entire body region. This was first defined in vertebrate embryos when the dorsal lips of blastopores of early amphibian embryos, when transplanted, could reprogram cells in the new host to form an entire body axis.[28] I discuss the Spemann Organizer in detail in chapter 12. The blastopore lip in the amphioxus also appears to have some properties of an organizer, at least inasmuch as it expresses some of the same genes, such as *Nodal*, *Lefty*, some genes in the BMP family, and so on, consistent with the presence of an organizer involved in the specification of the major body axes deep in chordate ancestry.[29]

Anterior-posterior patterning is also governed by *Wnt*/β-catenin and retinoic acid signaling,[30] a feature probably characteristic of chordates in general. *Wnt*/β-catenin expression specifies posterior fates for the tissues in which these genes are expressed; it is the suppression of their expression that helps define anterior fates. Cells along the anterior-posterior axis gauge their position according to a gradient of retinoic acid that regulates *Hox* genes and other genes involved in the fate of cells along this axis.

The stage following gastrulation is called neurulation, when the young animal hatches; the neural plate rolls up to form the neural tube; and the animal lengthens with the formation of somites. In the amphioxus, the somites start to form from the anterior by enterocoely until there are between eight and twelve pairs. After that, somites form from the tail-bud at the posterior end. These somites, unlike the ones that form at the anterior end, form by schizocoely. In vertebrates, by contrast, all somites emerge from the tail bud, probably representing the ancestral chordate condition,[31] and extend only as far forward as the hindbrain. Mesoderm in the vertebrate head remains unsegmented.[32]

The amphioxus also lacks the neural crest. This isn't to say that the roots of the neural crest cannot be found in the amphioxus; the neural plate border of the amphixous expresses genes central to neural crest migratory activity in vertebrates.[33] A key difference, however, concerns the expression of the gene *FoxD*. The amphioxus has a single *FoxD* gene: the series of genome duplications in the vertebrate ancestry created an entire *FoxD* family, one of which, *FoxD3*, became essential to specifying the properties of neural crest.[34] In addition to the neural crest, vertebrates have a number of sensory placodes—patches of superficial cells that migrate internally, emerging elsewhere on the surface where they develop as sensory structures and connect with the nervous system. The amphioxus appears to have sensory-cell precursors that behave in a way strongly reminiscent of vertebrate sensory placodes and they may well share an evolutionary origin.

The larval stage of the amphioxus is characterized by gross asymmetries that, for the most part, are corrected as the animal metamorphoses into the adult.[35] The larval stage begins when the mouth opens—not front and center, but on the left side of the animal. This opening is in fact the first left gill slit.[36] The second gill slit appears in the ventral midline but migrates to the right side, followed by 12–15 more, which do the same. Once these slits have formed, one at a time in due sequence, the entire array keel-hauls itself as a unit, moving ventrally and coming to rest on the left side. A new series of gill slits then develops, again as a body, on the right side, replacing the ones that have migrated; and the aberrant, large, and left-sided mouth becomes smaller and moves to its final position, at the front of the animal. Thus by these strange and still largely inexplicable evolutions two seemingly symmetrical series of gill slits are produced.

Another asymmetrically placed structure is the larval kidney or Hatschek's nephridium, found on the left between the ectoderm and the most anterior somite. Hatschek's nephridium is a serial homologue of more posterior, paired nephridia, each associated with a gill slit.[37] Also on the left side and opening to the environment is the ciliated pit. This fuses with Hatschek's diverticulum, a leftward outpocketing from the gut, an event reminiscent of the fusion of the developing neural tube and Rathke's Pouch—an outpocketing of the roof of the mouth—which, in vertebrates, contributes to the adenohypophysis, the anterior part of the pituitary gland. The homology of the ciliated pit with the pituitary is supported by the expression of the gene *Pax6* in the ciliated pit and pituitary alike.[38] It is interesting that the amphioxus equivalent of the anterior pituitary develops on the left, a feature held in common with the tunicate *Ciona* and connected with the expression of the gene *Pitx*, which as we have seen is involved in the specification of left-right asymmetry.[39]

The endostyle, in contrast, develops on the right side of the larva, moving to the floor of the pharynx at metamorphosis. As it develops it expresses genes characteristic of the developing vertebrate thyroid gland and produces iodine-containing proteins.[40] Behind the developing endostyle is the so-called club-shaped gland. This structure, found only in the larva, forms a tube connecting the pharynx with the external environment and, like the endostyle, it is involved in the production of the protein constituents of the mucus that, when spread across the pharyngeal slits, is used to trap food particles. The gland disappears completely at metamorphosis.[41]

A ventral contractile vessel serves at least some of the functions of a heart, and although the amphioxus version does not much resemble a vertebrate heart, with neither valves nor chambers, the expression of a gene involved in heart formation in animals as phylogenetically well-separated as *Drosophila* and vertebrates shows that it counts as such, at least at a deep level of homology.[42]

The larval stage concludes with metamorphosis, with the disappearance of the club-shaped gland; the migration of the first row of gill slits from the right to the left side; the appearance of the "new" right-side gill slits; the migration of the mouth and endostyle to their final position; the growth of cirri from the edges of the mouth; the movement of the larval anus from the right side to the left; and when folds from the body wall on either side grow outward and meet ventrally to form the atrium.

Other structures to disappear at metamorphosis include the orobranchial musculature: muscles supporting the large, lopsided larval mouth and developing gill slits.[43] These muscles are entirely replaced by adult musculature, and do not seem to be homologues of the vertebrate pharyngeal musculature. As in tunicates, movement of water through the adult pharynx is ciliary.

The amphioxus was once seen as the closest an animal could get to the vertebrate condition without actually becoming a vertebrate. The revelation that it is in fact a basal chordate, and that tunicates are more closely related to vertebrates,[44] should not be seen as a demotion—rather, that the amphioxus represents the closest we have to the common ancestor of all chordates. I present a diagrammatic representation of a basal chordate in fig. 8.1.

The closeness of the amphioxus body plan to vertebrates shows that vertebrates are in many ways rather conservative; that they perhaps owe many of their distinctive features to the two-fold duplication of the genome

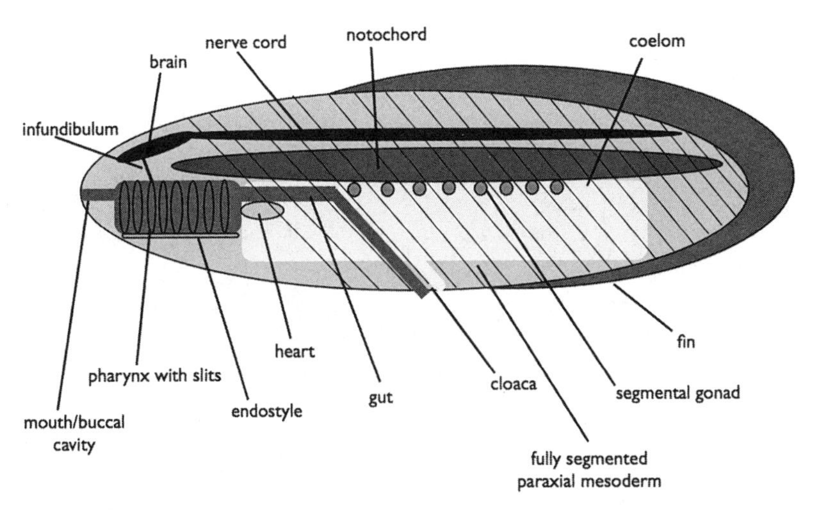

8.1 A cartoon showing the major features of a hypothetical basal chordate.

in their ancestry; and that tunicates, once thought primitive, have evolved in their own directions to achieve a sophistication in lifestyle and biology of a kind that vertebrates would find quite alien.

Is there any fossil evidence that might bear on the early evolution of the amphioxus? Perhaps the most famous early chordate fossil is *Pikaia gracilens*, from the Cambrian Burgess Shales of western Canada—famous because it featured strongly in Stephen Jay Gould's popular book *Wonderful Life* (1990), showing how a small, soft-bodied, and minor component of the fauna would, as a chordate, eventually inherit the Earth. However, detailed study of more than a hundred specimens of *Pikaia*[45] shows that whereas it is most likely to be a basal chordate, an alliance with the amphioxus is problematic. I discuss *Pikaia* and another fossil form, *Cathaymyrus*, in chapter 14.

CHAPTER 9

Tunicates

The 2,000 or so species of tunicates, also known as urochordates, constitute one of the most diverse and abundant elements of the marine fauna.[1] They are called "tunicates" from the extracellular sheath or "tunic," supported by an epidermal mantle, in which many individuals are enclosed. As adults they may be solitary or colonial, yet all filter seawater with a pharyngeal basket that dominates the adult body (fig. 9.1).

Aristotle originally classed tunicates as mollusks, a view endorsed by Haeckel. Given the shell-like tunic, occasionally enriched with minerals, enclosing a prominent filter-feeding apparatus, this is not as outlandish as it sounds. Lamarck coined the name "Tunicata" but classed them with sea-cucumbers. Thomas Henry Huxley added larvacea (also known as appendicularia) to Lamarck's Tunicata to produce the group we know today.[2] It was the discovery that some tunicate larvae have tadpole-like tails, each complete with a notochord and a dorsal, tubular nerve cord,[3] that brought them within the chordates and made them for many years the focus of discussion on vertebrate origins.[4] Long thought to be basal chordates,

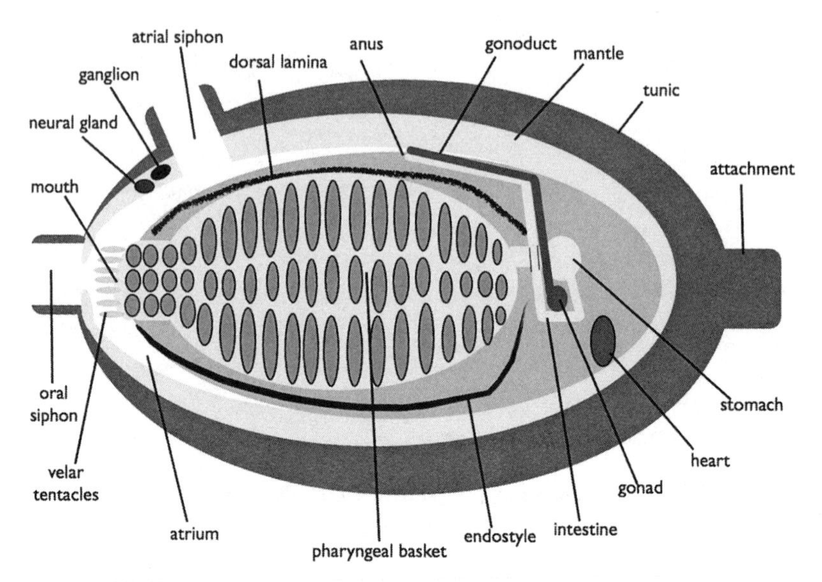

9.1 A cartoon showing the major features of adult ascidian anatomy. Heavily modified by the author, after Jefferies (1986) and Satoh (1994).

tunicates are shown by more recent genomic work to be the vertebrates' closest cousins in the invertebrate world.[5] This seems strange, given that the amphioxus looks very much like a vertebrate, whereas tunicates, with their lack of somites or indeed any obvious signs of segmentation, do not seem much like vertebrates at all.

The answer lies in the rate at which tunicates have evolved from their common ancestor with vertebrates. Whereas the amphioxus and vertebrates have evolved so slowly that their genomes preserve features presumably characteristic of ancestral chordates or even ancestral deuterostomes,[6] tunicates have evolved with such rapidity that many clues to their chordate heritage have been erased from their genomes as well as their external appearance.

So whereas tunicates are fascinating in their own right, they should offer us an object lesson in the waywardness of evolution, which does not preserve a neat, orderly progression of species from the lower to the higher, culminating in Man, as if the steady acquisition of vertebrate features—pharyngeal slits, then the notochord, then somites, a nerve cord with an ever-expanding brain—is the natural order of things. Until recently, the phylogeny of deuterostomes was organized in just this way,[7] perhaps overwelcoming of a simple and seemingly very natural notion of progression from simple to complex. Perhaps this is why the recent finding of our close kinship with tunicates seems so surprising. And yet tunicates have managed to become successful animals by molding the general features of chordates in ways entirely different from those on which their close relatives, the vertebrates, can claim their success.

Almost all the extant tunicates are the ascidians, or "sea squirts," and these are the tunicates that will be familiar to beach-combers. Ascidians are common worldwide from tide pools to the ocean depths[8] though they are not found in freshwater. The adult, solitary ascidian—the common species *Ciona intestinalis* is the best-studied example—spends all its adult life rooted to a single spot, filtering seawater through pharyngeal slits clothed with mucus from an endostyle, a glandular strip running along its ventral margin, much in the manner of the amphioxus. The pharynx is entirely surrounded by the atrium except along the ventral mid-line, and communicates with the external environment through two openings, the incurrent (or oral) and excurrent (or atrial) siphons. The pharyngeal basket is secured to the inside of the tunic by strands of tissue called trabeculae which function, in Jefferies' (1986) description, "like the guy ropes of a tent."

The bulk of the adult appears to consist of the pharyngeal basket. The viscera are confined to a relatively small volume at the base (posterior) of the adult. As water passes through the pharynx, sheets of mucus from the endostyle pass dorsally across the pharyngeal bars on either side, where they trap particles of detritus from the water current. The sheets meet at the dorsal margin where they are guided by a structure called the dorsal lamina and rolled into a rope, whence the mucus passes into the esophagus.

The circulatory system consists of a heart, capable of reversing the direction of blood passing through it, and a number of vessels. One end of the heart leads to a vessel serving the endostyle and thence a dorsal vessel by way of transverse vessels running through the gill bars; the other, to a network that spreads over the right-hand wall of the stomach and thence the viscera. The heart lacks any innervation and beats by virtue of spontaneous contraction of the muscles when enclosed in the pericardial fluid. The blood vessels are not lined with the tissue called endothelium, which appears to be unique to vertebrates.[9] The most remarkable of the eight different types of blood cells are the morulae, which contain large amounts of sulphur and also vanadium, sequestered from the seawater, which they transport to the tunic.

The nervous system consists of a ganglion located in the fork between the two siphons, from which four main nerve tracts run—a pair anteriorly and another posteriorly. All supply the body wall, especially the siphons. The right posterior nerve sends a trunk into the viscera and supplies a nerve plexus in the pericardium, gonad, and gut. Immediately ventral to the ganglion is the neural gland, the duct of which opens in the dorsal midline of the pharynx. It is possible that the neural gland is in some way homologous with all or part of the vertebrate pituitary gland.

The epidermal tissue that gives rise to the oral siphon is derived in part from cells forming the neuropore, the opening at the anterior end of the neural tube. This might reflect a very intimate connection between the mouth and the anterior central nervous system possibly common to all chordates.[10] The muscles that form the atrial siphon, in contrast, have more to do with the heart than the head. They are derived from the cardiopharyngeal mesoderm, the same pool of cells that gives rise to the heart and also the muscles lining the body wall, the same that put the squirt into "sea-squirt." The many similarities in development between cardiopharyngeal mesoderm in tunicates and vertebrates has led to proposals that some tunicate atrial siphon muscles are homologous with some craniofacial muscles in vertebrates.[11]

Many ascidians, such as *Ciona intestinalis*, are solitary, although others, such as *Botryllus schlosseri*, are colonial, and, uniquely in chordates, can reproduce asexually, by budding. A few ascidians are carnivores, the incurrent siphon forming a mouth that can capture small prey. Being sessile and soft-bodied exposes ascidians to predation themselves, so, as many plants do, they deter predators with an exotic biochemistry. Tunicates can sequester metals such as vanadium, manganese, and nickel from seawater[12] presumably to make them taste unpleasant, and create their tunics from cellulose, a highly unusual feature in any animal and possibly facilitated by gene transfer from bacteria.[13] The cellulose synthase enzyme does more than just make cellulose: it turns out that its activity is vital in coordinating the reorganization of the animal as it metamorphoses from tadpole larva to sessile adult.[14]

The tunics can become rather rock-like. The ascidian *Pyura chilensis*, for example, looks like a rock, which, when broken open, reveals a soft interior.[15]

Ascidians are generally hermaphroditic and fertilized eggs develop rapidly into an embryo. Cleavage is radial. With each cleavage, the important transcriptional activator β-catenin accumulates at the vegetal pole, prefatory to the formation of endoderm and, later, the notochord.[16] Vegetal expression of β-catenin is an ancient deuterostome feature found in echinoderms[17] but lost in vertebrates, where factors such as VegT—not found in ascidians—pattern the blastula.[18] Other transcriptional factors are important in the early development of ascidians. These include the FGF (Fibroblast Growth Factor) family, required for the formation of the notochord, the anterior parts of the nervous system, and the muscles in the larval tail tip. Transcription factors in the FGF family are activated, one way or another, by β-catenin, and their activities differ from their roles in vertebrates in important ways. In vertebrates, the transcription factor *Nodal* is important in the determination of mesodermal and endodermal cell fates, whereas in ascidians it plays no part in endodermal development and very little in that of mesoderm. In contrast, it helps direct the formation of the neural plate, something not seen in vertebrates. Transcription factors in the BMP (Bone Morphogenetic Protein) family likewise differ in detail in the activities between ascidians and vertebrates. In vertebrates, notochord and neural tube formation depend in part on BMP suppression, whereas in ascidians BMP promotes notochord formation and has no effect on neural tube development.[19]

In contrast, ascidian tissues tend to share expression patterns with those of vertebrates once the tissues themselves have been specified. *Brachyury* is expressed in the notochord, just as in all chordates, for example, and *Pitx* and *Nodal* are expressed in the epidermis of the left side (as in amphioxus and vertebrates, but not, as we have seen, in echinoderms), *Otx* in the most anterior parts of the central nervous system, and so on.

Gastrulation is typically deuterostome in that the archenteron obliterates the blastocoel and the blastopore forms the anus. Tunicate development is unusual, however, in that it is highly invariant, right down to the cellular level, even between distantly related species. That is, the fate of each cell in a tunicate embryo can be traced precisely, right back to the eight-cell stage or earlier. This reflects the fact that ascidian embryos and larvae make do with very few cells, each with a fate that is determined very early on.[20] There is some evidence that the precise position of a cell within the embryo determines its competence to respond to short-range molecular signals.[21] The nature of the signals themselves is much less important than the positioning of the cells, explaining, perhaps, why the choreography of cells in various ascidians is almost identical, but there is much less similarity between the signals at the genomic level[22]: all signs of the restless and tetchy genomes that lie within these apparently placid animals.

The ascidian embryo develops into a larva, if the species demands one, and if the larva has a tail (as in fig. 9.2), it can be seen as a kind of essence-of-chordate, in which all the features of a chordate are condensed to an irreducible minimum.

Indeed, it is at this stage that ascidian and vertebrate embryos resemble one another most closely. In general terms, gastrulation and neurulation in ascidians are very similar to the cognate processes in vertebrates. The ascidian larva looks a little like a tadpole in that it has distinct anterior and posterior regions.

The blob-like anterior part, known variously as the "head" or "trunk," contains much of the nervous system, including a balance organ and light-sensitive spot; the endoderm that will develop into the pharyngeal basket, endostyle, and digestive system; and the mesoderm that will eventually differentiate into the heart, blood, and muscles of the adult. The mesoderm appears restricted to the pericardium and appears to be made from cells sloughing off the inner surface of the archenteron.

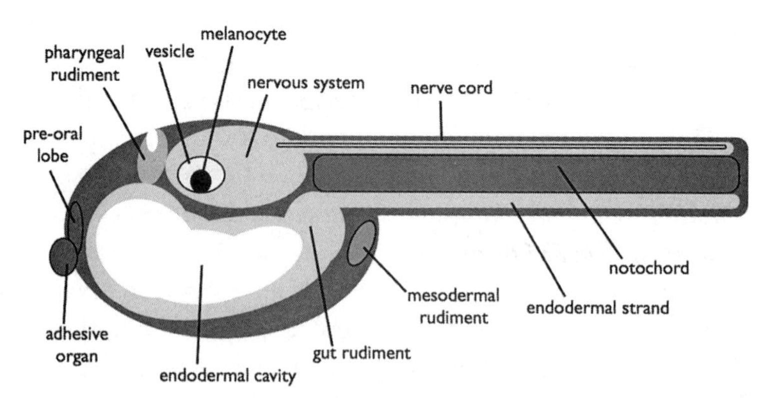

9.2 A cartoon showing the major features of an ascidian tadpole larva. Heavily modified by the author after Satoh (1994).

The extended posterior part contains the notochord; the dorsal tubular nerve cord; striated muscles derived from paraxial mesoderm; and a strand of endoderm that extends almost to the posterior end. The presence of this endoderm, which gives rise to the intestine of the adult, means that one cannot formally call the posterior part a "tail" in the vertebrate sense, as vertebrate tails are post-anal (and therefore post-endodermal) by definition. One might stretch a point by saying that as the ascidian larva does not feed and lacks an anus, we are entitled to call this structure a "tail" without too many pangs of conscience.[23]

The tail forms by a process of longitudinal extension in which the notochord buds off from the roof of the archenteron, the dorsal tubular nerve cord forming above it in characteristic chordate fashion, the observation of which caused Kowalevsky (1866b) to exclaim:

> *Nach allen diesen Gründen glaube ich mit vollem Rechte den Aschencylinder des Schwanzes der Ascidien mit der Chorda dorsalis des Amphioxus sowohl functionel, als auch genetisch vergleischen zu können.*[24]

which although functionally equivalent to "Eureka" does lack the Hellenic elegance (perhaps because, as far as we know, it was not uttered while in the bath).

In vertebrates, the extension of the notochord is largely the result of proliferation from a posterior growth zone, combined with a process known as convergence and extension, in which cells are drawn toward the midline and are then thrust backward into a narrower, even single file. Ascidians have the latter (convergence and extension) but not the former (proliferation). These cells of the presumptive notochord expand in size by a process of vacuolation (hollowing) and the developing notochord may expand further by the recruitment of extracellular material into the structure. As the notochord develops it stiffens into a hydrostatic skeleton, purchase against which swimming muscles can act.[25]

The ascidian neural tube is formed of just four rows of cells; one dorsal, one ventral, and two lateral, with a narrow lumen running down the center. A simpler, more abbreviated neural tube could hardly be imagined: indeed, the entire larval nervous system consists of around 100 neurons, compared with thousands in the amphioxus and millions in vertebrates. The tail muscles are similarly exiguous. With eighteen muscle cells per side in *Ciona intestinalis*, it's hard to say whether they can be further subdivided into somites or indeed segments of any kind.[26] Even though the notochord cells, their formation and elaboration, and associated striated muscles of tunicates resemble those of vertebrates, the notochord in tunicates is dramatically abbreviated by comparison. In *Ciona* it forms from a column of just 40 cells, development of which is triggered by the expression of the gene *brachyury*, the quintessential marker gene for the notochord in all developing chordates.[27]

Even at this small scale, however, the central nervous system is divided into distinct regions. At the anterior end are the sensory vesicle, which houses the gravity-sensor; and the light-sensitive spot with its associated pigment cells.[28] Progressively more posterior lies what passes for a brain; behind that, a mass of undifferentiated tissue that will form neurons in the adult[29]; the motor ganglion, containing the cell bodies of neurons driving the tail; and, finally, the nerve cord, consisting mainly of axons from those cell bodies in the motor ganglion.

Gene expression, too, varies from anterior to posterior. The gene *Otx*, well known as a marker for the anteriormost parts of the nervous system in vertebrates as well as fruit flies,[30] is expressed in the larval brain and neck, but not elsewhere, suggesting a fundamental homology between tunicate and vertebrate nervous systems, at least as regards the anterior-posterior axis. In addition, the expression of other genes suggests a rough correspondence between the "neck" and the vertebrate hindbrain.[31] Even in this extremely abbreviated state, the tunicate nervous system is cast along the same lines as that of the amphioxus and of vertebrates.

All in all, the tadpole larva of *Ciona intestinalis* consists of around 2500 cells, the fates of most of which can be traced back to the embryo. This abbreviation of development is matched by rapidity of life cycle. *Ciona* develops from an egg to a tadpole in about 18 hours, the entire life cycle taking less than three months. After hatching, the larva swims for a period of between hours and days, depending on species, while it seeks a permanent place to settle. The act of settlement triggers metamorphosis in which the tail is resorbed and the remainder of the animal undergoes dramatic remodeling that includes a ninety-degree rotation of the viscera. Coincident with the onset of metamorphosis is the expression of genes connected to the innate immune system,[32] and the triggering of a cascade of molecular interactions by the adhesion to a substrate of the the sensory papillae at the extreme anterior end of the larva.[33] Metamorphosis also depends on thyroid hormones,[34] well known to be involved in the regulation of metabolism in many vertebrates and to trigger metamorphosis in fish and amphibians as well as influence the metamorphosis of at least some echinoderms.[35] This suggests a role for thyroid hormones in the life cycle of the ancestral deuterostome.[36] In ascidians, however, as well as the amphioxus, thyroid hormone receptors are mainly triggered by a hormone called TRIAC (triiodothyroacetic acid) as distinct from the T3 (triiodothyronine) and T4 (thyroxine) found in other deuterostomes.

Tunicates do not, of course, have a thyroid gland. The homologous structure is the endostyle, formed during metamorphosis from an endodermal primordium in the larva. In the adult, the endostyle is a gutter that runs along the ventral edge of the pharynx and which is rich in gland cells. This sequesters iodine, and expresses a characteristic set of genes found in the very similar endostyle of the amphioxus, as well as the thyroid of vertebrates.[37] In tunicates, the amphioxus, and the ammocoete larva of lampreys, the endostyle has the additional function of synthesizing the mucoproteins that coat the pharyngeal bars.

There are distinct signs of evolutionary precursors of neural crest and placodes in tunicates. Melanin-containing pigment cells move away from the edges of the neural plate, and two of these cells are of particular interest in that they become sense organs: the gravity detector or statolith, and the light-detector, or ocellus. The development of these organs has much in common with that of neural crest cells at the molecular level[38] and it is very likely that the common ancestor of tunicates and vertebrates had some form of neural-crest-like tissue in which cells from the neural plate border expressed pigmentation and were capable of migration to some degree.[39] There is also evidence that tunicates have neurogenic placodes, if only in a limited sense.[40]

Some researchers wondered whether a creature such as an ascidian tadpole larva might have retained its shape throughout life, becoming sexually mature while still in the larval state, a process known as paedomorphosis,[41] providing a ground plan for the evolution of creatures such as the amphioxus or a primitive vertebrate. Given the extreme reduction and simplicity of tunicate tadpole larvae, this hypothesis, while popular, seems to rely on more than a soupçon of naively adaptationist wishful thinking. It has in any case been rendered moot by the discovery that tunicates are the invertebrates most closely related to vertebrates, suggesting that the rule in tunicate evolution has been one of loss and extreme modification, rather than ceaseless striving to some vertebrate apotheosis. Further, the trend in tunicates as a whole has been that the tadpole larva is reduced or lost even as a pelagic, swimming form has been retained in adults, contrary to what one might expect had the tadpole larva formed the ground plan for an adult that retained the swimming habit.

Nonetheless, the paedomorphosis idea might have yielded more traction by the twenty or so species of larvaceans or appendicularians. These tunicates are motile and retain a notochord-supported tail throughout life though, unlike with ascidian larvae, this is a true post-anal tail. Like ascidians they are solitary but mostly minute—of the order of a few millimeters in length—and float in the water, veiled in an elaborate construction called a "house," vaguely spherical in shape but containing complex passages and fine filters through which the animal draws particles to its mouth, driven inward by the motion of its tail.[42] The house is many times the size of the animal at its centre, and is made of cellulose, mucopolysaccharides, and at least eighty proteins called oikosins, half of which bear no similarity to any other known proteins, including those in ascidian tunics.[43] When the house gets clogged up with detritus the animal leaves it and inflates a new one, the germ of which has already been secreted by a specialized tissue called the oikoplast.

The best known larvacean, *Oikoplura dioica*, is a millimeter or so long and has a house as big as a walnut, though some larger species create mansions as big as soccer balls. Larvaceans are very abundant in the plankton: their discarded houses constitute an important source of food for other marine organisms, and contribute appreciably to the global carbon cycle when deposited on the ocean floor. All larvaceans except *Oikopleura dioica* are hermaphrodites (*O. dioica* has separate sexes).

Oikopleura lives its life in such a hurry that were it to get much faster it would die before it was born. As it is, the animal hatches within about three hours from fertilization, and runs its entire life cycle in just four days. Related to the rapidity of its life cycle is an exiguity of cells that is extreme even by the standards of tunicates. *Oikopleura* gastrulates at just 32 cells (compare with 110 cells in ascidians). By the end of embryogenesis, the notochord consists of 20 cells (40 in *Ciona*) and with just thirteen of the fifty downstream target genes of *brachyury* that, in *Ciona*, coordinate its formation.[44] That having been said, the notochord is elaborated greatly in later stages until it ends up with a prodigious one hundred cells. It also expresses Hox genes, something not found in the notochord of *Ciona*, and furthermore in a more or less collinear order. The tail muscles of *Oikopleura* consist of just twenty cells, ten on each side of the notochord. The animal deserts its own elaborately made house just two or three hours after building it, but constructs a new one within ten minutes.

Still more remarkable are the thaliaceans, the third group of tunicates. The seventy or so species are divided into three groups—the salps, doliolids, and pyrosomes. Of these, only the doliolids show any signs of having a notochord or nerve cord, and even then only in rudimentary form in some larvae. Thaliaceans are exotic creatures of the open ocean that will probably never have been encountered by a casual visitor to the sea shore.

Salps are generally found as single individuals or zooids, ranging between one and ten centimeters in length. Their incurrent and excurrent openings are at opposite ends, so the animal looks like an animated jet engine. Indeed, salps move through the water by a kind of jet propulsion, in which the action of circular muscles pulls water in at the front and expels it out the back. The circular muscles almost—but not quite—encircle the body. The single salp or "solitary" is the sexual stage in a complex life cycle. As it swims, it buds off a chain of asexual buds or "aggregates"—the eldest being the furthest away from the host solitary. Each aggregate produces a single egg, which is fertilized internally by sperm shed into the water from older aggregates. The embryo is nurtured internally and eventually grows into a mature solitary, and the cycle continues. Salps can live for a year or more, and, like larvaceans, form a considerable fraction of the "marine snow" that contributes to the ocean's carbon budget.

Like salps, doliolids exist as single individuals shaped like jet engines, though they tend to be smaller, around five centimeters in length; have muscles that entirely encircle the body, like hoops round a barrel, and, perhaps surprisingly given the musculature, move through the water by the action of the cilia that line the pharyngeal basket. The muscles are reserved for occasional, sudden leaps. The individual doliolid, a product of sexual reproduction, extends a stalk or stolon whence grow three rows of zooids; two rows of feeding zooids flanking a central row of phorozooids. These latter detach from the stolon and bud off a further generation, the so-called gonozooids. These are hermaphrodites and release eggs and (later) sperm, which meet in the water column. In some cases the embryo produces an ascidian-like tail, with a notochord and muscles but no nervous system, but this structure is resorbed even before the animal hatches.

Last but not least among the thaliaceans are the pyrosomes. Individual pyrosome zooids are tiny, but they form colonies many thousands strong. These colonies look like tubes with one sealed end: individual zooids are oriented such that the incurrent openings are on the outside, the excurrent on the inside, so the entire colony moves along by a stately kind of jet propulsion with the open end of the colony as the exhaust. Individual zooids are hermaphrodite and reproduce sexually—but, being tunicates, there is a twist. The embryo grows into a form called a cyathozooid, and this buds off four more zooids, either contributing to the extant colony or leaving to start a new one. Pyrosome colonies can be substantial, up to twenty meters in length with an opening up to two meters across, big enough for divers to swim around inside.

Understanding the evolutionary relationships of the various groups of tunicates with one another is complicated by the rapidity with which their genes and genomes have been evolving. Recent phylogenies suggest that thaliaceans have evolved from within the phlebobranch ascidians (the group that includes *Ciona*) or the aplousobranch ascidians (such as *Clavelina*). The position of larvaceans is less clear; they might either have evolved from stolidobranch ascidians (including such genera as *Botryllus*, *Styela*, and *Halocynthia*), or represent an offshoot basal to all other tunicates.[45] However, molecular phylogenies placing *Oikopleura* basal to the other tunicates might reflect that creature's extremely rapid evolution rather than its true phylogenetic position. In addition, evidence from the development of *Oikopleura dioica* suggests that beneath the extreme abbreviation of cell number and rapidity of development, various asymmetries in cleavage, gastrulation, and other aspects of development speak to an ascidian heritage.[46]

The first tunicate genome to be sequenced was that of *Ciona intestinalis*.[47] As expected from its exiguous cell count and rapid development time, it is small,[48] squeezing about 16,000 genes (more than half that of the human genome) into a genome of 160Mb (where Mb stands for "megabase," or million basepairs[49]). The genome of the related *Ciona savignyi*[50] (174Mb), the colonial ascidian *Botryllus schlosseri*[51] (580Mb), and the larvacean *Oikopleura dioica*[52] (70Mb) followed; the genomes of several other ascidians[53] and one salp[54] have been sequenced. It is clear that tunicate genomes are small. The largest known, that of *Botryllus schlosseri*, is comparable with that of the amphioxus (520Mb) but smaller than the sea urchin *Strongylocentrotus* (814Mb). The genome of *Oikopleura dioica* is by far the smallest known of any chordate, or indeed any animal. The genes are packed extraordinarily tightly and often organized into operons, that is, linked sets of genes of common function, a phenomenon more usually associated with the compact genomes of bacteria. The *Ciona* genome has operons, too, and there are negligible signs of synteny—that is, large regions of conserved gene order—between the genomes of *Oikopleura* and *Ciona*, a sign of extraordinarily rapid evolution. Contrast this with the extensive syntenies between hemichordates, the amphioxus, and vertebrates.[55]

Oikopleura has nine Hox genes, though not in a cluster—unlike any other known bilaterian genome.[56] They are, however, expressed in a more or less collinear way, just as if they had been in a cluster, mainly in the nerve cord, notochord, and tail muscle. In vertebrates, Hox genes play an important part in patterning the central nervous system. Whether they do the same in tunicates, however, is moot. Some Hox genes seem to be involved in the development of the larval tail and the dorsal visceral ganglion of the adult, but knockdown of five others had no discernible effects.[57] In general, the Hox and the associated ParaHox clusters in tunicates show significant signs of erosion, fragmentation, disorder, and wholesale loss.[58] The expression of those that remain is confined to the notochord, nerve cord, and muscles of larvae. Given the central importance accorded to Hox genes in the patterning of the body plans of animals in general, it is "stunning," as Stolfi and Brown put it in their 2015 review, "to see the rock stars of bilaterian embryonic patterning remain behind the scenes in tunicates." Lemaire et al. (2008) describe ascidians as "nature's extreme chordate experiment."

The genome of *Oikopleura* represents perhaps the extreme end in a tendency of shrinkage seen in tunicate genomes, together with the disintegration of chordate- or even bilaterian-like genome structure. This shrinkage is presumably connected with the general reduction in cell number, the rapidity of the life cycle, the generally deterministic development, and the speed of evolution seen more generally in tunicates.

Although tunicate genomes, morphology, and development have undoubtedly evolved rapidly, a theme that comes up repeatedly is the degree to which morphology is preserved despite radical genomic reorganization. Despite the use of very different patterning factors, for example, the fate maps of ascidian blastulae are very similar to those of vertebrates such as the frog *Xenopus* and basal ray-finned fishes such as sturgeon, but different from those of the amphioxus and non-chordate deuterostomes.[59] Detailed examination reveals further marked genetic differences. The genomes of *Ciona intestinalis* and *C. savignyi*—species in the same genus—differ markedly as regards sequences within chromosomes.[60] Cognate regulatory elements in closely related ascidians such as *Ciona intestinalis* and *Halocynthia roretzi* are so different that their sequences can no longer be aligned, yet the two animals are recognizably ascidians. The differences between the expression of notochord-specific genes in *Oikopleura* and *Ciona* have already been mentioned. Looking further afield, tunicates manage to form a tadpole larva without any assistance from any structure resembling the Spemann Organizer, essential in the development of vertebrates and possibly the amphioxus.[61] Further comparisons between gene expression profiles of *Ciona* or *Oikopleura* with vertebrates reveal marked differences[62] although conservation of expression is most likely in those tissues and organs most associated with the chordate body plan, such as the central nervous system, notochord, and associated muscle.[63]

The seeming discordance between divergent genes and similar morphologies may be resolved as follows: it seems to be that whereas the order, spacing, and orientation of transcription-factor binding sites in their DNA may vary, the transcription factors themselves do much the same things, species to species.[64] The connection of genes in regulatory networks[65] seems to be more important than gene order to development and morphology. Underlying similarity in the gene regulatory network of chordates presumably explains why despite their diversity, all chordates seem to have a tadpole-like stage in their development, and aspects of the more complex developmental routine of vertebrates can be discerned in the simpler tunicate form. These include a distinct boundary between the midbrain and hindbrain[66]; similar patterns of gene expression in the anterior central nervous system, in particular the sensory vesicles of tadpole larvae and the vertebrate hypothalamus[67]; molecular similarities linking the ascidian atrial and oral siphons with vertebrate cranial placodes[68]; the adult ascidian ganglion with the vertebrate hindbrain[69]; and substantial similarities between the molecular underpinnings of tunicate and vertebrate heart formation.[70] The presence of cells with some of the properties of vertebrate neural crest has been discussed above.

This is not to say that small changes in tunicate genes cannot produce radical transformations, for they patently can. Molgulids are solitary ascidians, many of which have tailless larvae, and yet hybridization is still possible between tailed *Molgula oculata* and tailless *Molgula occulta*.[71] The expression of chordate features in molgulids depends on the expression of a single gene, *Manx*.[72] This gene cannot be a master-switch for chordate features in general, as no detectable counterpart can be found even in *Ciona*, a relative of *Molgula*. Almost all thaliaceans have dispensed with the tadpole larva entirely, presumably by their own means.

Summarizing what is known about tunicate life history, Stolfi and Brown (2015) describe it as "biphasic." Tunicates contain all the major features of chordates—just not all of them together in a single life stage. The notochord, the nerve cord, and the paraxial mesoderm perhaps equivalent to somites are found in the larva, whereas the endostyle, heart, blood, digestive system, pharyngeal gill slits, and associated musculature are found only in the adult. This situation presumably evolved from a more integrated form by heterochrony (that is, differential speeds of development in different tissues or organ systems) and perhaps was accentuated by miniaturization in cell number and the underlying genome.

The fossil record of tunicates is as slender as one would expect for such soft-bodied animals. *Cheungkongella*, a fossil ascribed to tunicates and compared with the ascidian *Styela*, has been described from the Lower Cambrian Chengjiang Fauna of China,[73] though the affinities of this creature have been disputed by the describers of another Cambrian form, *Shankouclava*,[74] and when more specimens of *Cheungkongella* were discovered they were found to resemble another fossil, *Phlogites*, a creature with a pair of branching tentacles thought to be a lophotrochozoan[75] but recently rehabilitated as akin to cambroernids, an assemblage of soft-bodied fossil deuterostomes possibly related to ambulacraria,[76] and which I shall discuss further in chapter 14.

CHAPTER 10

Vertebrates

You'll have seen the briefest overview of the essentials of vertebrates at the start of chapter 1. Here I shall attempt to summarize their general features more in the style of chapters 6–9—that is, not as unique creatures, somehow set apart, but as just another deuterostome group (fig. 10.1).

The aim is to draw attention to those parts of vertebrate anatomy and development that have most in common with other deuterostomes. The following should also highlight those parts that are different. I'll explore these in more detail in chapters 12 and 13.

The vertebrates comprise the most successful group of deuterostomes in terms of numbers of species. There are at least 60,000 species of vertebrate alive today, including, in the whales, the largest animals ever known to have existed, and many spectacular extinct species, such as dinosaurs. Most vertebrates are primitively aquatic, mobile, and marine—we call them "fishes"—though uniquely among deuterostomes, vertebrates have invaded freshwater and, like the insects, have colonized the land and the air. Some vertebrates, such as whales, turtles, and the extinct ichthyosaurs,

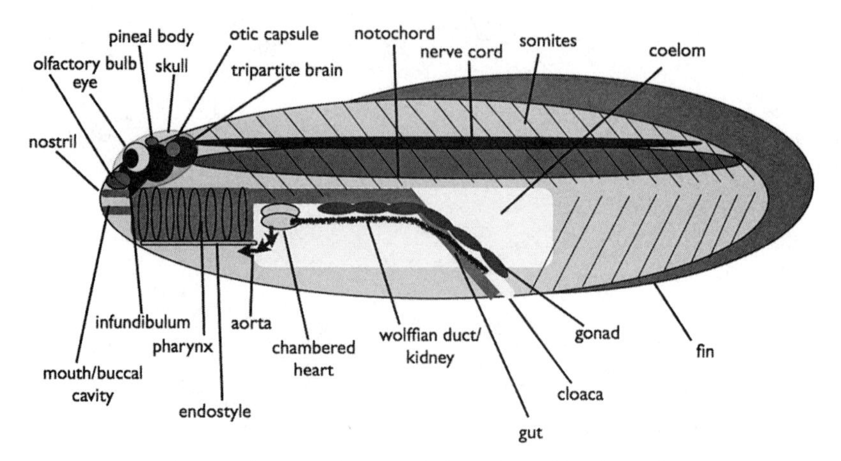

10.1 A cartoon showing the major features of a hypothetical basal vertebrate. Compare with the basal chordate in figure 8.1.

have re-colonized the sea. In contrast to most tunicates, vertebrates are active and mobile and have separate sexes. Although a few vertebrates are known to change sex during their lifetimes, there are no hermaphrodite vertebrates. Again, unlike tunicates, vertebrates cannot reproduce vegetatively, or by budding, and their capacity for regenerating tissues is relatively limited. Some vertebrates can reproduce asexually, by parthenogenesis, in which a female produces offspring without fertilization, but this is rare. The life cycle of vertebrates is very variable in length, from a few weeks to many decades, depending on the species, but in general it is much longer than is the case in tunicates.

Perhaps the most important difference between vertebrates and all other deuterostomes is that vertebrates are, in general, predatory. This is reflected in the marked elaboration in the musculoskeletal, sensory, and nervous systems of vertebrates compared with those in other deuterostomes.[1]

The most primitive vertebrate is the ammocoete, a tiny creature found buried in river gravels with its front end poking out into the water—a freshwater version of an amphioxus. Like the amphioxus, the ammocoete is a filter feeder. The mouth, ringed with tentacles, leads to the buccal cavity, curtained off from the pharynx by a pair of flaps called the velum, rather in the manner of the amphioxus. Water is drawn into the pharynx not by ciliary action but by the movement of the velum and muscular contractions of the pharynx. Openings on either side of the pharynx lead to seven pairs of gill pores that communicate with the outside by branchial slits. At the base of the pharynx is an endostyle, a rather more complex organ than the equivalent in the amphioxus or tunicates, but which serves precisely the same function and is presumably homologous with them. As in these other chordates, the endostyle produces mucus that serves as a trap for particulate matter. Like the amphioxus, ammocoetes have a notochord and body musculature organized into segments or somites. Unlike the amphioxus, however, the ammocoete has a much more developed head end with a distinct brain and paired eyes.

Is the ammocoete a kind of "missing link" between vertebrates and other chordates? Well, no. Although once thought to be distinct animals, ammocoetes were discovered to be the larval stages of lampreys that, on metamorphosis, lost their pharyngeal basket style of feeding and adopted a predatory lifestyle.[2]

All vertebrates alive today fall in to one of two major groups. The smaller of the two, known as agnathans or cyclostomes, comprises lampreys and hagfishes. These lack hinged jaws, paired fins, and any trace of hard tissue such as bone or enamel. The fossil record of agnathans is richer and very much more bony than one might guess from the extant representatives. Modern lampreys and hagfishes are important to zoologists as their anatomy and physiology might yield clues to the earliest stages of vertebrate evolution. Almost all modern vertebrates, however, belong to the gnathostomes, or jawed vertebrates. As the name suggests these have articulated jaws with teeth. They also have two sets of paired fins and many other refinements not seen in agnathans. In his verse *An Introduction To Dogs*, the humorist Ogden Nash gives a description of a gnathostome as succinct as one could hope to see (dogs, of course, being gnathostomes).

> The dog is man's best friend.
> He has a tail on one end.
> Up in front he has teeth.
> And four legs underneath.

There are 38 species of lamprey. They live in the sea as adults but migrate into rivers to spawn. It is in freshwater that the larval form of ammocoete lives, sometimes for several years, before metamorphosis. The newly adult lamprey migrates to the sea before returning to freshwater, breeding and dying. This lifecycle is reminiscent of salmon, but the exact opposite of eels, the fishes with which lampreys are most often compared.

The resemblance of lampreys to eels is, however, superficial, as true eels are gnathostomes. The lamprey skin bears no scales, and the mouth contains no jaws, but instead is surrounded by a circular oral funnel filled with horny teeth. The tongue, similarly armed with teeth, is attached to an elaborate arrangement of powerful muscles, and can be protruded from the mouth, all the better to grab onto any passing fish. On the top of the head is a single, median nostril, leading to the olfactory capsule at the front of the brain, and the hypophysis, beneath it. Behind the nostril is a pale spot corresponding to the pineal "eye," a light-sensitive patch that grows out of the midbrain. On each side of the head is an eye, fully developed and moveable. Behind the head are seven pairs of pharyngeal slits. About two-thirds of the way along the ventral margin is the cloaca, the combined opening for the urogenital system and the anus. The animal has median dorsal and caudal fins but no trace of paired fins.

Like the amphioxus, the lamprey contains a prominent notochord that extends from the posterior tip of the animal, though unlike with the amphioxus it terminates before it reaches the anterior end. The notochord is crowned at intervals with irregular cartilages or arcualia that correspond with parts of gnathostome vertebrae. The anterior tip of the notochord is at a level corresponding with the eyes. The brain and spinal cord are encased in a fatty meningeal body. The cartilages of the head are barely recognizable as a skull, but consist of a number of separate boxes to house the otic capsules and the olfactory capsule and provide a basicranial support for the brain, supports for the tongue, and so on. The branchial skeleton is a single, unjointed basket that supports the gills and extends posteriorly as far as the heart, partially enclosed in its own pericardial cartilage. The branchial skeleton is lateral to the gills, protecting them in internal pouches. This is in marked contrast to the gnathostomes, where the skeleton is medial to the gills, which are as a result more accessible to the outside.

The body cavity of lampreys is divided into two main coelomic compartments. These are, first, the pericardium, which surrounds the heart; and, second, the visceral coelom, in which the liver and urogenital organs are suspended. The two coeloms are not connected to each other.

However, the visceral coelom is connected to the cloaca by a pair of pores, through which sperm and ova pass.

The animal is heavily muscled: the most prominent muscles are the segmented somitic muscles used in swimming, and which work against the notochord. Other muscles include the massive apparatus that operates the tongue; the muscles of the gill pouches; and the muscles that move the large, well-developed, mobile, image-forming eyes.

The presence of eyes and other elaborate organs of special sense is another major difference between vertebrates and other deuterostomes. The single nostril of the lamprey is the external opening of the blind-ended nasohypophysial duct that passes from the exterior and comes to an end between the roof of the mouth cavity and the base of the brain. The olfactory capsule, at the front of the brain, lies atop the nasohypophysial duct and opens into it dorsally.

In addition to eyes, lampreys also have ears, in the form of semicircular canals, the organs that in gnathostomes constitute the inner ear. The displacement of otoliths by moving fluid within the semicircular canals give the animal information on its position in space. Lampreys have two semicircular canals (per side), unlike the three found in gnathostomes, and yet are quite capable of distinguishing between roll, pitch, and yaw. The ear is developmentally and evolutionarily related to the lateral line system, a network of superficial canals on the head and trunk that also contains organs sensitive to the displacement of water. The ear and the lateral line system both develop from an acoustic placode, more deeply invaginated and developed in the head (forming the ear within its otic capsule) than in the rest of the body. The lateral line system is innervated by a branch of the auditory nerve.

The lamprey brain is constructed along the same lines as that of any vertebrate, but is large and elaborate compared with that of any other deuterostome.

At the very front of the lamprey brain (fig. 10.2) is the single olfactory capsule, the primary organ for the sense of smell. There would be a pair of these in gnathostomes. Nonetheless, even in lampreys, the olfactory capsule sends not one but a pair of olfactory nerves to the telencephalon, which with the diencephalon constitutes the forebrain. The diencephalon in its turn receives the optic nerves, one on each side, although these cross

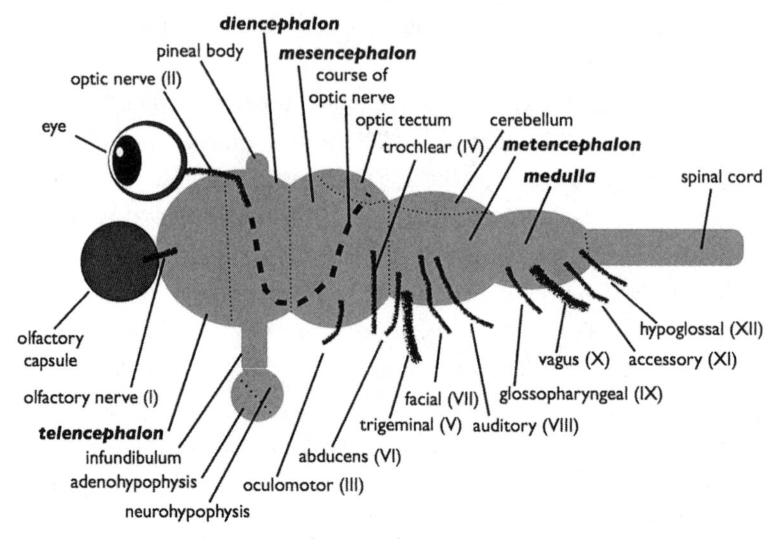

10.2 A cartoon showing the major features of the vertebrate brain.

each other on the floor of the diencephalon, then pass dorsally and posteriorly to terminate in the midbrain (mesencephalon) on the opposite side. The ventral part of the diencephalon is extended into the infundibulum, terminating in the neurohypophysis. This meets the adenohypophysis, a pocket upthrust from the anterior gut. These structures together constitute the pituitary.

The dorsal part of the mesencephalon, called the optic tectum, receives the optic nerves, as described, as the ventral part receives the oculomotor nerves, which move the eye muscles, along with the trochlear nerves, which also move the eye muscles, but whose roots are posterior and more dorsal. Behind the mesencephalon is the hindbrain, or rhombencephalon, divided into the metencephalon (the chief part of this being the cerebellum) and the medulla oblongata, the hindmost part of the brain. The hindbrain receives quite a few nerve roots: both major parts of the trigeminal; the abducens; the facial; the auditory (innervating the ear and lateral-line canals); the glossopharyngeal and vagus (these innervate the gills); and finally the roots of the spino-occipital (spinal accessory and hypoglossal) nerves that innervate the muscles in the base of the pharynx.

The hindbrain grades into the spinal cord, which receives segmental motor and sensory nerves. As in the amphioxus these join the spinal cord in separate roots, in contrast to the situation in gnathostomes (and hagfishes) when they join up into ganglia before entering the spinal cord.

At this point I should say a few words about these various nerves, which together constitute the system of cranial nerves, a major feature of vertebrate anatomy taught in classical zoology and medicine, and which was taught to me. The cranial nerves are useful landmarks for the anatomy of the vertebrate brain and the much-debated segmental nature of the head, and they are worth getting to know. They can be set out as a chart, with numerals and names, listing them from anterior to posterior, like this:

I. Olfactory
II. Optic
III. Oculomotor
IV. Trochlear
V. Trigeminal
VI. Abducens
VII. Facial
VIII. Auditory
IX. Glossopharyngeal
X. Vagus
XI. (Spinal) Accessory
XII. Hypoglossal

I have a mnemonic for these which, though mild in comparison with many, is perhaps not as politically correct as today's standards might demand. It has, however, served me well for more than thirty-five years. Perhaps you can think of one for yourself. The functions of the olfactory, optic, facial and auditory nerves more or less speak for themselves. The oculomotor, trochlear, and abducens nerves innervate the muscles that control the eyeballs. The glossopharyngeal and hypoglossal nerves innervate the tongue and parts of the pharynx. Perhaps the most important cranial nerves in terms of size and function are the trigeminal and vagus, which innervate various parts of the body. The vagus is notable in that it sends nerve tracts to the heart and beyond, deep into the viscera, far from the brain. Together, the cranial nerves are found, more or less like this, in all vertebrates from hagfish to humans, and in no other animals.

Another key feature of the vertebrate body is the kidney, a system for filtering excess water from the coeloms directly, or from the blood. Similar organs are found in other deuterostomes—one thinks of the glomerular complex attached to the heart in the prosome of enteropneusts. The kidney has taken on a much greater role in vertebrates, however, and life outside the marine environment would be impossible without it. The kidney exists to regulate the balance of salts in the body. All life on Earth functions by virtue of the properties of a water, in which salts of various kinds are dissolved. Life evolved in the sea, which is why it's no accident that the concentration of salts in the water sloshing around inside every animal is more or less the same as in seawater. And that's fine if you are an animal that lives its whole life in the sea. The problems start when animals adapted for marine life move into freshwater, which contains fewer dissolved salts than seawater. An animal evolved for sea life that moves into freshwater will find it has saltier body fluids than its new environment, and will therefore take on excess water, by osmosis. It can protect itself to some extent by waterproofing its skin, but this won't work for tissues such as gills, which require intimate contact with the water in order to facilitate the uptake of oxygen and the removal of carbon dioxide. Animals living in freshwater need to be able to remove excess water, while retaining salts. This is where the kidneys come into play.

In terms of development, the kidney forms as a series of small tubes, the Malpighian tubules, one in each somite, draining the coelom into a pair of vessels, the Wolffian ducts, suspended in mesenteries in the roof of the main, visceral coelom. Without going into enormous detail, the tubules are arranged in such a way that they expel water but retain salt. The waste is a dilute solution we call urine. In addition to excess water, urine may contain nitrogen-rich products left over from the breakdown of proteins. In most fishes, this is ammonia, which dissolves readily in water and requires no special means of expulsion. But ammonia is highly toxic and cannot be allowed to accumulate in tissues in any quantity, so it is often metabolized into a soluble by-product, urea, and this is the major nitrogenous constituent of urine in most vertebrates. This is the substance responsible for the yellow color and distinctive smell of urine. Nitrogenous waste can be metabolized into uric acid, which is highly insoluble in water. Uric acid crystallizing out in the urine is the cause of bladder stones: if it escapes the urinary tract and gets into joints, it causes gout. The Wolffian ducts drain into the cloaca, and, by a quirk of development, are intimately associated with the reproductive system, which exploits Wolffian plumbing as a way of getting gametes to the outside. I discuss the anatomy and development of the kidney in greater detail in chapter 13.

The vertebrate heart has two main chambers, the atrium and ventricle, which in lampreys are closely appressed together as in gnathostomes, and the heart is innervated by a branch of the vagus nerve. The main output of the heart is the ventral aorta, which passes anteriorly until it reaches the pharynx, where it branches dorsally, supplying the tongue and the gills. These various branches collect oxygen and merge into the dorsal aorta, which runs backward just beneath the notochord, giving off arteries to each segment; to the kidneys; and the mesenteric artery supplying the liver and intestine. The veins collect blood from various parts of the body and debouch into the sinus venosus, the main input into the heart. In broad terms, the circulatory system in the lamprey is cast on the standard vertebrate model, leaving aside substantial alterations with the evolution of lungs in the bony fishes and the subsequent separation of the system into pulmonary and systemic branches, with the consequent division of the heart into left and right halves.

Most organisms have a kind of immunity. That is, they have mechanisms to tell the difference between tissues that are part of their own bodies, and material that has come in from somewhere else. Such self-nonself recognition systems are known in tunicates, especially hermaphrodites, where they are used to prevent an organism's sperm fertilizing its own eggs or, if the species is colonial, for allowing clones to tell the difference between their own colony and another.[3] Vertebrates have a very distinct system of adaptive immunity, in which the immune system can be trained to recognize foreign matter very specifically. This is the basis of vaccination, in which the immune system is taught to recognize a threat and to remember it in case it comes across it again. The number of possible threats is infinite: the genetic material, much less so. So how can this possibly work? Gnathostome genomes contain many genes that, when expressed as proteins called immunoglobulins, can be joined together to form antibodies in a very large number of combinations. It turns out that lampreys (hagfishes, too) are also able to juggle proteins to create a so-called adaptive immune system, but the proteins they use are different from those in gnathostomes.[4]

Lamprey development is classically deuterostome. Cleavage is radial and gastrulation is by enterocoely. Neurulation is similar to the process in the amphioxus. At some point a lip appears at the presumptive posterior end of the blastula and grows downward until the yolkier cells are encompassed and form the base of the archenteron, the primary gut cavity. The opening beneath this lip is the blastopore, which will eventually form the anus. By the end of gastrulation, the germ of the notochord is present in the dorsal wall of the archenteron, the presumptive somitic mesoderm on either side, except at the anterior end, where it gives way to the mesodermal prechordal plate—the foundations for the head. The appearance of a deepening groove in the upper ectoderm signals the beginning of neurulation, that is, the appearance of the brain and nerve cord. This is roofed over by ectoderm growing forward from near the posterior end, leaving an open end at the front called the neuropore, and a connection with the archenteron at the posterior, the neurenteric canal, both of which eventually close up. The archenteron wall grows inwardly to separate the notochord (but not the prechordal plate) from the archenteron itself, and the somites begin to form in the paraxial mesoderm.

Apart from lampreys, the extant agnathans include the hagfishes, which are by any stretch of the imagination among the most disgusting of all vertebrates, as well as among the most fascinating. They are most notorious for being able to secrete, with the slightest provocation and in almost no time at all, incredible quantities of very sticky slime: a feat which led one scientist, no doubt with a twinkle in his eye and wry nod to experience, to suggest that bare-handed hagfish-catching might make for an interesting Olympic sport.[5] There are approximately sixty species of hagfish, and they tend to live in the chillier ocean depths in both hemispheres, although there are some tropical species. They spend much time buried in mud where they feed on worms, and also emerge to feed on a wide variety of prey. It's long been assumed that hagfish are largely scavengers feeding on carrion, but recent work reveals them as active predators.[6]

Like lampreys, hagfishes are long and superficially eel-like, with a single medial nostril; a feeding arrangement of sucker-like mouthparts and rasping tongue supported by powerful muscles; pharyngeal pouches on either side; an unconstricted notochord; and a skull consisting of a somewhat loose collection of cartilages.

There are, however, many important differences between lampreys and hagfishes. External features of hagfishes include four pairs of tentacles on the snout; the branchial openings set well posterior to the head; and a large opening, the esophagocutaneous duct, on the left side behind the branchial openings. Uniquely for vertebrates, the body fluids of hagfishes are isotonic with seawater. The heart does not receive any innervation, being "pumped" by the activity of the pericardium; and there are several other "accessory" hearts in various other parts of the circulatory system. The atrium and ventricle of the main heart are connected but set apart from one another, quite unlike a heart in a lamprey or gnathostome. The nasohypophysial duct is not blind-ended, but continues into the roof of the pharynx. The two coeloms—the pericardium and the abdominal cavity—are connected by a large internal duct on the right hand side; and there seems to be a complete absence of vertebrae, even to the primitive extent seen in lampreys. Hagfishes are blind, or nearly so, and even though their eyes have lenses they are immobile, as there are neither muscles to move them nor oculomotor, trochlear, or abducens nerves to innervate them. The brain is compact and has no cerebellum: the ear comprises just one single semicircular canal per side, and the lateral line system is either fragmentary (in *Epatratus*) or absent (*Myxine*). The cartilages in the head bear little relationship with those in lampreys or gnathostomes. There is neither a branchial skeleton nor a distinct braincase, the brain and spinal cord being enclosed in an unbroken fibrous sheath like the skin of a sausage.

The relationships between hagfishes, lampreys, and gnathostomes have been contentious. The classical view was that lampreys and hagfish belonged in a distinct and separate group, the Cyclostomata. However, the many differences between hagfishes and lampreys, and the perceived primitiveness of hagfishes, led to the view that hagfishes were the most primitive of all vertebrates, and that lampreys stood closer to gnathostomes. In addition, researchers suggested that lampreys and hagfish could not legitimately be grouped together based on shared primitive characters notably the absence of characters seen in gnathostomes, such as jaws, paired limbs, paired nostrils, myelin sheathing of nerves,[7] and so on; and that there were positive reasons for separating lampreys from hagfishes based on features shared by lampreys and gnathostomes, but absent in hagfishes. Lampreys and gnathostomes were called "myopterygia" in recognition that in both groups, but not hagfishes, the rays of the median fins can be moved by muscles; conversely, hagfishes were excluded from vertebrates sensu stricto but included in a larger group, the Craniata, as they had a cranium (of sorts) but no vertebrae.[8]

The "cyclostome" view has, however, enjoyed a recent resurgence. The presence of very similar and somewhat specialized feeding mechanisms in both groups[9] speaks to a shared common ancestry that excludes gnathostomes, and there is no evidence for such apparatus in any of the many fossil agnathan groups. Yet more recent work has suggested that many features of hagfish craniofacial development thought to be primitive might actually represent evolution away from a cyclostome norm[10] and also that the absence of vertebrae in hagfishes might represent a secondary loss, as traces of vestigial vertebrae have been found in some hagfish, instantly invalidating the idea of hagfishes as invertebrate craniates.[11] The weak, buried, and unmuscled eyeballs of hagfish also represent a secondary loss.[12] Further evidence for the monophyly of the cyclostome clade comes from the very distinct adaptive immune system shared by lampreys and hagfishes, as discussed above (and in chapter 13) and also from molecular phylogeny.[13] The evidence for the cyclostome group now seems compelling.

Hagfishes are peculiarly hard to study. Unlike lampreys they generally live and breed in deep water. Relatively little is known about their development, and nothing at all about their reproductive habits in the wild.[14] They are fussy spawners, and eggs and embryos are hard to obtain and culture. In the past few years, Shigeru Kuratani, Kinya Ota, and their colleagues have managed to persuade captive hagfishes to spawn and produce fertilized eggs in the laboratory. Their study on hagfish embryology, confirming that the development of neural crest was essentially the same as that in other vertebrates,[15] was the first significant study on hagfish development for more than a century. A more recent study reveals marked similarities between hagfish and lamprey cranial development, indicative of a fundamental cyclostome bauplan.[16]

Lampreys are known as fossils from the Devonian Period, and both lampreys and hagfishes from the subsequent Carboniferous Period, showing that the lineages have been distinct for at least 360 million years.[17] The seas and rivers of the Cretaceous Period, 125 million years ago, contained lampreys all but identical to modern forms.[18]

By the same token, there has been time enough for morphologically diverse lineages to have evolved from within the cyclostomes that have subsequently become extinct, leaving no easy correspondence in the extant fauna. Fossils of such creatures will be hard to recognize as such, given the rather limited range of forms among modern cyclostomes with which they might be compared. A case in point is *Tullimonstrum*, from the same Carboniferous rocks in Illinois that have yielded a variety of soft-bodied agnathan vertebrates, including fossil lampreys and hagfish, and whose affinities have defied interpretation despite its popularity—it is the State Fossil of Illinois—and the existence of thousands of specimens. *Tullimonstrum* is a segmented and vaguely fish-like creature with eyes on the ends of long stalks, and a peculiar articulated snout terminating in what look like snapping jaws. Recent work has identified *Tullimonstrum* as a vertebrate, possibly on the stem-lineage of lampreys,[19] though this is likely to remain controversial. Another form is *Pipiscius*, a fossil metazoan with a feeding arrangement of two concentric circles of teeth leading into a large buccal chamber. This has been interpreted as a lamprey-like agnathan, but similar forms from China have been used to argue against this assignment.[20]

The genome of the sea lamprey *Petromyzon marinus* has been sequenced,[21] offering a new view of both the origin of vertebrates, and of the differences between cyclostomes and gnathostomes. It contains 26,046 genes, comparable in number with the genomes of other vertebrates. Study of the genome supports the view that some time after the vertebrate lineage diverged from that of the tunicates, the entire genome was duplicated, and duplicated again.[22] There must be some truth in this given that that vertebrates have (at least) four clusters of Hox genes where the amphioxus and tunicates have just one. However, evolution following the duplication events makes their reconstruction much less clear-cut than one might like, especially as many of the copied genes have since been lost in each lineage,[23] and some authors have suggested that one of the whole-genome duplication events happened before the lamprey-gnathostome split, the other in the ancestry of gnathostomes.[24] Conversely, there are signs of whole-genome duplication in the lamprey lineage alone, suggested by the discovery that the Japanese lamprey *Lethenteron japonicum* has six Hox clusters.[25]

Whatever its particulars, consequences, and order of events, it is certain that whole-genome duplication of some degree did happen during the earliest history of vertebrates, offering at least the potential for the evolution of genes with new functions not seen before. The lamprey genome study identifies 224 families of genes characteristic of vertebrates, and not known in other animals. Many of these are concerned with the organization of the nervous system, specifically neurohormone signaling, and myelination—surprisingly so, given that cyclostomes lack myelin. This suggests that the molecular machinery for producing myelin was in place before the appearance of myelin itself.

Ninety-nine percent of all vertebrates are gnathostomes, that is, vertebrates with moveable jaws articulated to a distinct braincase into which the olfactory, optic, and otic capsules, as well as the more anterior spinal nerve roots, have become integrated. The ear contains three distinct semicircular canals, and in many primarily aquatic vertebrates (that is, fishes) there is a distinct and extensive lateral line system. Gnathostomes also have two pairs of paired fins—pectoral and pelvic—rooted in girdles of cartilage or bone anchored securely in the body wall. The notochord is surrounded and even obliterated by vertebrae.

Gnathostomes also show extensive development of various kinds of hard tissue, including bone, dentine, enamel, and various forms of mineralized cartilage. The mineral fraction is hydroxyapatite, a form of calcium phosphate. The first gnathostomes were fishes called placoderms in which the head and trunk were encased in stout bony armor. Even today, the skins of sharks are adorned with arrays of small, sharp tooth-like structures or "denticles." This is why sharkskin feels rough to the touch and was once used as an abrasive, like sandpaper. Other fishes may have scales more or less clothed in hard bony or enamel-like tissue. Only later in gnathostome evolution, especially with the evolution of tetrapods, was the dermal armor restricted to the head, and enamel and dentine to the teeth.

There are two extant groups of gnathostomes. They are the cartilaginous fishes or chondrichthyans, comprising sharks, rays, skates, and their allies; and the osteichthyans, or bony fishes, comprising everything else, including tetrapods. Two other groups of gnathostome are wholly extinct. These are the acanthodians or spiny sharks, which became extinct in the Permian period (some 250 million years ago), now thought to be early offshoots of chondrichthyans; and the aforementioned placoderms, now regarded as stem-gnathostomes, more or less closely related to all other gnathostomes, and known from the Late Silurian and Devonian periods.[26] Given the increasing evidence for internal fertilization and live birth in a variety of placoderms,[27] it is possible that this represents the primitive condition in gnathostomes.

Generations of students, myself included, have been raised on dissections of sharks as model gnathostomes. The spiny dogfish *Squalus acanthias* is a typical example and this is the one Jefferies (1986) chooses as his model.[28]

Like all gnathostomes including you and me and Ogden Nash's dog, the spiny dogfish has teeth at the front, and these are typically gnathostome in that they are made of dentine and covered with enamel. The rest of the internal skeleton is, however, made of cartilage. The vertebrae are much more defined than the arcualia of lampreys, and articulate with ribs—elements not seen in cyclostomes. The cranium completely encloses the brain and is attached to a complicated branchial skeleton defining seven branchial slits. In contrast to cyclostomes, the branchial skeleton is medial (internal) to the gill lamellae, and it is likely that it is not homologous with the branchial skeleton of cyclostomes. Likewise, the muscles of the branchial arches are external to the somitic muscles, and not internal as in cyclostomes.

What would be the most anterior gill slit, between the mandibular and hyoidean gill arch, is reduced to a spiracle, possibly a consequence of the presence of the jaws. Most of the branchial skeleton (all except for unpaired ventral elements) and the part of the skull anterior to the notochord arise from neural crest: the rest is mesodermal in origin. The fins—median and paired—are also supported by cartilaginous skeletal elements.

The mouth opens into the pharynx, which is continuous with the gut, recalling the simplicity of the ammocoete larva of lampreys, with none of the complicated tongue and associated musculature seen in adult cyclostomes. Like cyclostomes, the spiny dogfish has two well-defined coelomic compartments, the pericardium and visceral cavity, connected by a duct.

The somitic musculature is very like that of lampreys, but with a gnathostome refinement: the division into dorsal (epaxial) and ventral (hypaxial) groups by a horizontal septum co-planar with the vertebral column.

The embryo of the spiny dogfish starts as a tiny dot appressed to the surface of a much larger yolk. It soon develops into a disk or blastoderm of two layers of cells with a space between, like a squashed blastula. The cytoplasm of the lower layer of cells is continuous with the yolk mass. Gastrulation starts when one end of the blastoderm becomes thickened into a "lip" and folds inward, taking parts of the external surface with it, obliterating the original cavity and forming the archenteron. As with the lamprey embryo, the lip forms the presumptive rear end of the animal. Not long after this event, the notochord forms in the roof of the archenteron, with rods of paraxial mesoderm on either side destined to form the somites of the trunk.

After gastrulation comes neurulation in very much the usual chordate pattern. A ridge in the blastoderm extends forward from the blastopore lip. The ectodermal covering of this ridge folds downward and inward, deepening until its right and left edges meet at the top and fuse, at first in the center, and then zipping up before and behind, eventually closing the anterior neuropore, and the posterior neurenteric canal connecting the nervous system with the gut.

The mesoderm on either side of the notochord divides into somites, but the division is not complete. Although the dorsal parts become segmented, the ventral mesoderm remains a continuous sheet, the lateral plate.

As the embryo grows and the yolk shrinks, the head end curls downward and neural crest begins to form, at first in the head region, spreading rearward to emcompass the ear region, the hindbrain, and the nerve cord. The optic vesicles start to appear, as do the presumptive ears—the otic placodes—just as in lampreys.

The neural crest exhibits a marked difference in behavior depending on whether it ends up in front or behind the posterior boundary of the branchial region. Anterior to that point, neural crest gives rise to the cartilages of the skull and branchial skeleton, as well as the trigeminal, facial, auditory, glossopharyngeal, and vagus nerves. Posterior to that point, neural crest gives rise to the dorsal nerve roots and the hypoglossal and spinal nerve ganglia.

The embryology of human beings is very like that of the shark, although without the yolk mass. For we, too, start as a disk, rather than a lamprey-like ball, even though there is no yolk at all, presumably an evolutionary vestige of the yolky eggs of our reptilian ancestors.[29]

More than 120 gnathostome genomes have been sequenced. Perhaps the most interesting for the purposes of this book is that of the elephant shark *Callorhinchus milii*,[30] mainly because it has evolved relatively slowly and shows many similarities with the genomes of quite distantly related gnathostomes such as humans. As anatomy of the spiny dogfish stands for that of gnathostomes in general, the genome of the elephant shark stands for gnathostome DNA.

CHAPTER 11

Some Non-deuterostomes

Over the years several animal groups besides ambulacrarians and chordates have been proposed as deuterostomes on the basis of embryological characters or adult anatomy. Notable candidates have included brachiopods, phoronids and bryozoa (ectoprocta), pogonophores and sipunculids. Molecular evidence has without exception removed these from the deuterostome picture. Two other groups have exerted more persistent claims. Although neither is now seen as belonging to the deuterostomes, their tenacity perhaps exposes the limits of what is possible with either molecular phylogenetics or comparative anatomy.

One case concerns the marine worm *Xenoturbella* and its allies, in which molecular solutions have been set in complete defiance of morphology, or lack of it. The other, the chaetognaths, shows how hard it is to place a group whose members appear to have evolved rapidly into more or less their extant forms, covering their evolutionary tracks.

Debate has rumbled about the affinities of the marine worm *Xenoturbella* since its scientific description in 1949. It is free-living and found in sea-bed mud off the coasts of Scotland and Scandinavia. It is vaguely lentil-shaped and up to four centimeters long, covered in cilia (the length is very variable, as the worm readily changes its shape). It has no brain, a blind gut (that is, with no anus), no coeloms, and a simple net-like nervous system. Many researchers considered it as a simple flatworm similar to acoels[1] based on a number of features including development.[2]

Placing *Xenoturbella* with acoels as branching off close to the root of the bilateria seemed the most sensible hypothesis[3] based on their anatomy, supported by some ultrastructural work[4] and molecular phylogenetic studies.[5] Other researchers, however, saw reminiscences of hemichordates in immunocytochemistry[6] and the structure of the extracellular matrix,[7] although Pedersen and Pedersen (1986) concluded that any resemblance between the "complex" epidermis of *Xenoturbella* and that of hemichordates was "superficial" and hinted that *Xenoturbella* was perhaps a more complicated creature than it appeared at first.[8]

But that wasn't the end of the tale: further molecular work turned up the surprising news that *Xenoturbella* was neither acoel nor hemichordate but—of all things—a bivalve mollusk[9]—based on the presence of mollusk DNA and mollusk-like eggs and larvae inside specimens.[10] Perhaps *Xenoturbella* started as a regular mollusk in its early stages, and degenerated to its simple, wormlike adulthood, leaving no trace of its affinities. Later work on DNA overturned this idea. Any mollusk DNA or eggs in *Xenoturbella* came from a mollusk-heavy diet (cue cheesy headlines such as "A Diet of Worms"). Whatever *Xenoturbella* was, it made its living by feeding on mollusks and their eggs and young.[11]

Once authentic *Xenoturbella* DNA had been identified, showing that it was not a mollusk, some researchers revived the idea that it should move in with the deuterostomes,[12] in particular the ambulacraria.[13] Not only that, but it should bring its friends the acoels along, too.[14]

What are we to make of this? Perhaps the best guides are phylogenomic studies using a large number of species and a wealth of new genomic information, made possible thanks to powerful algorithms and heavy duty computing. Even here, though, results have hardly been unanimous. One such study placed *Xenoturbella* as the sister taxon to ambulacraria, although the study did not include any acoels.[15] Another that included various acoelomorphs recovered *Xenoturbella* as sister to the ambulacraria—

but split the acoelomorphs, placing them variously as early offshoots of the Bilateria.[16] A third found that acoelomorphs were basal bilaterians, and that *Xenoturbella* was best placed within this group.[17] A recent review of acoels[18] is agnostic on the placement of *Xenoturbella* among the acoels and considers the implications of the placement of these various worms among deuterostomes, suggesting that if the latter is supported, then one might have to redefine what is meant by "deuterostome." Another review frames the problem more starkly—are acoels (with or without *Xenoturbella*) closer to the stem of Bilateria, or to deuterostomes? Either answer has profound implications for our understanding of the evolution of animal life.[19]

If *Xenoturbella* (with or without acoels) is a sister taxon to the ambulacraria, the implications are drastic. They are that either these very simple worms lost a large number of features seen as characteristic of deuterostomes and even bilaterians, including pharyngeal slits, coeloms, and a through gut; or that these complex features evolved convergently in ambulacraria and chordates. Although we know that convergence is pervasive, and that loss of structures is perhaps as common as increases in complexity, the loss of structures in acoels and *Xenoturbella*—if they are deuterostomes—has been so thorough as to mask any ancestry almost completely, leaving only some enigmatic signs in the genes and various details of physiology, all of which are open to interpretation.

Alternatively, some or all of these worms might be very early offshoots of deuterostomes, from a time when the distinction between deuterostomes and protostomes was moot, one looking very much like another, and before the acquisition of such specialized features as a through gut or pharyngeal slits.

This possibility has implications for the form of the earliest bilaterians, which some suggest is a more complex affair than acoels or *Xenoturbella* can offer—having, perhaps, coeloms, segmentation, and a sophisticated nervous system.[20] If so, then acoels are both primitive (in that they have a simple body plan) and derived (in that this simplicity evolved from earlier complexity, though perhaps not as complex as implied by an ambulacrarian).

The final possibility is that acoels and *Xenoturbella* are neither more nor less complex than they appear, representing the unadorned yet undecayed offshoots of the earliest bilaterians,[21] before the evolution of the more complex common ancestor of protostomes and deuterostomes.

The problem of deciding the interrelationships of *Xenoturbella* and acoels is the same as that which plagues the reconstruction of branching events in evolution that happened a very long time ago, as reckoned from the use of modern-day relatives whose genes and morphology have been evolving more or less rapidly from an unknown ancestor that lived more than half a billion years ago.

The very latest findings cast considerable doubt on the association of *Xenoturbella* with deuterostomes. Analysis of two hemichordate genomes[22] places acoels (without *Xenoturbella*) as a sister-group to the bilateria, well outside deuterostomes, the position of *Xenoturbella* itself being more equivocal. Although *Xenoturbella* mitochondrial DNA bears some comparison with that of hemichordates, transcriptome data from nuclear DNA places it firmly among the acoels and nemertodermatids, and indicates that both *Xenoturbella* and acoelomorpha form a group at the very base of the bilateria, branching off after ctenophores, cnidaria, and sponges, and before the nephrozoa (deuterostomes plus protostomes[23]). The conclusions follow discovery of more species of *Xenoturbella*, some quite large, between 10 and 20 cm long,[24] and phylogenomic work on a broader range of acoels and nemertodermatids than had been attempted hitherto.[25]

These results are satisfying in that they do not require multiple loss of important and hitherto fairly definitive deuterostome characters, and also in that they break the long branch between bilateria and other animals, illustrating that bilateral symmetry evolved before a through gut, metameric segmentation, and a complex, centralized nervous system. On the other hand, one cannot simply dismiss the acoelomorpha (including *Xenoturbella*) as primitive relics and leave it at that—there is substantial diversity within the group,[26] and many acoelomorphs contain features not seen outside the group, such as a distinctive muscular pharynx and so on. The last word might not have been said on the phylogenetic position of *Xenoturbella*.[27]

Like *Xenoturbella*, the position of the chaetognaths (arrow-worms) within the tree of life has been a persistent problem, and, like *Xenoturbella*, they have intermittently nipped at the heels of the deuterostomes. The chaetognaths form a group of about 100 species of specialized predators on copepods, mostly found in the plankton, although they have also been found in benthic and deep water habitats.[28] They are bilaterally symmetrical, small (a few millimeters to around 12 centimeters in length), and fierce, but their resemblance to Jack Russell terriers ends there. Distinguished by a set of grasping spines at the front, their rigid bodies are bounded by fins, making them look very like the fletchings of arrows, hence their vernacular name. The number of coelomic compartments inside any given chaetognath is debatable, and the development of the mesoderm and the body cavities is peculiar, obscuring any obvious resemblances with deuterostomes or even coelomates.[29]

The link with deuterostomes comes from observations of early embryos, which appear to undergo radial cleavage, with the blastopore forming the anus. The hindparts of the adult are post-anal, reminiscent of the post-anal tail of vertebrates. However, molecular phylogenetic work puts chaetognaths well outside the deuterostomes.[30] Mitochondrial DNA flags them as protostomes[31] perhaps close to lophotrochozoa such as annelids or mollusks.[32] Other studies, in contrast, ally them with ecdysozoans such as nematodes, though there might be long-branch effects[33]; or, yet, close to priapulids, though placement is likewise uncertain.[34] More cautiously still, some work has placed them as basal protostomes, outside both lophotrochozoa and ecdysozoa, with many unique features of their genetic architecture.[35] Hox gene structure contains features that are reminiscent of lophotrochozoa and ecdysozoa as well as unique to chaetognaths.[36] Such a uniqueness could suggest branching from bilateria before the deuterostome-protostome split.[37]

Chaetognaths have a long if spotty fossil record. Phosphatic spines from various Palaeozoic settings known as protoconodonts are very similar to the grasping spines of chaetognaths.[38] Body fossils are known from the Lower Cambrian of China,[39] the Middle Cambrian Burgess Shales of British Columbia, as well as the Carboniferous Mazon Creek fauna of Illinois. These fossils look essentially modern so shed no light on chaetognath interrelationships. The fossils *Amiskwia sagittiformis* and *Oesia disjuncta* from the Burgess Shales have been compared with chaetognaths, but detailed work has challenged these interpretations[40]: *Oesia* is now interpreted as an enteropneust.[41]

The consensus, then, is that there is no consensus. The phylogenetic placement of chaetognaths remains one of the great unresolved mysteries of animal phylogeny. The lesson of this protracted indecision seems to be that the chaetognaths appeared a long time ago and, having evolved, came to adopt more or less their present form rather quickly. It is possible, even likely, that chaetognaths represent a good example of a long branch, unbroken by intermediate forms that have become extinct and have not yet been discovered either as fossils or in the extant biota.

CHAPTER 12

Vertebrates from the Outside, In

12.1 INTRODUCTION

In chapters 6–10 I set out the basic biology of deuterostomes, group by group. A number of themes have emerged. First, each group has its own idiosyncrasies, from which one can learn very little about vertebrate origins. Apart from vertebrates, the most successful deuterostomes, in terms of numbers of species, are echinoderms and tunicates. The biology of echinoderms is dominated by its peculiar scheme of biomineralisation, which obscures rather than enhances relationships with other deuterostomes. Tunicates, on the other hand, have evolved so rapidly and have diverged so far from their common ancestry with vertebrates that very little can be discerned about the nature of that legacy. The other groups—hemichordates and amphioxus—seem to have evolved more slowly and might shed more light on deuterostome evolution. However, these groups are highly depauperate so it is hard to gauge how representative the present members of those groups are of the groups as a whole, over the course of an evolutionary history that goes back to the Cambrian. More positively, we've seen that quite a few of the features seen in vertebrates are seen, in one form or another, in other groups. Overall, deuterostomes do provide a guide to the stepwise accumulation of characteristics in vertebrates, even if that guide is subject, in places, to a great deal of interpretation.

In this chapter and the next, I shall review the same material but from a different angle—that of embryonic features, organs, and systems. In this chapter I shall look at all those features most commonly discussed as setting vertebrates apart—things like the notochord, the head, segmentation, and so on. In the next I shall look at some equally important and characteristic features of vertebrates that don't usually get the same billing, because they are less visible: features like the heart, blood, immune system, the various parts of the gut, the kidneys, reproductive system, and so on. In all cases I shall examine their possible antecedents and equivalents in other deuterostomes and animals more generally. The aim will be to start to fill in the gap between vertebrates and other animals.

The topics in this chapter are, in general, arranged in the order in which they are met during the development of a typical vertebrate. This seems like a sensible way to proceed as some features are contingent on their predecessors. The first is the so-called "organizer," that special region of the dorsal lip of the blastopore that, at gastrulation, lays down the body axis and without which many of the features of vertebrates and other chordates cannot develop.

Gastrulation is accompanied by the emergence of the notochord, that eponymous characteristic of all chordates. In chordates the development of the notochord is tied to the activities of the organizer, and it becomes a potent organizer in its own right. The emerging notochord is flanked by two rods of mesoderm. This, the paraxial mesoderm, becomes divided into somites. So much is true in vertebrates and the amphioxus. But there are important differences between the organization and fate of somites between amphioxus and vertebrates.

Signals produced by the notochord trigger neurulation, that is, the formation of the distinctive chordate nervous system, with its dorsal, hollow nerve cord and, in vertebrates, the enlarged brain. Once the main features of the central nervous system have been formed, cells from either side of the developing neural tube migrate outward and transform other tissues into various structures. This population of migratory cells, the neural crest, is responsible for many features we instinctively see as vertebrate, particularly the structure of the head and face.

I shall discuss the extent to which the vertebrate head is an entirely new structure or an elaboration on ancient foundations. This discussion touches on a larger, historical question, between the "segmentationists," who thought that the vertebrate body could be divided into segments from stem to stern; and the "antisegmentationists," who view the vertebrate body as fundamentally bipartite.

Finally in this chapter I shall look at hard tissues. Many deuterostomes have cartilage, or a tissue very like it, supporting either the branchial skeleton (in enteropneusts and the amphioxus), or much else (in vertebrates). Cartilage is a tissue with a deep and ancient heritage: the innovation seen in vertebrates is its mineralization into a variety of distinctive hard tissues such as bone, dentine, and enamel.

12.2 THE ORGANIZER

Just before gastrulation, the early embryos of chordates develop a thickening, overhang, or "lip" at one end (fig. 3.1). Tissue flows or involutes beneath this lip, creating a pocket that pushes into the embryo, obliterating the blastocoel and replacing it with the archenteron (that is, the presumptive gut) bounded by a two-layered structure: the gastrula. The opening beneath the lip (nowadays called the dorsal lip) is the blastopore and forms the presumptive anus (fig. 3.3).

In a now-historic series of experiments, a German doctoral student named Hilde Mangold discovered that when this lip was taken from the gastrula of a lightly pigmented newt (*Triton cristatis*) and transplanted onto the gastrula of a related, darkly pigmented species (*Triton taeniatius*), it developed into an entire body axis, including a notochord, neural tube, and somites, persuading the surrounding tissue of the host to take part.[1] The lip was somehow special, as other pieces of transplanted tissue simply merged into the host milieu, without exerting any influence at all.[2]

Mangold used two differently pigmented newts so she could tell the difference between the fates of the grafted and the host tissue. The grafted lip always differentiated and elongated into a notochord and sometimes parts of the floor of the neural tube and the occasional somite. The surrounding host tissue contributed most of the rest of the neural tube as well as most of the somites, the kidney, and the gut. Neither graft nor host produced anything resembling a head. Nevertheless, the transplanted lip had persuaded (Spemann's word was "induced") the otherwise indifferent neighboring tissue to differentiate into a variety of tissues it would not otherwise do, and organize them relative to one another into a recognizable chordate body axis.

Although the organizer clearly formed a second body axis when transplanted, Spemann (after Mangold's death) remained unable to perform the obvious follow-up experiment. That is, what happens to embryos when the organizer is excised? How would organizer-less embryos fare? His failure had less to do with his lack of expertise than the particularities of newts. Another amphibian, the African frog *Xenopus laevis*, proved a more than adequate replacement and has since been the model organism of choice for this kind of work. As Gerhart describes in his review (2001), a frog embryo from which the organizer has been removed develops into what Spemann called a *bauchstück* ("belly piece"). Gastrulation of a sort calmly proceeds; the blastopore closes to a point. The embryo does not elongate by convergent extension (which, as we have seen with tunicate tadpole larvae, is a distinctive feature of notochord formation), and there is no notochord, nerve cord, or pharyngeal slits. There is, however, coelomic mesoderm, blood cells, and posterior gut, and the embryo can live as long as there is still yolk to digest. Without the organizer, however, the embryo simply does not develop the features we think of as distinctively chordate, in particular the notochord and dorsal tubular nerve cord.

The organizer is not some formless magic blob of potency. Spemann suspected that it had some kind of internal architecture, and so it has proved. Removal of progressively greater amounts of organizer tissue results in embryos that look more and more like a *bauchstück*. Conversely, encouragement of greater-than-usual development of organizer tissue results in more head-dominated little monsters.[3] Clearly, the proportion of embryonic tissue in an organizer is important for normal, balanced development.

How does the organizer work? Spemann thought that it contained substances that induced the naive tissues round about to develop. There was an alternative model, however, in which tissues would prepare themselves to develop but would be blocked from doing so until the restraint was lifted by the organizer: like actors, fully cognizant of their lines, but waiting for the appropriate cue to deliver them. Evidence for the latter view came from experiments showing that excised ectoderm from gastrulae could develop into eyes, or pieces of brain tissue, when subject to shocks such as sudden changes in acidity or salt concentration. The shock would jolt any restraints from their moorings.

This is, essentially, the current model. Signaling by proteins of the Bone Morphogenetic Protein (BMP) and Wingless (*Wnt*) families from the "ventral zone" at the ventral side of the embryo, antipodal to the dorsal lip, blocks the development of cells otherwise competent to do so unless the siege is relieved by one or more of a large number of inducers from the organizer in the dorsal lip itself. These substances are evoked by the activities of other factors, notably *Nodal* and β-catenin,[4] and there are a formidable array of them including (but not limited to) *goosecoid*, *noggin*, *chordin*, *cerberus*, and *dickkopf*—the functions of some of which can be guessed from their names. The organizer can also regulate itself by negative feedback, like the governor on a steam engine. One participant in the system is a factor called Anti-Dorsalizing Morphogenetic Protein, or *Admp*.[5] This is usually repressed by BMPs. When BMP activity fades, *Admp* awakes and binds to *chordin* in the organizer, and the bound pair shuttle to the ventral zone where they promote BMP signaling, therefore shutting down further *Admp* and *chordin* synthesis.[6]

If you look at a genetic circuit diagram in which the interactions of all these players are set down, it looks like a Mexican stand-off in which everything appears to repress everything else. But the secret of development, as with comedy, lies in the timing. The factors vary in concentration from one part of the embryo to the other, and these relationships change as the embryo grows and develops. The result is that the interaction of all these factors does not result in stasis, but sets up gradients in concentration whose values allow the cells to develop in ways appropriate to their station.

The precise details of the molecular interactions between the organizer and the ventral zone are still being worked out and are in any case beyond the scope of this book. Suffice to say that they have been dissected by the careful study of mutations, some of which have dramatic and somewhat alarming results. When left unchecked, the gene *cerberus* produces frogs with multiple heads[7]; the deletion in mice of *Lim1*, a transcription factor associated with the organizer, results in a creature of perfectly normal appearance apart from the total absence of a head anterior to the ears.[8]

Does the organizer have antecedents and parallels in other deuterostomes? Back in the 1960s, Tung and colleagues managed to reproduce Mangold's experiments with the amphioxus, showing that the dorsal lip of an amphioxus gastrula, when transplanted to another part of another amphioxus gastrula, induced the development of axial structures.[9] This suggested that the amphioxus had something like an organizer. It took another half-century to demonstrate that the organizer in the amphioxus works much as it does in vertebrates, that is, by a kind of balanced antagonism between BMPs and *Wnt*-mediated signaling and a host of dorsalizing factors such as *chordin*, although there are some differences in detail.[10] This was more than mere confirmation, as there had been considerable doubt that the organizer was anything more than a vertebrate-specific innovation.[11]

Tunicates, in contrast, seem to have dispensed with the organizer as they have with much else, raising the question of how they produce chordate-specific structures such as the notochord. Although induction by the notochord is necessary to produce the overlying nerve cord, as in chordates generally, the expression of organizer-related genes such as *chordin* is rather different, and not associated with anything like an organizer.[12] It could be that rapid development and the extreme exiguity in cell numbers in tunicates has dispensed with the need for the inductive mode of development otherwise required to synchronize the behavior of hundreds or even thousands of cells.

Hemichordates do not have an organizer, though work on *Saccoglossus* shows that patterning of the dorsoventral axis is governed by a balance between BMPs expressed on the dorsal midline and antagonists such as *chordin* and *Admp* expressed ventrally.[13] This suggests that the interplay between BMPs and their antagonists represents an ancient patterning system co-opted into the chordate organizer, along with mechanisms for anterior-posterior patterning, which are separate in *Saccoglossus*, as in protostomes such as *Drosophila*. In contrast, the development of neural tissue in *Saccoglossus* is not repressed by BMPs, in contrast to both chordates and *Drosophila*, and might be a reflection of the diffuse rather than centralized nervous system in hemichordates. It is possible that this particular role of BMPs was acquired independently in insects and chordates.

There is, however, intriguing evidence for something like an organizer in echinoderms.[14] The ventral ectoderm of a sea urchin (corresponding with the dorsal side of a chordate) expresses *Nodal*, as does the organizer in chordates. Knocking out *Nodal* results in completely dorsalized embryos that are fully radial and lack a mouth. Injection of *Nodal* into such embryos restores them to normality, showing that *Nodal* confers the properties of an organizer. Crucially, embryos doubly injected with *Nodal* produce a pair of perfectly formed conjoined larvae, fused back to back. As in chordates, *Nodal* expression prompts the expression of BMPs and *chordin* that, between them, set up suitable conditions for the determination of the body axis.

Other factors found associated with the chordate organizer are also found in echinoderms and contribute to the specification of the dorsoventral axis, including *goosecoid*, *Lim1*, and *lefty*. There is, however, not just one kind of *Admp*, but two, each expressed at the opposite pole relative to the other, and subject to different transcriptional control. This may be a feature specific to echinoderm development.

As Lapraz et al. (2015) suggest, a genetic circuit involving the interplay between repressive BMPs or *Wnts* and molecules such as *chordin* is likely to have a very long pedigree, where it is put to work specifying various body axes. In the ancestry of deuterostomes it seems to have become very much associated with the dorsoventral axis where, in that form, it assumed a degree of potency that one could call an organizer. The form of the organizer observed in echinoderms seems too much like that of chordates to be a parallel development, in which case the system in hemichordates seems to be secondarily simplified. However, it could be that we simply do not know enough about hemichordates at this point.

What seems to be special about the chordate organizer is that it's more than just the mechanism for specifying the dorsoventral axis. With the formation of the notochord, it specifies the anterior-posterior axis, too. The emerging, converging, extending notochord acts as a powerful signaling source, way beyond the confines of the dorsal lip. It's to the notochord that I now turn.

12.3 THE NOTOCHORD

The notochord is the product of molecular activity within the organizer during gastrulation.[15] As discussed above, a mass of cells converges on the dorsal midline and is then extruded out backward to form a single file. Work on zebrafish *Danio rerio* has identified the genes without which this process cannot occur.[16] The resulting structure is the notochord, the defining axial structure of the chordates. The formation of the notochord radically redefines the axes of the embryo and is the reason why, in chordates, the formation of the anterior-posterior axis, determined by the notochord and its signaling activities, is so intertwined with the formation of the dorso-ventral axis, whose signaling helps determine the structure of the notochord itself and the position of the overlying central nervous system. The notochord also helps specify left-right asymmetry, demonstrating its central role in chordate development as well as structure. Perhaps surprisingly it is also involved in the development of the heart and aorta,[17] as well as endoderm derivatives such as the liver and pancreas.[18]

The structure and fate of the notochord differ widely among chordates. In the amphioxus it is present throughout life and becomes somewhat muscular. In vertebrates, and those tunicates that retain a notochord, the cells become vacuolated and expanded, creating a structure rather like a string of water-filled plastic bags bounded by a tough external sheath, whose turgidity provides support for the tail muscles. Among vertebrates, hagfishes and some gnathostomes such as sturgeons and coelacanths have unconstricted notochords of a similar sort. In most vertebrates, however, the notochord is a transient structure, soon replaced by the vertebral column. Before it disappears, though, it acts as a source of *Sonic hedgehog* and BMP antagonists that help pattern the overlying neural plate.

As a tissue, the notochord has much in common with cartilage. Notochord cells express various types of collagen. The essential difference, though, is that cartilage derives many of its structural properties from the extracellular matrix secreted by its cells, whereas in the notochord the substances characteristic of such a matrix are retained in the vacuoles of its constituent cells. Notochord tissue is like cartilage turned inside out. When bones form from cartilage, extracellular matrix rich in type II collagen becomes infused with a different form of collagen, type X, a harbinger of bone. There are clear parallels with the replacement of notochord by vertebrae: notochord running through each vertebra expresses type X collagen before becoming ossified.

Notochord tissue between the vertebrae does not express type X collagen and so does not become bone. Instead it ends up as the *nuclei pulposi*, slugs of gelatinous matter in the centers of intervertebral discs. When, once upon a time, I lifted a heavy weight and thereby ruptured an intervertebral disc, squeezing out some of this matter, this was just the hem of my chordate heritage showing through. Another such manifestation is chordoma, a rare cancer of parts of the spine and skull base. As Nibu and colleagues (2013) describe, this originates in the remnants of the notochord and is related to spontaneous or hereditary duplication of *brachyury*, the gene most associated with notochord identity.

Brachyury comes from the Greek for "short tail." The gene is also known as *T*—standing for "tail"—and thereby, if you'll excuse me, hangs a tale. The name *brachyury* was originally applied to mutant mice with vertebral abnormalities and shorter tails than normal. The mutation was found to be lethal when homozygous, with embryos failing to form a notochord; the extra-embryonic sac called the allantois; and much of the posterior of the animal, alongside additional defects in the neural tube and somites. Not surprisingly the embryos died after about ten days of gestation. The *brachyury* gene encodes a transcription factor absolutely essential for the formation and development of the mesoderm, in particular the notochord and the structures that depend on the notochord's prior formation.[19] The importance of *brachyury* to the formation of the notochord is underscored by the discovery of its notochord-related expression in other vertebrates, tunicates, and the amphioxus.[20] In ascidians, *brachyury* is expressed exclusively in the notochord[21] though in larvaceans it is also expressed in posterior endoderm and hindgut.[22]

Brachyury is not alone—it is the founding member of a whole family of transcription factors known collectively as T-box (*Tbx*) genes.[23] Homologues of *brachyury* have been found in all multicellular animals so far examined, including those without notochords or even mesoderm, such as placozoa, cnidaria, and ctenophores, as well as protists considered close relatives of metazoa, such as the filasterian amoeba *Capsaspora owczarzaki* and a fungus, *Spizellomyces punctatus*. Only plants seem to be without it,[24] suggesting an origin in the common ancestry of opisthokonts—that is, the group that includes multicellular animals, fungi, and their protist relatives (fig. 3.4). Its primary function—at least in metazoa—seems to be the regulation of cell adhesion and cell movements, especially in gastrulation.

In the ctenophore *Mnemiopsis leidyi* and the sea anemone *Nematostella vectensis* it is expressed in cells around the blastopore.[25] In protostomes—or at least insects—*brachyury* is required for the formation of endodermal derivatives of the mid- and hindgut,[26] recalling its role in the development of the liver and pancreas in the mouse. From these results, Kispert et al. (1994) explicitly suggested homology at some level of notochord and insect hindgut.

The homologue of *brachyury* in the sea urchin *Lytechinus variegatus* is expressed in a ring around the blastopore and is involved in cell movement associated with gastrulation[27]: in the enteropneust *Ptychodera flava* it is similarly expressed around the blastopore[28] but not in the stomochord, thus weakening any claim that this structure is related to the notochord. At later embryonic stages in both echinoderms and enteropneusts, *brachyury* is variously expressed in the coeloms,[29] suggesting an ancestral deuterostome role in foregut and especially hindgut specification.

In summary, then, *brachyury* seems to have an ancestral role in the control of cell adhesion and movement. In the evolution of the metazoa, its expression became concentrated around the blastopore, whence, during gastrulation, it could direct the patterning of mesoderm, especially in the posterior parts of the animal, as well as parts of the endoderm. The blastopore is, of course, the site of the vertebrate organizer, and in that context *brachyury* helped direct the distinctive cellular choreography of convergence and extension required in vertebrates and other chordates to create the notochord. So although the notochord is a distinctive feature of chordates, one can see clear antecedents in regard to its formation and function.

Some recent work has suggested a connection—whether a homology or a parallelism—between the notochord and a mesodermal, midline structure in the annelid worm *Platynereis* dubbed the axochord.[30] This stems from work on the homologue of *brachyury* in *Platynereis*[31] suggesting that the common ancestor of protostomes and deuterostomes had a tripartite, tube-shaped gut with distinct mouth and anus. Whereas it is likely that all metazoa share some similarities in expression and even function of *brachyury*, it is more likely that the axochord and the notochord represent independent expressions of the development of a mesodermal midline structure in which *brachyury* would naturally be involved. The debate ties into a larger question, that is, the state of the last common ancestor of bilaterian animals. Was the last common ancestor a simple, unsegmented worm-like creature with a sac-like gut, similar to an acoel? Or was it a more complex creature with a through gut, some form of segmentation, and presumably some kind of skeleton—hydrostatic or muscular—against which the segments could gain purchase during locomotion? That is a question beyond the scope of this book. Segmentation, however, is well within our purview, and forms the next segment, if you will, of this chapter.

12.4 SOMITOGENESIS

We last looked at our archetypal vertebrate embryo during gastrulation, as tissue flowed inward through the blastopore lip and extended forward internally. The notochord forms in the roof of the resulting cavity, the archenteron, and strips of paraxial mesoderm form on either side.[32] These strips become segmented in pairs, one of each pair on either side of the notochord.

As the embryo lengthens, the blastopore lip, still a potent organizing center, becomes known as the tail bud. Segmentation starts at a remote remove from the tail bud, however, at the anterior ends of the mesoderm strips, at the level of what will become the otic vesicles. The pairs of somites form from the front, with each successive pair forming posterior to the one before. The somites start from mesenchyme—basically, stuffing—but as each pair pinches off the somites become more organized and epithelial, that is, boxes of layered cells each bound by a basement membrane and surrounding a hollow coelomic space. The entire process is, however, controlled from the tail bud.

Perhaps the best model so far advanced to explain somitogenesis is the "clock and wavefront."[33] In this model, biochemical changes in the cells turn each cell into a clock, so arranged that the clocks in adjacent cells are in phase—that is, locked to one another. At the same time, a wavefront of slow, biochemical change travels along the mesoderm from the anterior toward the posterior, as the body as a whole elongates. The timing of the clock and the progress of the wavefront are so arranged that when the wavefront meets a group of cells in a permissive phase of their cycle, they undergo a series of rapid changes that results in their budding off to form a somite.

Cooke and Zeeman (1976) proposed that the basic parameters of clock and wavefront were characteristic of a given species, thus explaining both the variety in the numbers of somites in vertebrates, and how this number remained consistent within species—from 30 pairs in some fish to several hundred in snakes. In all vertebrates so far examined, the clock-

and-wavefront works through the paraxial mesoderm until it is exhausted: the vastly greater number of somites in snakes relative to other vertebrates is a function of the segmentation clock of snakes ticking especially quickly, producing a larger number of smaller somites.[34] In mice, a pair of somites buds off every two hours to form 65 pairs: in humans, a new somite pair forms every four to five hours to give a total[35] of between 38 and 44.

The precise details of the movement within the segmentation clock vary from one vertebrate to another, but in general they involve three signaling pathways encountered elsewhere: *Notch*, *Wnt*, and FGF. Precisely what it is about these three pathways that makes the pace of each distinctive pulse is elusive. It is likely a result of interactions between members of all three pathways. However, there is much evidence that the pacemaker involves the expression of a one or more genes in a family called, for various arcane reasons I needn't detail here, the *hairy-enhancer-of-split-related* (*HER*) genes. Expression of *HER* genes tends to start in a broad region in the posterior end of the presomitic mesoderm, the domain moving forward, narrowing and slowing as it goes.

HER genes encode proteins that interact with *Notch* proteins. These latter are receptors sitting on the surfaces of cells. Once activated, they transmit signals to the cellular interior, triggering a cascade of further gene expression. Many genes in the segmentation clock mechanism operate by negative feedback. That is, when the production of messenger RNA or protein from the gene exceeds a certain threshold, it represses its own further production. When the concentration falls below a certain level, production resumes. This seems to be true of the *HER* proteins: the delay between their initial production and their subsequent self-repression seems to correspond with the period of somitogenesis.

So much for the clock. What of the wavefront? This is set up by opposing concentration gradients of various molecules along the strips of paraxial mesoderm. The sweet spot—the crest of the wave—moves posteriorly along the mesoderm as it lengthens, somites budding off at the front and more mesoderm being added at the tail bud behind. The gradient is set up, in part, by various members of the FGF family (and possibly *Wnt* family). The genes for these are expressed in the paraxial mesoderm in the tail bud but expression tapers off, so concentration of the gene products (messenger RNA and protein) decays as one moves anteriorly. The FGF/*Wnt* gradient is reinforced by an opposing gradient of retinoic acid, an important morphogen. This is concentrated near the front of the presomitic mesoderm and tapers posteriorly.

Retinoic acid has another key role, and that is ensuring that the paired somites line up symmetrically on the right and left. In embryos in which the effects of retinoic acid are blocked, somite formation becomes somewhat haphazard, as the embryo responds to the very strong left-right asymmetry coordinated by the *Nodal* pathway. The activity of retinoic acid in buffering the presomitic mesoderm against this asymmetry explain why somites (and their mature derivatives, such as vertebrae, ribs, and bilateral musculature) are arranged symmetrically, in marked contrast to the asymmetrically arranged heart and viscera.[36]

Once the wavefront has passed, a genetic program in the maturing somite is activated to shape it such that it has a distinct anterior and posterior region. This anterior-posterior division is important: it ensures that neural crest cells and nerve axons penetrate the anterior parts of somites, and governs the formation of vertebrae, in which a vertebra forms from the fusion of the posterior part of a somite to the anterior of the one behind.

In addition to a sharp anterior-posterior division, mature somites divide up dorsoventrally into three components. These are the dorsal dermatome, forming the dermis of the back; the myotome, forming the segmental trunk musculature; and the ventralmost sclerotome, which forms vertebrae. Each somite is different from the other, its position in the trunk being specified by Hox genes that control regional differentiation.[37] It is the *Hox* genes that are largely responsible for the regional variation of vertebrae in our spinal column, for example.

Do the somites of vertebrates have antecedents? The amphioxus has somites, whose development uses many of the same molecular factors as found in the development of somites in vertebrates.[38] In the amphioxus, though, the somites run—as the notochord—all the way to the anterior tip of the animal. In addition, they develop rather differently from vertebrate somites. The rostral somites (those closest to the front end) are generated by enterocoely from the archenteron. FGF signaling is restricted to these rostral somites.[39] The others form successively by schizocoely from the tail bud, forward; in contrast to the vertebrate system in which they form from the front, backward.[40] If their formation is governed by any kind of clock, there seems to be no wavefront.[41] The clock-and-wavefront mechanism seems to be a uniquely vertebrate innovation. So, too, is the system whereby retinoic acid buffers the somites against prompts toward left-right asymmetry. No such system is found in the amphioxus[42] and the somites aren't arranged in a neat two-by-two array, but staggered in a "herringbone" pattern, with the left somite being placed slightly forward with respect to that on the right.

For all these differences, the somites of vertebrates and the amphioxus are sufficiently similar to be considered homologous. The common ancestor of amphioxus and vertebrates would have had somites, in which case one would expect them to have been present in the common ancestor of vertebrates and tunicates. But if tunicates ever had somites, they show no signs of them today. The muscles on either side of the notochord in a tunicate tadpole's tail do not develop from the segmentation of paraxial mesoderm, but directly from muscle cells that arise early in development and are subsequently arranged on either side of the tail.[43]

Despite the many differences between the generation of somites in the amphioxus and those in vertebrates, it is notable that in both systems there is a clear disparity between the development of mesoderm in the anterior of the animal compared with that of the posterior. The rostral somites of amphioxus form by different rules from those that govern the remainder, so one is entitled to wonder whether these rostral somites are homologous, at least in a very general sense, with the mesoderm in the vertebrate head that otherwise plays no part in vertebrate somitogenesis.

12.5 SEGMENTATION AND THE HEAD PROBLEM

So we come to perhaps the most intriguing and difficult problem, the one at the heart of any discussion of vertebrate origins—that is, the evolution of that most distinctive structure, the vertebrate head.

But first, we must ask ourselves, What do we mean by segmentation? Many animals are divided into a number of similar sections strung together like carriages in a train. In many cases these are the germs of individuals, each one eventually going its own way as a complete animal. This can be seen, for example, in the strobilation of tapeworms, where each segment is in fact an individual; and in the formation of zooids in some tunicates. In the present context, however, segments are serially repeated sections in a single individual, each section homologous to another, and which together form an integrated structure. This is so-called metameric segmentation, and occurs widely in the animal kingdom, most notably in annelid worms, arthropods, and of course vertebrates. Somites represent one sort of metameric segmentation.

Metameric segments can form in a number of different ways. In the fruit fly, *Drosophila*, the first creature in which the molecular underpinnings of segmentation were understood, all the segments coalesce at once from un unsegmented, syncytial primordium. In other insects, segments are formed serially, as somites are in vertebrates and the amphioxus. As we have seen with somites in the amphioxus, segments that appear serially homologous can form in different ways even in the same animal.

In vertebrates there are, in fact, three quite separate segmental systems. These are the pharyngeal arches, ultimately derived from endoderm; the somites, from paraxial mesoderm; and the rhombomeres in the embryonic hindbrain, derived from ectoderm.

The degree to which vertebrates appear segmented as a whole depends on the interplay between these separate systems.[44] As I have shown, pharyngeal arches are an ancient feature found in all deuterostomes, at least primitively. Hemichordates have pharyngeal arches substantially homologous with those in chordates, even though these creatures lack a sophisticated nervous system or segmented mesoderm. Somites are the segments that form in the twin strips of paraxial mesoderm that form alongside the notochord in the trunk in the amphioxus and vertebrates, and presumably in the ancestry of tunicates. Rhombomeres are segmental divisions of the embryonic hindbrain in vertebrates, and I shall discuss these later in connection with the neural crest. Even though these three segmental systems appear wholly integrated in any particular animal, each one evolved in its own evolutionary context.

The reason I described somitogenesis before discussing segmentation more generally is that the segmentation of paraxial mesoderm in vertebrates is entirely obvious for all to see without much in the way of interpretation. Moreover, the mechanisms of somitogenesis can be unpicked and understood more or less without reference to other systems. At the same time, we can see how segmental systems can become integrated and impose patterning on other tissues, even other segmented systems. A good example is the division of vertebrate somites into anterior and posterior compartments such that only the anterior compartment of each somite is innervated, each by a spinal nerve from the nerve cord.

Understanding the role of segmentation in the origin of the vertebrate head offers complexity of an entirely different order. There are two principal causes of confusion. In vertebrates, but not in the amphioxus, much of the head is formed by the interaction of ectodermally derived neural crest with prechordal mesenchyme in the head.[45] The neural crest is entrained into a number of distinct streams, each associated with a rhombomere, and to make matters still more complex the neural crest interacts with the pharyngeal arches, the result being the creation of the dermal skeleton—that is, much of the head and face—as well as parts of the visceral (that is, pharyngeal) skeleton. Worse still, the degree to which the neural crest dictates the formation of certain tissues does vary somewhat between species. Although the rostral region of the amphioxus has its own peculiarities not seen in vertebrates, it is, as a whole, relatively uncomplicated, with a neat array of somites extending all the way to the anterior tip of the animal; each somite innervated by a spinal nerve; the whole entirely unobscured by a skull, eyes, ears, or any other neural-crest-derived impedimenta. One is entitled to ask whether germs of a head exist in the amphioxus in some occult form, or if the animal is essentially a headless trunk.

Ideas about the evolution of vertebrates have tended to fall into one of two camps, known as "segmentationist" and "antisegmentationist."[46] In the first view, vertebrates are thought to have evolved from a highly segmented creature similar to the amphioxus, and vertebrate structures such as the head are elaborations on structures already present in that ancestor. The antisegmentationist view, in contrast, holds that although various organs and systems of vertebrates are indeed segmented, this need not say much about the condition of the vertebrate ancestor, especially if it was essentially unsegmented, like a tunicate; bipartite, like a tunicate tadpole, with a muscular and possibly segmented tail portion and a blob-like head retaining a pharynx segmented according to its own rules; or headless, as the amphioxus appears to be, with its notochord extending all the way to the anterior tip.

The segmentationist view goes back to nature-philosophers such as Goethe who imagined that the entire vertebrate skeleton could be read as a series of more or less modified vertebrae. The skull of vertebrates was essentially a series of vertebrae fused together, somewhat like the sacrum only more extreme, such that discerning the precise number was a matter of debate. The modern era of segmentation owes much to the studies of the young zoologist Francis Maitland Balfour (1851–1882),[47] who worked on sharks, and his successors, notably Edwin Stephen Goodrich (1868–1946). In their view, the head of a shark could be sliced into segments, each one having its own ectodermal (cranial nerve), mesodermal (somitic), and endodermal (pharyngeal) derivative, and that these segments in the head formed a continuous, serially homologous series with the somites of the trunk.[48] Deviations from this essentially segmental pattern could be put down to secondary modifications or losses. The amphioxus, with its very clear segmental structure, could be cast as an ancestor, despite the fact that its gill slits develop in a decidedly odd and asymmetric way without reference to the organization of the somites.

If the amphioxus was the poster child for the segmentationists, the tunicate tadpole larva was for the antisegmentationists, especially the paleontologist Alfred Sherwood Romer (1894–1973). He interpreted the vertebrate body as a kind of fusion of two entirely discrete sections: a "somatic," comprising the central nervous system, the axial skeleton, and the somitic mesoderm, and the "visceral," comprising the pharynx and other essentially endodermal structures.[49] Romer's "somatico-visceral animal" speaks to a tunicate heritage as Goodrich's idealized diagrams of vertebrate head structure owe much to the amphioxus. Perhaps the most extreme antisegmentationist stance of recent times was taken by Gans and Northcutt (1983), who held that the greater part of the vertebrate head was, essentially, a neomorph, without antecedents. Their argument was based on the fact that much of the vertebrate head is a result of the activities of neural crest, an embryonic tissue absent from the amphioxus, and equally unknown (at the time) in tunicates.

Here I shall concentrate on mesoderm. No serious researcher nowadays would adopt either an extreme segmentationist or antisegmentationist view. Goodrich-style segmentationism is exploded by the demonstration that the segmentation seen in cranial nerves, mesoderm, and pharynx each has its own evolutionary history.[50] On the other hand, the mesoderm of the vertebrate head appears to be governed by different laws from those that rule the somites of the trunk, yet there are clear correspondences to be drawn between the disposition of unsegmented mesoderm in the vertebrate head and the seemingly more structured somitic arrangement of the amphioxus. Is the mesoderm of the vertebrate head primitively unsegmented (the antisegmentationist view) or did it once have segments that correspond with amphioxus rostral somites (the segmentationist view), those once-clear divisions, but now lost?

Linda Holland and colleagues (2008) adopt an unashamedly segmentationist stance, and it is perhaps no accident that much of their evidence comes from the amphioxus. Lampreys contain three pairs of mesodermally derived head cavities, which can be considered as homologues of the 8–10 most rostral somites of the amphioxus, particularly as they form by enterocoely, just as they do in the amphioxus, and in contrast to the remaining somites. The eye and jaw muscles of gnathostomes derive from unsegmented head mesoderm and appear homologous with mesodermal head cavities in sharks, which form from solid blocks of tissue (that is, a form of schizocoely). The evidence is consistent with the evolution of the head mesoderm from a segmented, amphioxus-like ancestor, with the progressive loss of overt segmentation in lampreys and sharks, and the transformation of the once-segmented mesoderm into eye and jaw muscles. From these observations one can construct a scenario in which the mesoderm of the vertebrate head, including the muscles of the eyes you are using to read this, are evolutionary derivatives of the anteriormost somites of an amphioxus-like ancestor. This scheme has substantial support from genetic and ultrastructural studies showing in particular that the anterior central nervous system of the amphioxus is organized into regions in very much the same way as a vertebrate brain. The cerebral vesicle corresponds to the vertebrate diencephalon (though the telencephalon is lacking); there is a distinct midbrain-hindbrain boundary, and the eye-spot at the anteriormost tip of the nervous system is homologous with the paired eyes of vertebrates.[51] Admittedly, the organization of the nervous system is only tangentially related to the homologies between mesodermal compartments in the amphioxus and vertebrates, but it does establish that at least part of the anterior end of the amphioxus corresponds with cognate parts of the vertebrate head.

In the antisegmentationist corner we find Shigeru Kuratani and his associates,[52] who draw a distinction between the development of mesoderm in amphioxus and that in the vertebrate head. In particular they wonder whether the head cavities of sharks are not in fact homologous with the head cavities in lampreys and, by extension, the amphioxus, and that at least some of the mesodermal features seen in the heads of gnathostomes are peculiar to that clade. If there are any mesodermal segments in the heads of vertebrates that are in any way comparable with the egregiously visible somites of the trunk, they are exceedingly hard to see.[53]

For example, Kuratani and colleagues (1999) failed to find anything equivalent to myotomes in their studies on the head of the lamprey, the preotic mesoderm appearing essentially as a continuous mass until secondarily "regionalized" by the development of the pharyngeal pouches and the otic vesicle. The preotic mesoderm does develop by enterocoely, as it does in the amphioxus, but its segmentation into discrete vesicles is more likely to be a function of external constraints applied to the whole mass than any inherent tendency toward segmentation in the mesoderm itself—in marked contrast to the somites of the trunk. Kuratani (2008) makes the point that the spinal cord is not inherently segmented: the segmental pattern of spinal nerves is dictated by the somites of the trunk. In the head, in contrast, it is the segmental nature of the rhombomeres, not the mesoderm, that dictates the patterning of the spinal nerves.

There are, nonetheless, head cavities in sharks, which Balfour termed premandibular, mandibular, and hyoid, and which had the distinction of somites thrust upon them, with the implication that they were serial homologues of trunk somites. Kuratani (2008) argues that when those in the Balfour school looked on lampreys, they naturally interpreted any mesodermal masses they saw through shark-shaped lenses. Later, molecular work has served to draw further distinctions between the head and trunk mesoderm of vertebrates.[54] For example, expression of a T-box gene in amphioxus marks the anteriormost three somites early in development, but is restricted to more posterior somites later.[55] T-box genes are expressed in the head in vertebrates, but not, as far as is known, in patterns that mark somites.[56]

As noted above, there do seem to be close correspondences between the molecular patterning of the anterior nervous system in amphioxus and vertebrates, but this need not imply that the mesoderm is similarly segmented in each. Besides, this neuronal patterning is very ancient and is presaged by similar expression patterns in the ectoderm of enteropneusts, which lack somites altogether and have nothing remotely resembling a chordate nervous system.[57]

Kuratani (2008) concludes that mesodermal head cavities are absent in lampreys, and are therefore a feature of gnathostomes. Although seen in gnathostomes to a greater or lesser extent, they are most prominent in sharks. Because Balfour and his school just happened to work on sharks, they got themselves caught up in the coat-tails of a remnant Goethean idealistic morphology; assumed that the anatomy of sharks was archetypically vertebrate; and that their head cavities were serially homologous with somites in the trunk.

What is one to make of this? First, one has to take into account that each researcher brings to the table their own impressions gathered from the material with which they are most familiar. Holland works mainly on the amphioxus, and so is more inclined to interpret any cranial mesoderm segmentation confidently in the light of somites. Kuratani, for his part, works on the lamprey, where mesodermal segmentation in the head is extremely hard to make out, and, if it exists at all, is heavily modified by the presence and activities of structures not seen in the amphioxus, notably the otic vesicle and the rhombomeres of the hindbrain. All must labor in the long shadows of Balfour and Goodrich, who pulled together what we now know as three disparate segmental systems (neural, somitic, and pharyngeal) and shoehorned them into a single system in true Goethean style.

Whether or not any cavities one sees in the mesoderm of lampreys are somites or not, there does seem to be a clear difference, developmentally, between the preotic mesoderm in the vertebrate head and the postotic mesoderm, that is, the paraxial mesoderm that is divided into the somites of the trunk. There is, for example, no trace in the head of the clock-and-wavefront model of mesodermal segmentation, so characteristic of the trunk. If the head mesoderm is segmented, it must be along different lines from that of the trunk mesoderm, in which case head segments need be no more serial homologues of trunk somites than are the gill arches or rhombomeres.

Does this mean that the amphioxus has no "head"? Clearly this cannot be so. First, the most rostral somites of the amphioxus develop differently from those in the trunk. In Holland's view, the enterocoelous development of these somites is homologous with the enterocoelous development of somites in the head mesoderm of lampreys (if these are observed at all) and the (more definitive, if schizocoelous) head cavities of sharks. This suggests that the amphioxus has a region that corresponds with the head of vertebrates.

Kuratani would demur, noting the difficulty of observing somites in the head of lampreys; advancing the possibility that head somites in vertebrates are a gnathostome innovation (and thus amphioxus rostral somites are irrelevant to the discussion); and in any case that differences in development need not say much about homology. After all, no one doubts that the segments of flies, which develop all of a piece, are homologues of those of (say) locusts, which develop one at a time, so whether a somite develops by enterocoely or schizocoely is not relevant.

On the other hand, all agree that the various components of the anterior parts of the amphioxus central nervous system do seem to have clear homologues with those of vertebrates. Whether or not the mesoderm in the rostral part of the amphioxus corresponds with the head mesoderm in vertebrates, the amphioxus nerve cord shows detailed homologies with the front end of the vertebrate central nervous system.

To me, the big difference between the amphioxus and vertebrates is not the head mesoderm, or the nervous system, but the notochord. In the amphioxus this extends all the way to the front. In vertebrates, however, it extends forward only to a point in the middle of the head where its progress is halted by the development of an intervening structure. A finger-like extrusion called the infundibulum grows downward from the midline of the base of the diencephalon, a part of the developing forebrain, to meet Rathke's Pouch, an upward extension of the foregut. The result is the hypophysis, whose development represents a clear interruption to the forward progress of the notochord. A possible homologue of this in the amphioxus is the so-called ciliated pit, an epidermal opening on the left side that fuses with a structure called Hatschek's diverticulum, a left-leaning outpocketing from the gut. Earlier, I contrasted the asymmetrical development of this structure in amphioxus with its midline situation in vertebrates—an asymmetry that allows the notochord of the amphioxus to extend rostrally.

Does the forward extension of the notochord in amphioxus mean that the animal simply does not possess any structure homologous with any prechordal part of the vertebrate head? Well, no—as I have discussed, there are close, detailed correspondences between the anterior nervous systems of the amphioxus and vertebrates. These correspondences are initially very hard to make out from the apparently very simple anterior nerve cord of the amphioxus, with its modest cerebral vesicle, compared with the much larger and more developed brain of vertebrates, with its pre-chordal cephalic flexure and clear divisions into fore-, mid-, and hindbrain. We know that the notochord is a potent signaling center and exerts a profound influence on the nerve cord as the two structures develop. Could the presence of a notochord in the rostral part of the amphioxus act to inhibit the development of a brain that, in vertebrates not so encumbered, expands greatly in the prechordal region? Could the notochord also act on the mesoderm of the anterior end of the amphioxus, prompting it to develop into somites that are serial homologues of those in the trunk?

If so, then Holland and Kuratani are both partly right. The amphioxus does indeed have a head. The mesoderm in the head develops into somites. However, the vertebrate prechordal mesoderm, because it lacks the notochord, does *not* develop into somites. In which case, the somites of the amphioxus need not be homologues of any divisions in the prechordal mesoderm that one might come across in the vertebrate head.[58]

The problem now becomes one of phylogenetic polarity. Did vertebrates evolve from an amphioxus-like ancestor with a full-length notochord that pulled back, allowing the prechordal mesoderm to develop in a new way? Or did the amphioxus evolve from an ancestor with a prominent prechordal region, suppressed by the anterior extension of the notochord? The peculiar asymmetries of amphioxus development suggest the latter, in which case the amphioxus becomes less suitable as the model of a vertebrate ancestor than many would like. After all, the amphioxus lineage has been evolving away from its common ancestor with vertebrates for precisely the same length of time that vertebrates have been evolving from that same point. Even though the amphioxus appears to have evolved extraordinarily slowly compared with vertebrates,[59] much can happen in half a billion years. At no time in that unimaginable stretch of time has the amphioxus been obliged to remain in its unadorned ancestral state simply for our convenience. What we should dearly love to have is information about tunicates, but mesodermal segmentation in these animals, if they had any at all, appears to have been lost.

12.6 THE NERVOUS SYSTEM

The capacity to move electrical charge from one place to another seems to be a fundamental property of cells. Even the simplest are able to receive and respond to stimuli. From this fact, it might seem only a small step for some cells to become specialists in moving electrical charge, using it to communicate information such that the organism can formulate a response. Such cells are the neurons, the basic building blocks of nervous systems. Neurons receive information through projections from the body of the cell called dendrites: they relay it either through dendrites or much longer, tubular extensions known as axons. Some axons may be very long indeed. The longest axons in the human body are those bundled together into the sciatic nerve, carrying signals from the spinal cord to the toes of each foot.

Neurons rarely form a complete circuit. A neuron receives information from some organ or receptor, and once the signal has passed along the axon or dendrite, it must convey that information to the effector. Sometimes the signal must be patched through several neurons, across cell-to-cell junctions called synapses, before it reaches its final destination. In other words, the electrical signal has to be converted into chemical form, shuttled across the gap, and converted back into electrical form for onward transmission.

Not surprisingly the molecular machinery of synapses is formidably complex, but the basic building blocks of many molecular components of the synapse are found in single-celled organisms, and multicellular organisms such as sponges that lack nervous systems. It might be said that the synapse antedated the neuron.[60] These molecular components include "channels" that allow ions to pass across membranes, and proteins that allow cells to stick together and communicate with one another.

Not all cell-to-cell communication needs to be conducted through neurons. We've met neuron-free cell-to-cell communication before. It's the process that coordinates the wavefront in the clock-and-wavefront model for somite formation; the convergence-and-extension process that creates the notochord; the influx of cells through the blastopore during gastrulation; and indeed every instance of cell movement as cells divide and organize themselves during development. Many other instances of motion happen by simple cell-to-cell communication. Sponges, single-celled organisms, fungi, and plants can move without any neurons at all.[61]

Neurons are ectodermal in origin, and their beginnings can be traced to a process called neurogenic induction, in which the ectoderm of an animal is partitioned into neurogenic and non-neurogenic fields. This partitioning involves the expression of Bone Morphogenetic Proteins (BMPs) such as *decapentaplegic* (*dpp*) in non-neurogenic ectoderm, and the antagonists of

BMPs in neurogenic fields.[62] These antagonists include *short-gastrulation* (*sog*) and *chordin*. If these names sound familiar, they should, for we have already met BMPs and their antagonists in another context, that of the establishment of dorsoventral patterning. The role of BMPs in neural induction is to suppress the activities of neurogenic genes, ensuring that the nervous system forms in the right place. BMPs are sufficiently potent that they can direct neural patterning all on their own, without any other cues. After the primary division of the ectoderm into neurogenic and non-neurogenic fields, the neural ectoderm is divided further into three longitudinal stripes, each expressing its own suite of homeobox-containing transcription factors.

It seems that neurogenic induction directed by BMPs is closely tied to the establishment of the dorsoventral axis, and the process appears to have been highly conserved in all bilaterians. The systems are closely similar in vertebrates, fruit flies, and annelid worms—each representing one of the three major divisions of the Bilateria. Its roots lie even further back in evolution, as parts of the BMP system are found in cnidaria such as sea anemones and jellyfishes, where they direct the formation of a body axis that, though not strictly dorsovental, is orthogonal to the main body axis.

The simplest nervous systems consist of neurons strung together in a net-like structure, an irregular yet planar arrangement of neurons connected to an internal contractile layer, external sensory cells, and one another.[63] In some animals such as cnidarian polyps and the acoelomorph *Xenoturbella*, the nervous system consists almost entirely of a nerve net, though polyps such as *Hydra* have a denser ring of neurons around the mouth, and some of *Xenoturbella*'s acoel relatives have simple brains, showing how nerve nets can be concentrated into a ganglion or plexus in which neurons may interact directly with one another, to catalogue and store information. The converse is also true. Animals such as vertebrates, with otherwise sophisticated central nervous systems, still have nerve nets that innervate the gut and which coordinate the wave-like motion or "peristalsis" that propels food along it. As anyone suddenly gripped with a fit of vomiting or diarrhea knows only too well, such gut movement is entirely outside voluntary control and is testament to the awesome power of a brainless nerve net to respond to stimuli with rapid, coordinated motion.

Although nerve nets might seem simpler than more concentrated aggregations of neurons into ganglia, nerve cords, rings round the mouth, and brains, one should not assume that they are always more primitive. Many animals have both cords and nets, and the broad conservation of the BMP system of neural induction mechanisms in bilaterians means that it is possible that the simple nervous systems we see in some animals result from the secondary disaggregation of once complex, centralized nervous systems.

For all that the process of neural induction appears closely similar in all bilateria so far studied, the central nervous system—that is, the brain and any associated sensory organs and major nerve trunks—may take many different forms. In protostomes in general, the brain tends to be formed of ganglia (that is, aggregations of neurons) in a ring surrounding the anterior gut, and the nerve cord (or cords) is ventral, and solid. In chordates, in contrast, the brain lies entirely dorsal to the gut and is hollow, and the single nerve cord is likewise dorsal and hollow. As I noted in chapter 4, the switch between ventral and dorsal was first discussed by Geoffroy in the early nineteenth century, and does seem to be connected with a genuine inversion in dorsoventral patterning in the chordate lineage.[64] Understanding the evolutionary origin of this switch, and the reason for the hollowness of the chordate central nervous system, is made more difficult by the peculiarities of the nervous systems in echinoderms and hemichordates, of which more below.

The formation of the central nervous system in chordates is known as neurulation (in contrast to neural induction, which is common to bilaterians generally). During the formation of somites from paraxial mesoderm, the area of ectoderm on the dorsal midline known as neural plate is reshaped so that it is broader in the cranial region than in the parts destined to form the spinal cord.[65] As with the formation of the notochord, this reshaping involves that species of cellular choreography known as convergence and extension. A furrow forms along the midline, largely due to a hinge-like movement in the midline that prompts the more lateral parts of the neural plate to fold up. This hinge mechanism is directed by the ubiquitous morphogen *Sonic hedgehog* (*Shh*) secreted by the underlying notochord. Indeed, the presence of the notochord as a potent organizing center is essential for the direction of neurulation. The edges of the furrow grow upward as crests and flip toward each other by virtue of bilateral hinge mechanisms (this time inhibited by *Shh*). The cells at the leading edges extend protrusions called lamellipodia toward one another, allowing the two edges, once met, to knit together. So much is clear for the neural tube from the base of the brain posterior to the level of the sacrum: the hinder parts of the spinal cord are actually formed from a process called secondary neurulation governed by the tail bud, the remnant of Spemann's organizer.

In vertebrates, neural tube formation and closure in the cranial region are a little more complex than in the trunk due to the expanding volume of the presumptive brain, and (presumably) the absence of the notochord. In molecular terms, cranial neural tube closure involves a different set of proteins from those involved in spinal cord closure. The final closure of the neural tube is a delicate process, especially in the cranial region. So-called neural-tube defects (NTDs) afflict around one in every 1,000 human pregnancies, and constitute the second most common form of congenital problem after heart defects. NTDs range from the embryonically lethal anenkephaly, where the anterior neural tube fails to close and the brain does not form, to a range of mild-to-severe problems collectively known as spina bifida.

Neurulation also occurs in the amphioxus and in tunicates, though with idiosyncrasies in each case. In the amphioxus, the neural plate is first roofed over with non-neural ectoderm before the neural tube itself rolls up.[66] In tunicates, the process is more similar to vertebrates in that it is neural tissue that does all the rolling up—except that the process is reduced to a very strict choreography of a small number of precisely organized cells.[67]

The amphioxus central nervous system seems very simple when compared with that of vertebrates. At the anterior end it swells slightly to form the cerebral vesicle, and there is a light-sensitive spot at the front. Studies of neuroanatomy and gene expression show close correspondences between the amphioxus and vertebrate anterior central nervous system.[68] However, there is no clear evidence in the amphioxus for the telencephalon,[69] and in particular nothing corresponding to olfactory bulbs, but there are regions corresponding to the diencephalon; the light-sensitive spot is a homologue of the paired vertebrate eyes[70]; and a structure called the lamellar body might be a homologue of the pineal.[71] There is a distinct midbrain and hindbrain. In vertebrates, the midbrain-hindbrain boundary (MHB) is an important landmark and has organizer-like properties. Other organizers in the forebrain include the anterior neural ridge (ANR) and zona limitans intrathalamica (ZLI). The cognate regions in the amphioxus do not have organizer properties[72] and may be differently arranged in their details,[73] making direct comparisons between vertebrates and the amphioxus less secure than they might be. Nevertheless, the wealth of cryptic detail in the outwardly modest amphioxus nervous system suggests that the ancestral chordate had a rather sophisticated central nervous system, secondarily reduced in the modern amphioxus. It is possible that this reduction is connected with the rostral extension of the notochord as suggested above.

The tunicate central nervous system shows traces of organization more vertebrate-like than those in the amphioxus, for all that the adult nervous system is in essence a ganglion in the cleft between the two siphons, and reduction in the more vertebrate-like larval central nervous system (CNS) is extreme. At the anterior end of the larval CNS is a sensory vesicle, which in some forms may house a light-sensitive organ and a gravity sensor. Posterior to that is, in order, the brain; a mass of cells that will contribute to the adult CNS; the motor ganglion; and finally, the nerve cord.[74] Detailed mapping of the connections between neurons in the brain of the larva of *Ciona intestinalis* reveals that whereas the numbers of neurons on each side of the brain are approximately the same, there are pronounced lateral differences in connection patterns.[75] However, gene expression shows some correspondence between the anterior-posterior organization of tunicate and vertebrate central nervous systems.[76]

The brains of extant vertebrates are qualitatively and quantitatively larger and more complex than those in the extant amphioxus and tunicates. Although the amphioxus genome has evolved extremely slowly, it is nonetheless hard to know whether the modest compass of the anterior end of its central nervous system is primitive, or the result of the diminution of something more complex. The genomes of tunicates, on the other hand, have evolved extremely quickly, making the question even harder to answer. It is possible, however, that at least some of the complexity of vertebrate brains results from the whole-genome duplications specifically in the vertebrate lineage with subsequent elaboration.[77] The telencephalon, in particular, seems to be a specifically vertebrate innovation.[78]

The earliest chordate, then, had a central nervous system something like an amphioxus. It developed by neurulation and had a diencephalic forebrain, midbrain, and hindbrain and tubular nerve cords, and some signs of landmarks such as the ANR, ZLI, and MHB, although the properties of these regions as organizers appears restricted to vertebrates.[79] The region corresponding to the vertebrate hindbrain would have been patterned by Hox gene expression. It would, however, have lacked any structure corresponding to a telencephalon, and possibly also olfactory bulbs. A consequence of this is that the eyes, if paired, would have been much closer to the anterior end of the animal than we are used to seeing in vertebrates. It may or may not be significant that reconstructions of early vertebrates such as *Metaspriggina*[80] or of conodont animals, whose status with respect to vertebrates has been a matter of dispute, inevitably put the eyes at the extreme anterior end.

From this one is entitled to ask whether the singleton status of the amphioxus terminal eyespot is primitive or derived, and if the latter, whether the eyespot represents the left eye, the right eye, or both. I am inclined to think that a single eyespot is primitive, truly median, and homologous with the pair of eyes that evolved with vertebrates, along with the telencephalon and the tendency for the forebrain to separate into hemispheres. This evolutionary step would have "opened up" the mid-face, creating space for the nose and forcing the eyes apart. One can draw an analogy with the evolution of gnathostomes, with their paired nostrils, from the single, median nostril of agnathans, in a further episode of face-creation that ultimately resulted in the evolution of jaws.[81]

A clue to the evolution of paired eyes might come from a rare congenital deformity called cyclopia, in which a fetus develops with a single eye in the middle of the forehead. The nose is either absent entirely or present as a kind of proboscis just above the single median eye, or on the back of the head. Cyclopia is a manifestation of a brain defect called holoprosencephaly, in which the forebrain fails to split along the midline into two hemispheres, thus failing to create the space for both eyes in what is originally a single eye field at the most anterior tip of the neural plate. The causes of holoprosencephaly are varied but some result from spontaneous mutations in *Sonic hedgehog*.[82] It is perhaps notable that the part of the amphioxus brain corresponding to the forebrain is not conspicuously divided into hemispheres, and that the animal lacks olfactory bulbs.

It is well to be skeptical of drawing evolutionary conclusions from rare teratologies, that is, to cast them as throwbacks or atavisms. However, it is possible that the one-eyed condition of the amphioxus is a product of the same regulatory landscape that produces holoprosencephaly in humans, and both speak to a ground plan of anterior brain specification in chordates.

Tracing the roots of the chordate nervous system more deeply into the phylogeny of deuterostomes immediately runs into a thicket of problems. The nervous systems of echinoderm larvae do not carry over into the adults after metamorphosis.[83] This finding falsifies at a stroke the popular ideas of Garstang that the adult nervous systems of chordates could have evolved from the convergence and involution of neurogenic ciliary bands in an echinoderm-like larva.[84] In general, the nervous systems of echinoderms take the form of a superficial net, and the various nerve rings and radial nerve cords of adult echinoderms do not express Hox genes.[85] This is a crucial difference between echinoderm and chordate nervous systems. Whatever nervous systems were present in the most primitive echinoderms, and whether they had any genetic or anatomical relationship with chordate nervous systems, will probably remain forever unknown.

Hemichordates (specifically, enteropneusts), in contrast, display a wealth of information on nervous systems, but as discussed previously it remains likewise difficult to relate much of it to nervous systems in chordates. As with echinoderms, the nervous systems of larval hemichordates (in indirect-developing species) are not retained into adulthood. Adult enteropneusts have a superficial nerve net that is particularly dense in the proboscis. In addition they have not one but two nerve cords—a dorsal and a ventral. Part of the dorsal cord is hollow and appears to develop by a process very like neurulation.[86] Genes expressed along the anterior-posterior axis in the chordate central nervous system are expressed in a more-or-less collinear way in enteropneusts—but in rings, over the entire ectoderm. It's as if deuterostomes once had a condensed chordate brain and central nervous system that in enteropneusts has become smeared all over the ectoderm. Either that, or it is a relic of a time in the evolution of deuterostomes before brains condensed from more diffuse nets.[87] In any event, it shows that the spatial organization of genes now expressed in the vertebrate central nervous system have very deep roots among deuterostomes.

There is another possibility, however. This is that enteropneusts have lost an ability seen in many other animals—that is, the use of BMP signaling to pattern ectoderm into neurogenic and non-neurogenic regions.[88] This could suggest that the ancestor of enteropneusts, deuterostomes, and possibly bilaterians in general, had a more centralized central nervous system that enteropneusts have lost, even if they have retained the pattern, imprinted on their ectoderm.[89]

Much remains to be learned about how the nervous system in enteropneust larvae develops into that of the adult. Studies on *Saccoglossus kowalevskii* show that some genes related to neuronal development are expressed all over the ectoderm but become more restricted to nerve cords in the adult—but whether this means that the nerve cords condense from a nerve net, or develop from an entirely different population of cells, is not clear.[90] A study on *Balanoglossus simodensis* suggests that larval nerves do not persist into the adult.[91] This parallels what is known in echinoderms, and is another nail in the coffin of Garstang's ideas that the adult central nervous system of chordates evolved from a fusion of ciliated bands in an echinoderm- or hemichordate-like larva.

The genetic patterning of the dorsoventral axis in enteropneusts involves BMP and its antagonists, as in chordates, but in the reversed sense, showing that chordates are inverted with respect to other deuterostomes as well as other protostomes. This casts further doubt on the homology of the collar cord with the dorsal, hollow nerve cord of chordates.

Although a great deal remains to be learned about the nervous systems of enteropneusts, it appears that there is no clear homology between any particular part of the system, such as the ventral or dorsal cord, or the collar cord, with any part of the chordate nervous system. There does remain the very general correspondence between the anterior-posterior patterning of the system, and further correspondence between enteropneust and chordate neurulation might be revealing. The nerve cord of chordate requires signals from the underlying notochord to develop, in particular *Sonic hedgehog* (*Shh*). The stomochord of enteropneusts is a forward extension of the gut and thus endodermal, and not a homologue of the notochord as once thought. However, it does express genes in the *hedgehog* family and it is possible that the collar cord can respond to this.[92] It could be that neurulation precedes the wholesale dorsoventral inversion in the ancestry of chordates.

12.7 NEURAL CREST AND CRANIAL PLACODES

Perhaps the single most important evolutionary novelty in the evolution of vertebrates is the neural crest: so important, in fact, that vertebrates might even be defined by it.[93]

Neural crest is a transient, embryonic population of cells that arises during development from the lateral border of the neural plate. The cells share a number of distinctive properties. First, they are migratory. That is, they move from their site of origin to roam throughout the body. Second, they are multipotent: "stem" cells able to develop into a number of different cell types, depending on where they end up. Finally, although ectodermal in origin, they undergo a process of reprogramming into multi-purpose mesenchyme, capable of being further molded into a wide range of different forms. Neural crest cells contribute to various structures including much of the dermal skeleton of the head and face; cartilage in the pharynx; many neurons and their supporting cells or glia, including the sympathetic nervous system; the entire enteric nerve net; the chromaffin cells in the adrenal glands that secrete various important hormones; and pigment cells in the skin. Although various cell types are known that have some of these properties, only neural crest cells are simultaneously multipotent, migratory, and capable of undergoing the ectoderm-to-mesenchyme transition.

More speculatively, the stem-cell-like properties of neural crest cells might have had a role in allowing vertebrates to achieve larger sizes than is typical for invertebrates. Most invertebrates are rather modest in size, and certainly in bulk, whereas vertebrates include the largest animals that have ever lived. We human beings hardly match whales or sauropod dinosaurs, but we are as individuals much more massive than all but a few invertebrates, none of them deuterostomes.

As a consequence, the internal spaces of vertebrates are far more remote from the environment than the few cell diameters typical of invertebrates. This has had implications for the range of environments available for vertebrates to colonize. Life originated in the sea, and most animals are exclusively marine, including all non-vertebrate deuterostomes. Vertebrates belong to a select group to have colonized freshwater; an even more exclusive band to have made a success of life on land and in the air. Another consequence is the extensive elaboration of many internal organs and organ-systems, a theme I'll discuss in the next chapter.

Both cyclostomes and gnathostomes alike have neural crest, but gnathostomes enjoy a number of additional, specific neural-crest-influenced features such as jaws and mineralized hard tissues. These differences aside, the gene-regulatory network that underlies the formation, migration, and transformation of neural crest cells is highly conserved among vertebrates and includes such genes as *SoxE* and *FoxD* as part of its regulatory machinery.

Neural crest cells do not create new structures out of nothing, but interact with existing structures to make new, improved versions, sometimes co-opting evolutionary older cell types and tissues. Cartilage, for example, is a tissue found throughout the Bilateria, and is used in the construction of the pharyngeal skeleton of hemichordates and the amphioxus.[94] The presence of neural crest in vertebrates did not conspire to replace this cartilage, but co-opted its development and use in the vertebrate skull.[95] This suggests that the neural crest itself cannot have appeared from nowhere, but must have had identifiable precursors among the invertebrates.

Cells have been variously identified in ascidians that originate near the neural tube; are capable of migration; and give rise to pigmented cells, including sensory cells in the otolith and ocellus.[96] Misexpression of a gene called *twist* in some of these cells produces mesenchyme-like cells capable of longer-range migration.[97] Tunicates, therefore, have cells with properties very like those in vertebrate neural crest cells in that they are capable of conversion to mesenchyme, are migratory, and produce pigment. The wholesale remodeling of the body wrought by neural crest in vertebrates, especially the anterior end, is, however, conspicuously absent in tunicates, suggesting that the equivalents of neural crest cells in tunicates, if that is what they are, lack the multipotency characteristic of neural crest cells in vertebrates.

Hints of neural crest in the amphioxus are even harder to find. One hypothesis suggests that the origin of neural crest might have some connection with the pigmented cells in the amphioxus eye-spot.[98] The amphioxus does, however, have many of the components of the gene network that in vertebrates constructs the neural crest, although they do different things: the origin of neural crest presumably involved the co-option of preexisting genes and regulatory modules.[99] As far as we can tell from the extant species, there are no cells in amphioxus that have the multipotency or migratory potential of neural crest cells in vertebrates.

Earlier I discussed Hox-cluster genes, and the remarkable discovery that the order in which these genes is ranged along the chromosome specifies the segmental identity of structures in the body in the same linear order, a property known as collinearity.[100] In the fruit fly, *Drosophila*, a given Hox gene tends to specify structures in a specific segment of the body, notably excluding the head. In vertebrates, however, the expression domains of Hox-cluster genes within the neural tube extend from the posterior end up to well-defined, successively more anterior boundaries in the hindbrain. The further forward one goes, the fewer Hox genes are expressed until, eventually, there are no more: as in *Drosophila*, Hox-cluster genes play no part in the specification of structures in the head.[101] The anterior boundaries of Hox expression correspond with quite distinct segmental compartments in the hindbrain known as rhombomeres. Hox genes are also expressed in a segmental and collinear way in the amphioxus, although expression is confined to the neural tube.[102] Collinearity of Hox expression is maintained to some extent in tunicates even though the cluster itself has either partially (*Ciona*) or completely (*Oikopleura*) disintegrated.[103]

The expression of Hox genes in vertebrates has important consequences for the disposition of neural crest, as it is from the borders of the rhombomeres that neural crest flows ventrally and anteriorly to populate the pharyngeal arches and thus form the head and face.[104] To be more specific, neural crest comes from the even-numbered rhombomeres. Crest from rhombomeres 2, 4, and 6 respectively populate the first, second, and third pharyngeal arches, taking their distinctive patterns of Hox expression with them. The first pharyngeal arch eventually develops into the upper and lower jaws, the palate, and, through Meckel's cartilage (evolutionarily a part of the lower jaw), the hammer and anvil bones of the mammalian middle ear. The second develops into the stapes in the middle ear, parts of the hyoid bone that supports the tongue, and many facial muscles; the third, much of the rest of the hyoid. The neural crest also contributes to the skull vault.[105]

In vertebrates, of course, there is not one Hox cluster but four, a result of the two episodes of whole-genome duplication some time during vertebrate ancestry. It has been argued that the efflorescence of neural crest in vertebrates was connected with this whole-genome duplication. The study of the genomes of extant chordates leave us in no doubt that such whole-genome duplications happened, though as I discussed earlier there remains some doubt as to whether both duplications were complete by the time the cyclostome lineage had diverged from that of gnathostomes. The elaboration of gnathostome features unseen in cyclostomes, such as jaws, paired limbs, enamel, and dermal bone might suggest the latter. However, as Green et al. (2015) note, the evolution of neural crest might have been facilitated not so much by the fact of whole-genome duplication but for the increased opportunities such duplication offered for a more nuanced regulation of genes involved in the specification and behavior of neural crest.

Whole-genome duplication, however, might have left tell-tale signs that the anterior end of the vertebrate body is profoundly distinct from the more posterior parts. As mentioned above, regional specification of neural segmental identity (and thus neural crest) by Hox-cluster genes does not extend more anteriorly than the hindbrain.[106] Second, it seems that the anterior boundary of expression of any vertebrate Hox gene tends to match with those of its paralogues. In the trunk, in contrast, paralogues have different domains of expression, as if they had had more time to evolve from a common ancestor. Third, the organization of neural crest into various streams from the rhombomeres appears to be a property conferred by neural tissue, and thus intrinsic to it. In other words, neural crest from rhombomeres makes its mark on the various tissues it meets. In the trunk, in contrast, the fate of neural crest tends to be defined by the milieu in which it arrives.[107] Lineage tracing in vertebrates reveals a cryptic but distinct posterior boundary of cranial neural crest activity, in the anterior parts of the shoulder girdle[108]: morphologically, at the posterior margin of the pharynx. The reasons for this seeming difference between the behavior of trunk and cranial neural crest are obscure. Green et al. (2015) discuss a number of possible scenarios that need not detain us here. Suffice to say that there is a difference, and if the evidence of Hox gene expression is brought in, there seems to be a "newness" about cranial neural crest compared with trunk neural crest. Hox-gene paralogue boundaries are clearer; cranial neural crest exerts far more influence on the underlying tissue than trunk neural crest; and the results, in terms of the transformation of structures, are more startling. It could have been that the antecedents of neural crest cells in the ancestors of chordates, and possibly tunicates too, performed various neural-crest-like functions before the evolution of the head as a distinct, integrated structure. The appearance of the head and

Table 12.1. Vertebrate cranial placodes, derivatives, and roles

Anterior	
Adenohypophyseal	Endocrine cells of the anterior pituitary
Olfactory	Sensory and neuronal cells in nose and vomeronasal organ
Lens	Forms the lens of the eye
Posterior	
Otic	Hair cells and nerves in the ear
Lateral Line	Sensory cells in the lateral line in aquatic vertebrates
Trigeminal	Ganglia of the trigeminal nerve
Epibranchial	Viscerosensory neurons in tongue, lung, and gut
Paratympanic	Hair cells in paratympanic organ in middle ear
Hypobranchial	Associated with epibranchials, role unknown

the co-option of the neural crest into the cranial region, together with the duplication and reduplication of the genome, might have reinforced one another to produce what we now think of as the vertebrate head.

Allied to neural crest cells are the epidermal cranial placodes,[109] thickenings of the epidermis, especially in the head, which in vertebrates form dense concentrations of the sensory and neuronal components of many sense organs and important hormone-secreting cells and tissues.

Patthey et al. (2014) list the various placodes, their derivatives, and their roles. I have tabulated these in Table 12.1.

The alliance between placodes and neural crest is the product of the fact that neural crest and placodes are always found together: no animal exists that has neural crest without placodes, or vice versa. Like neural crest cells, the cells associated with developing placodes migrate from their epidermal sources to their final destinations, a process helped by adjacent neural crest cells. However, it is thought that neural crest and placodes had distinct origins, originating from (respectively) the neural and non-neural sides of the neural plate border.[110]

Cyclostomes and gnathostomes all share essentially the same suite of placodes, suggesting that these structures evolved in the common ancestor of vertebrates. Can we discern any traces of placodes in tunicates or the amphioxus? The answer is a qualified yes: the tunicate *Ciona* has been shown to have patches of epidermis expressing genes similar to vertebrate placodes and with properties reminiscent of olfactory and adenohypophyseal placodes.[111] There might also be some connection between the secretory properties of Hatschek's pit, a structure in the amphioxus with structural homologies with the pituitary as we have seen, though structural homology and neurosecretory properties, even together, need not be evidence of the existence of a placode, as Patthey et al. (2014) note. Some placodes, notably the lens, have absolutely no counterparts outside vertebrates. It is possible that the elaboration of preexisting neural and secretory properties of various cells were pulled together into distinct placodes, as with the neural crest, during, and perhaps as a consequence of, the two whole-genome duplications in the early history of vertebrates.

Embryologically, placodes develop from the preplacodal ectoderm, a crescent of ectoderm surrounding the anterior neural plate, and anterior to it, the horns of the crescent running posteriorly on either side. Front and central is the region that will develop into the adenohypophysial placode, flanked on either side, and progressively more posteriorly, by paired presumptive olfactory, lens, and more posterior placodes. Each individual placode develops according to a specific molecular mechanism, although the transcription factors *Six1* and *Eya* are involved in placodal specification more generally.[112] The development of placodes is also influenced by signals from other regions of the developing central nervous system. For example, the adenohypophysial and olfactory placodes respond to signals from the anterior neural ridge (ANR), a potent organizing center in the forebrain; the development of the lens placodes is influenced by the large, median "eye field" just posterior to the ANR, whereas the more posterior placodes respond to signals from the midbrain-hindbrain boundary and the rhombomeres (see fig. 2 in Schlosser et al. 2014).

The dorsal neural plate is a defining feature of all chordates, and the division into neural and non-neural ectoderm is already well specified in the amphioxus lineage. However, the elaboration of the placodal domain anterior to the neural plate is a refinement seen only in tunicates and vertebrates. There is no evidence in the amphioxus for any ectodermal territories homologous to that which might give rise to the posterior cranial placodes. That having been said, some of the genes expressed in the posterior placodes do turn up in the pharyngeal pouches, suggesting that these genes were co-opted by vertebrates into placode specification.[113] Tunicates show more specific signs of the development of homologues to the posterior placodes. For example, *Pax2/5/8*, a marker for the posterior preplacodal area giving rise to otic, lateral line, and epibranchial placodes, is expressed in the neck region of the larval neural tube as well as in the siphon primordia.[114]

Schlosser and colleagues (2014) suggest that the earliest chordates had a "protoplacodal domain," a region of non-neural ectoderm in which neurosecretory cells tended to become concentrated. In the ancestry of tunicates and vertebrates this protoplacodal domain divided into anterior and posterior subdomains. The anterior subdomain differentiated into the primordia of the oral siphon in tunicates, or the lens, olfactory, and adenohypophysial placodes in vertebrates. The posterior subdomain in contrast produced atrial siphon primordia in tunicates, and the otic, lateral line, and epibranchial placodes in vertebrates. In addition, many genes expressed in the atrial primordia in tunicates are the same as those associated with deuterostome gill slit formation more generally, suggesting that an ancestrally pharyngeal expression was co-opted in overlying ectoderm.[115] Taken together, it seems as if the common ancestor of tunicates and vertebrates had placodal domains already divided into anterior and posterior subdomains.[116]

12.8 THE SKELETON

Most deuterostomes are rather small, and, being immersed in seawater for their whole lives, have little need for additional support. The only exceptions are echinoderms and vertebrates, both of which have extensive skeletons of connective tissue that have become mineralized. Although echinoderm and vertebrate hard tissues are very different, there are signs in deuterostome ancestry that genetic programs exist that predispose toward the mineralization of connective tissue.[117]

There are essentially four kinds of skeletal tissue in extant vertebrates, each associated with its own kind of secretory cell. They are cartilage, produced by chondrocytes; bones, by osteoblasts; dentine, by odontoblasts; and enamel, by ameloblasts.

Cartilage is the base material on which the skeleton is founded. Secreted by chondrocytes, it is a tough, rubbery material variously rich in the fibrous proteins elastin and collagen, as well as proteins called proteoglycans—that is, proteins to which various carbohydrate moieties are chemically combined. Genetic programs for chondrocyte and thus cartilage development run deep in bilaterians[118] and there are signs of the transient expression of vertebrate-like cellular cartilage in the amphioxus.[119] Cartilage is the only skeletal material found in cyclostomes, where it is of a peculiar form that contains little or no collagen, and there is no sign that this was ever mineralized in cyclostome ancestry. Strictly speaking, then, mineralization is a feature of gnathostomes, rather than vertebrates, though it was found in a wide range of fossil jawless forms, the "ostracoderms," considered more closely related to gnathostomes than to cyclostomes; as well as in primitive jawed vertebrates called placoderms.[120]

Bone consists mainly of collagen that is enriched with crystalline calcium phosphate in the form of hydroxyapatite. As a tissue, bone contains various kinds of cell: mainly the osteoblasts that produce the bone, and the osteoclasts that digest it. It is easy to forget, when one looks at a dry skeleton, that bone is very much a living tissue, constantly being remodeled by a balance of forces. Bone growth depends on the coordinate interplay of osteoblasts and osteoclasts.

Dentine is another mineralized tissue in which collagen fibers become infused with hydroxyapatite. It has a characteristic structure in which microscopic tubules run through the matrix. In mammals it is found only in the teeth, as the mineral support for enamel, but it has a wider distribution in vertebrates.

Enamel (called enameloid in fishes) is the most heavily mineralized vertebrate tissue and indeed the hardest substance produced by living things. Consisting almost entirely of crystalline hydroxyapatite, it contains no living tissue when mature. We tend to think of enamel as solely found on the tooth crowns, but it is found more widely in vertebrates, associated with dentine and bone in structures called odontodes, of which teeth are a specialized variety. Crystalline enamel forms on very specific proteins called amelogenins, in contrast to the collagen found in other skeletal tissues.

The genes that encode many proteins essential for the mineralization of enamel and bone are believed to have evolved from the duplication and reduplication of genes in the Secretory Calcium-Binding Phosphoprotein (SCPP) family.[121] These genes, which form a cluster, in turn arose from a gene for bone proteins called *SPARC* (Secreted Protein Acidic and Rich in Cysteine) and a close relative, *SPARCL1* (*SPARC*-Like 1). It is possible that the duplication of *SPARCL1* genes happened early in vertebrate evolution:

it is also interesting that genes for various types of collagen are linked to each of the four Hox clusters found in vertebrates—in which case the vertebrate skeleton and its mineralization were facilitated by the large-scale genome duplication early in vertebrate evolution.

As with neural crest and placodes, there are hints of *SPARC* in nonvertebrate deuterostomes. Sea urchins have a version of *SPARC*, but it does not appear to be involved in the very particular mineralization style of echinoderms.[122] In contrast, the *SPARC* found in the ascidian *Ciona* is more suitable for mineralization: the tests of ascidians contain spicules composed of various calcium-based minerals.[123]

The skeleton itself is not a single entity. The most visible part, the dermoskeleton, comprises the external parts of much of the head and face, and is derived from neural crest. Dermal skeletal elements posterior to the head are not of neural crest origin. The endoskeleton comprises the braincase or neurocranium; the jaws and gill arches, or viscerocranium; the vertebral column; and limb elements.

The presence of collagen-stiffened gill bars throughout deuterostomes shows that the viscerocranium was the earliest part of the skeleton to form. The first signs of hard tissues, however, are in the bony head shields of early jawless fishes known as heterostraci. These tissues are assembled into a characteristic organ called the odontode—a sandwich of enameloid on dentine on a base of bone—supported by a mineralized dermis. *SPARC* is expressed in the basal lamina of the dermis: it is perhaps no coincidence therefore that the earliest mineralized skeletons started as dermal reinforcement.[124]

12.9 Summary

In this chapter I have looked at some distinctive features of vertebrates and examined possible antecedents in chordates, deuterostomes, and occasionally animals more widely. Some of these features, such as the organizer, notochord, and segmentation, are found in all chordates, but vary between the groups in idiosyncratic ways. The generation of somites in the amphioxus, for example, is different from the corresponding process in vertebrates, and is all but lost in tunicates. Other features that at first seem unique to vertebrates appear to have echoes in other animals. One thinks in particular of the neural crest and placodes, highly elaborated in vertebrates but seen to only a limited extent in tunicates, and not at all in the amphioxus. Leaving aside the possibility that tunicates have lost some neural-crest-like functionality, it seems that in neural crest, vertebrates have taken preexisting genetic programs and co-opted them for new uses. This process has been both enhanced and complicated by the two whole-genome duplications that occurred at some time in vertebrate ancestry. One might say much the same for the elaboration of the head and the development of the skeleton.

In their influential paper in 1983, Gans and Northcutt floated the idea that much of the vertebrate head was a neomorph—a new structure not seen in either tunicates or the amphioxus. The question is without doubt simplistic, but it was a question that had to be asked as it focused attention on an abiding problem, and that is what we mean by the term "head." It undoubtedly refers to the anterior part of the body, sometimes highly specialized and housing the mouth, brain, and organs of special sense, but such might be used to refer to the anterior portion of an insect as much as a vertebrate.

In the particular context of vertebrates, the term has at various times been used to refer to regions of substantially different character. It could mean, for example, the region anterior to the anterior end of the notochord, or the anterior extent of Hox-cluster expression; the posterior margin of the hindbrain, or of the pharynx; or, the domain in which neural-crest modifies the fate of tissues in which it comes into contact. If the last, the head includes various parts of the viscera including parts of the heart, the adrenal glands, and the nervous system that lines the gut. If this seems odd, consider the "head" of the tunicate tadpole larva, which includes not only the central nervous system and presumptive pharynx but almost all the viscera.

The problem arises, in part, because the vertebrate body contains three entirely separate segmental systems—the central nervous system, the somitic mesoderm, and the pharyngeal slits—not all of which are required to be in register in the same way, either with one another or with other structures such as the notochord. Hagfish pharyngeal slits, for example, migrate very far posteriorly during development. At the other extreme, some of the extinct jawless fishes known as cephalaspids have a pharyngeal cavity that extends substantially anterior to the brain,[125] a condition not seen in any extant vertebrate. By the same token, the anterior extension of the notochord in the amphioxus is probably a peculiarity of that lineage and needn't reflect the common ancestral condition of chordates. The relationships between these organs and systems are clearly fluid, and if this notion seems surprising, our astonishment is a testament to the influence of researchers such as Balfour and Goodrich, whose work relied on the idea that the segmental systems seen in chordates were always in register, and in which the parochial anatomy of sharks was made to stand as representative of vertebrates as a whole.[126]

Were one to put aside this strictly Goodrichian view of anatomy, relieving the various segmental systems of vertebrates from having to evolve in precise segmental lockstep with one another, it might be possible to work out the order in which each evolved as a way of crossing the bridge between invertebrate and vertebrate.

CHAPTER 13

How Many Sides Has a Chicken?

13.1 INTRODUCTION

One of the less appreciated features of vertebrates is that they are generally rather large animals, perhaps a consequence of the evolution of neural crest.[1] Most animals are only a few cellular diameters across, so that no part of their interior is very far from the environment. Vertebrates are sufficiently large to require sophisticated nervous, endocrine, and vascular systems to ensure that all parts of the animal are integrated; a skeletal system to support them; and, pertinent to this chapter, a distinct visceral compartment, far from the external environment and to an extent insulated from it.

More than any other animals, vertebrates have elaborate viscera, including the heart, the liver, and the kidneys—to mention just some of an extensive mixed grill—all connected with a primary circulatory system of arteries and veins, and a secondary circulatory system of lymphatic vessels. In most animals, blood vessels are merely spaces between other tissues. Vertebrate blood vessels are unique in being lined with a distinctive tissue, the endothelium. The blood coursing along these vessels is likewise elaborate, containing a sophisticated system of adaptive immunity. As well as ensuring that oxygen permeates the remotest corners of the vertebrate body, the blood also carries hormones, the products of diverse glands, which influence the affairs of parts of the body far from their genesis.

The title of this chapter comes from a playground joke, which like many of the genre is more profound than it first appears. The answer to the question posed is this: a chicken has *two* sides: an outside, and an inside. In the last chapter I looked very much at the outside. Now it's the inside's turn.

Although the day-to-day doings of the insides are of great interest to physicians (and their patients), they are perhaps less often the focus of studies of evolution, being less obvious than features such as the brain, skull, and central nervous system. I think, however, that the reason for their neglect lies with a deep-seated notion of manifest destiny, such that vertebrate evolution is seen as a progressive elaboration toward the human estate. This is why it is often cast, obviously or tacitly, as the evolution of structures such as the skull and especially the brain. In this short chapter I shall attempt to cast a little light into that void, and show that the evolution of the insides might have things to tell us about vertebrate evolution as interesting as those of the more frequently studied parts described in the last chapter.

This progressive stance is evident in the highly influential paper by Romer (1972) on the vertebrate as a "dual animal—somatic and visceral." Romer proposed that the vertebrate body was constructed very much along the lines of our playground chicken, with a distinct outside and inside.[2] The outside, or somatic animal, comprising the musculature (much of it in the form of striated muscle), axial skeleton, body wall, central nervous system, and organs of special sense, is concerned with "external affairs"—that is, monitoring the external environment and responding to it. The inside, or visceral animal, in contrast, comprises the gut and its appendages, lined with largely smooth muscle,[3] together with the pharyngeal skeleton and supplied by its own nervous system—the enteric nervous system—only loosely connected with the central nervous system. Romer speculated that the deuterostome ancestor was a primitive filter-feeder consisting entirely of visceral components, and followed Garstang[4] in supposing that the somatic component arose as a larval dispersal device seen today as the tunicate tadpole larva. Paedomorphosis produced the ancestors of vertebrates and the amphioxus.

Romer's ideas were already losing their attraction with the advent of cladistics in the late 1970s, in which such progressivist thinking was frowned upon as unfalsifiable. The realization, later still, that tunicates are more closely related than the amphioxus to vertebrates[5] makes it look quaint and antique. Furthermore, Romer's view of the vertebrate body as an uneasy alliance of two different kinds of creature—somatic and visceral—tends to downplay the many feats of integration seen in vertebrates, notably in the circulatory, immune, and endocrine systems.

13.2 THE ENTERIC NERVOUS SYSTEM

Romer showed that the pharyngeal skeleton in vertebrates has very old roots, being found, as we have seen, in the amphioxus, tunicates, and enteropneusts. The surprise is that in vertebrates it is constructed in large part by neural crest.[6] The same might be said of the enteric nervous system (ENS). Arranged as two interconnected plexuses, one in smooth muscle of the gut, the other in the mucosa of the gut wall, the ENS is reminiscent of the nervous systems found in the simplest animals such as cnidarians, the body walls of echinoderms, and so on, perhaps representing the ancient, decentralized nervous systems of animals in general that existed long before any tendency toward centralized structures such as brains and spinal cords. With only loose connections to the central and autonomic nervous systems, the ENS is the only part of the vertebrate nervous system that is able to function entirely independently of the central nervous system (CNS). It looks very much a part of Romer's "visceral animal," subsumed within the somatic animal yet not entirely tamed by it. And yet, like the pharyngeal skeleton, the ENS of vertebrates is very largely a derivative of the neural crest.[7]

Although adopting a somewhat lower profile than the CNS, the human ENS has a total number of neurons of around 500 million, approximately equal to that of the spinal cord. In life, the ENS controls the movement of most of the gut, in particular the small and large intestines, propelling food along its length. It is also responsible for modulating the passage of fluids across the intestinal wall. Not all the gut, however, is controlled by the ENS. The esophagus and stomach receive substantial input from the hindbrain, and control of defecation in the rectum is regulated by the lumbar and sacral regions of the spinal cord. The small and large intestines, however, are the kingdom of the ENS.

Given its importance, one would expect that developmental or pathological problems with the ENS would be serious, and indeed they are. Hirschsprung's Disease, which occurs in approximately one in every 5000 live births, is a pathology in which neural crest cells fail to colonize the hindgut, which is therefore unable to move material along its length and results in constipation and blockage, and can be lethal if the affected region of the gut is not excised. Infectious disease can also produce fatal outcomes when it involves the ENS. Neurotoxins secreted by pathogens such as *Vibrio cholerae* (the agent of cholera) paralyze the ENS, disabling its ability to control fluid flux. The result is severe diarrhea and life-threatening dehydration.

The precursors of the ENS originate in the neural crest of the hindbrain, particularly in the region associated with the vagus nerve (cranial nerve X), and migrate along the gut in a wave from anterior to posterior. Some parts of the ENS, however, derive from trunk neural crest in the sacrum, and these colonize the gut from the posterior end forward, as far anterior as the umbilicus. ENS precursors of sacral origin start their journey somewhat later than their vagal colleagues, and augment rather than replace the vagally derived ENS already in place.[8] Other parts of the trunk neural crest form ganglia of the sympathetic (that is, autonomic) nervous system through which nerves from the ENS communicate with the central nervous system. Derivatives of the vagal neural crest also contribute to the heart, and the vagal neural crest has been described as a transitional zone between the head and the trunk.[9] It is to the heart that I now turn.

13.3 THE HEAD AND THE HEART

I've already discussed another influential paper that divided the vertebrate body into two parts: the head versus the rest. In that paper, Gans and Northcutt (1983) suggested that the advent of neural crest and placodes to create a new head represented a clear dividing line between vertebrates and other animals. That distinction, like Romer's, has also since blurred. As I discussed in chapter 12, tunicates share with vertebrates at least some of the molecular groundwork for neural crest and placodes, features not seen (or not seen as extensively) in the amphioxus, a finding that sits well with our current understanding of chordate phylogeny, namely, that the common ancestor of tunicates and vertebrates shared developmental features that the amphioxus primitively lacks.

Gans and Northcutt's emphasis on neural crest and placodes—the structures seen in the head—also diverts attention away from other features perhaps shared between vertebrates and tunicates; subjects more in tune with this chapter. I refer in particular to the heart. The tunicate heart is an altogether more elaborate structure than the pulsatile ventral vesicle seen in the amphioxus. Recent work shows that the evolution of the vertebrate heart goes with that of the head, in particular the facial and branchiomeric musculature. As Diogo and colleagues show,[10] the development of the musculature of the head also involves the heart. Investigations into this development reveal themes that tunicates and vertebrates hold in common.

A central concept in the argument set out by Diogo and colleagues is that of the cardiopharyngeal field (CPF), a developmental domain from which muscles of the head and heart both arise. The human head includes (at least) six distinct groups of muscles, mainly related evolutionarily to the branchiomeric muscles, and some of which have derivatives found in the heart. For example, derivatives from the first or mandibular arch give rise to to masticatory muscles in the head, and parts of the right ventricle in the heart. Similarly, derivatives from the second or hyoid arch give rise to facial and hyoid muscles as well as the base of the pulmonary trunk (if on the left) and the base of the aorta (if on the right). The neck—a kind of transitional zone between the somitic and nonsomitic parts of the animal—is also drawn into the picture. Nonsomitic muscles of the neck (presumably associated more with the head than the trunk) share a clonal relationship with muscle cells in the atria.[11]

The message here is that the evolution of major elements of the head cannot be discussed without consideration of the heart, suggesting that when we talk of the development of the head, what we mean is the development of the cardiopharyngeal field, including the head, pharynx (now the neck region in humans), and parts of the anterior viscera including parts of the heart and major blood vessels. This might seem odd in the context of human anatomy, but when one looks at primitive vertebrates it makes perfect sense. In the lamprey, the brain, gills, and heart are protected by a single, confluent basket of cartilage; in various extinct, armored jawless fishes, the armor protecting the head protects not just the brain but the pharynx and the anterior viscera including the heart. Any evolutionary concept of the head, therefore, must extend more posteriorly than we might think.

Invertebrates have anterior ends, and pulsatile vesicles that move blood from place to place, but neither is in general as complex as the vertebrate head or the vertebrate heart; that both structures appear together at the base of the vertebrate family tree suggests a common developmental origin, in the CPF.

The CPF is not a single unit, but is divided into the first heart field (FHF) that gives rise to the tubular structure that will develop into the left ventricle and parts of the atria; and, developing later from progenitors in pharyngeal mesoderm,[12] the second heart field (SHF) that produces parts of the right ventricle and additional parts of the atria. Muscles from the SHF have potential to develop as heart or skeletal muscles depending on their exposure to other tissues, notably neural crest. Some of the same molecular developmental signals that determine cardiac muscle also act on craniofacial mesoderm to specify head musculature[13] and, significantly, differ from those operating in the trunk.[14]

The vertebrate heart has a distinctive S-shaped structure and is divided into distinct compartments or chambers, each specialized either to receive blood or to pump it out again.[15] Blood from the posterior end of the animal flows anteriorly into the sinus venosus and the atrium. These are situated dorsally to the chambers concerned with outflow—the ventricle and conus arteriosus—and the blood therefore flows into these from above. The blood is thence pumped forward to be oxygenated in the gills that line the pharyngeal arches. Blood, flowing dorsally in the pharyngeal vessels, collects dorsally in the aorta and then flows posteriorly to perfuse the organs before collection and return to the sinus venosus and atria. This is the archetypal, primitive system, and there are many variations. In land vertebrates that primarily ventilate their tissues from the lungs, the system has been radically remodeled: in mammals, the heart is divided such that the atria and ventricles are divided to produce completely separate pulmonary (lung) and systemic circulations, the details of which lie well beyond the scope of this book.

The tunicate heart starts with just two cells that divide into four ventral cells in the trunk. These migrate to the pharyngeal endoderm, dividing again to produce heart precursor cells (the FHF, essentially) and secondary trunk ventral cells; these latter divide further to produce secondary heart precursors (homologues of the SHF) and muscle precursors. These migrate to the atrial siphon placode (possible homologues of the otic, lateral line, and epibranchial placodes in vertebrates, as discussed in chapter 12). These atrial siphon muscles are, therefore, homologues of many branchiomeric skeletal muscles in vertebrates; the trunk ventral cells that produce them are likewise true cardiopharyngeal progenitors as seen in the heart and pharyngeal musculature of vertebrates, and express the same molecular markers as are found in the vertebrate SHF.

The development of the tunicate heart, therefore, is rather like the development of the vertebrate heart and head, in cartoon microcosm: so much so that it is beginning to be used as a model system for investigating aspects of vertebrate heart development such as tissue regeneration and the circumstances of congenital heart defects.[16]

The amphioxus, in contrast, does not seem to have a proper heart, in the sense of any one contractile vessel that has a clear homology with the heart observed in vertebrates and tunicates, and it is likely that a heart never evolved in its lineage. Rather, the developing amphioxus has what one might term a "hematopoietic domain," a decentralized region defined more by gene expression than anatomy, that might stand as the root from which the heart, circulatory system, and blood of tunicates and vertebrates eventually sprang. Amoebocytes in amphioxus hemal vessels could be the evolutionary progenitors of vertebrate endothelium. Pascual-Anaya and colleagues (2013) compare this hematopoietic domain with another such domain described in vertebrates, the "aorta-gonads-mesonephros" (AGM) region, in which the development of blood vessels is linked with that of another vertebrate organ system, the urogenital system. Indeed, the pericardium that surrounds the heart has many developmental and genetic similarities with the kidney, and parts of it might have originated, in evolution, as part of the excretory system.[17]

13.4 THE UROGENITAL SYSTEM

It might seem a mystery, if not a source of either revulsion or fascination, why our organs of generation are so closely connected with those of excretion. Why are those body parts concerned with the act of sex also involved with urination? The answer lies deep in our evolutionary past.

Animals with internal coelomic compartments invariably have some way of regulating their contents, and connecting them with the outside world. They do this with specialized structures that filter coelomic fluid, retaining macromolecules but voiding excess water and nitrogenous wastes produced by the breakdown of spent proteins. In vertebrates, knots of capillaries, known as glomeruli, project from the walls of the aorta and into the coelom. Here they are invested or enclosed by structures containing specialized cells called podocytes. These podocytes filter the blood to produce urine that is passed down a tubule, or nephrostome, to be collected in a central repository, such as the bladder, or voided directly from the coelom through the body wall.[18] But the gonads of many animals also hang inside coelomic compartments, and release gametes (that is, sex cells) into them. These, too, use coelomoducts as a means of reaching the outside world. In this way, two different activities of animals—reproduction and excretion—have become irreversibly linked, by virtue of sharing the same transport system.

As described earlier, the mesoderm in vertebrates divides longitudinally into a number of different regions including paraxial mesoderm that divides to become somites, as well as the unsegmented lateral plate mesoderm, which divides further into the somatic mesoderm that forms the inner lining of the body wall, and the splanchnic mesoderm that constitutes the outside surface of the gut, leaving the coelomic space in between. After these various domains of mesoderm are formed, the paraxial and lateral plate mesoderm meet once more, between the paraxial muscles and the dorsal edge of the coelom. The rendezvous is in the form of two longitudinal strips, one each side of the body axis, running posteriorly from approximately the position of the heart. It is from these strips of so-called intermediate mesoderm that the genital ridges form, developing into gonads, eventually to be populated by the germ cells precursors (these form in the endoderm and migrate to the gonads as they form). But the intermediate mesoderm also develops into the Wolffian ducts, fated to develop into the system that conveys the germ cells to the outside, and, while they are about it, the kidneys that filter coelomic fluid and void nitrogenous waste.

Wolffian ducts, kidneys, and their formation are unique to vertebrates. The mesoderm of the amphioxus is more or less completely divided into somites, without the formation of the continuous mesodermal sheet that is the lateral plate. Perhaps for this reason, the gonads of the amphioxus are rather simple, typically invertebrate structures, and the excretory system consists of segmentally arranged tubules or nephrons reminiscent of those in flatworms.

The vertebrate kidney is divided, developmentally, into three.[19] The most "primitive"—that is, the first to form, is an elaboration of the most anterior parts of the system, closest to the heart, known as the pronephros. This is a small bundle of tubules that drains the pericardium—reminiscent, at least functionally, of the heart and associated glomerular complex in enteropneusts.[20] The pronephros is functional in the ammocoete larva of the lamprey, and the hagfish is the only vertebrate known to retain a functional pronephros throughout life. This retention is perhaps related to another unique hagfish feature, that its body fluids are isotonic with seawater.

In lampreys, the pronephros is replaced by the mesonephros, a more or less segmentally arranged system in which body fluids are squeezed from bundles of blood vessels, or glomeruli, into cup-like arrangements, collectively the Bowman's capsules, that drain into the renal tubules, and these fluids are collected, eventually, into the renal ducts that are developmental descendants of the Wolffian system.

The mesonephros is divided into a posterior portion, the opisthonephros, which does the actual excretion; and an anterior portion closely involved in reproduction. In female vertebrates, the anterior portion forms from the embryonic Wolffian ducts and develop into the oviducts, each one connected to an ovary hanging from the ceiling of the visceral coelom. The testes in males are found in a similar position, and the non-excretory part of the mesonephros serves to transport sperm. Thus the intimate connection in vertebrates, and especially gnathostomes, between the kidney and the reproductive systems.

In the amniotes—that is, obligately terrestrial vertebrates that lay shelled eggs—the mesonephros develops into the third and final kind of excretory structure, the metanephros. In male human embryonic life, the testis develops from the receding mesonephros as the metanephric kidney takes over renal function (though the mesonephros recedes entirely in females).

The metanephric kidney is a sophisticated organ in which all the renal apparatus—glomeruli, podocytes, nephrostomes, and all—is collected together into a single structure, with accessory tubules that retain as much body fluid as possible while excreting nitrogenous waste, as well as managing salt balance. These features of the metanephric kidney are clearly adaptations to that highly specialized niche of life spent entirely away from water.

13.5 THE GUT AND ITS APPENDAGES

The head and kidneys are not the only parts of the body whose development relies, to some extent, on the heart. The gut and its appendages, such as the liver and pancreas, are patterned by an anterior-posterior arrangement of expression domains of transcription factors, closely dependent on mesoderm. Gut development is synchronized with the formation of the somites and requires a constant interplay between the splanchnic mesoderm and the endoderm of the gut wall. Indeed, the endoderm of the gut cannot survive long in culture without signals from mesoderm to maintain it. The liver, in particular, requires the presence of cardiac mesoderm to develop; the development of the pancreas and the lungs, too, relies to an extent on signals from cardiac mesoderm.[21]

It's not just the lateral plate mesoderm that conducts the endodermal developmental ensemble: the notochord too, plays its part, particularly in the development of the pancreas.[22] The pancreas, a complex organ that produces hormones such as insulin as well as enzymes for secretion into the gut, is shaped by other tissues, such as blood vessel endothelium, and the mesoderm that forms the aorta. However, the role of the notochord elsewhere in gut development has yet to be established. Although the possible homologies between the pancreas and the tunicate pyloric gland have been noted elsewhere, it is a surprise to learn that the notochord plays a part in its development in vertebrates. The lesson seems to be that if vertebrates started as two distinct animals—somatic and visceral—that fused together, this fusion is now such that one part cannot function without the other. Although the enteric nervous system can operate without the governance of the central nervous system, the gut as a whole—the core of the "visceral" animal—cannot develop without the somites and other elements such as the notochord, perhaps the most visible structure of the "somatic" animal.

13.6 IMMUNITY

All organisms have ways to defend themselves against microscopic and parasitic interlopers. They also need a way to recognize their own tissues as such, and distinguish these from those of other organisms. Vertebrates have a system of immunity that is both unique and uniquely complicated, so much so that biologists have tended to assume that it developed all of a piece in a kind of "big bang."[23]

That statement requires some clarification and context. Organisms, even plants, have systems of what is known as "innate" immunity. That is, they have a standing repertoire of genetic variation that creates molecules capable of interacting with infectious threats.

Vertebrates, however, have a conceptually different system known as "adaptive" immunity. This means that a specific immune response can be mounted against a threat of which the organism in question had no previous experience, and—if the organism survives—memory of that threat is stored against possible reoccurrence. This is the basis of vaccination, and relies on a kind of evolution and natural selection in microcosm.

In vertebrates, the detectors of external threat are Y-shaped protein macromolecules called immunoglobulins that are either bound to the surfaces of immune-system cells called lymphocytes, or released free into the blood as antibodies. The protein macromolecules are shaped to recognize any threat, or antigen, by an intense episode of DNA rearrangement in a battery of already highly variable genes. Gene segments known as V (for "variable"), D (for "diversity"), and J (for "joining") are cut, pasted, and shuffled together under the supervision of two enzymes, *RAG1* and *RAG2* (where "RAG" is short for "Recombination-Activating Gene"), to create an inexhaustible variety of antibodies. The key, though, is the system in which the single variety of antibody required to combat the current infection, out of all the astronomical number of possibilities, is selected, cloned, and propagated. As one might imagine, a mechanism that tampers with an organism's DNA—its source code—is complex and very highly regulated. And only vertebrates appear to have it.[24]

A further qualification is necessary. Among extant vertebrates, only jawed vertebrates have the system of adaptive immunity based on the V-D-J recombination system. Lampreys and hagfish also have a system of adaptive immunity, but based on an entirely different substrate: somatically derived Variable Lymphocyte Receptors (VLRs), variants of so-called leucine-rich repeats.[25] Although VLRs are structurally and chemically completely unrelated to immunoglobulins generated by V-D-J recombination, they also propagate clonally in response to external threat. The cyclostome system is conceptually similar to that found in gnathostomes even if different in all the details.

The fact that only vertebrates have adaptive immunity does not mean that systems of innate immunity found elsewhere are neither sophisticated nor complex.[26] Echinoderms, the amphioxus, and tunicates all have elaborate systems of innate immunity, such that the line between innate and adaptive is becoming hard to draw. It might be possible to break down the evolution of vertebrate adaptive immunity into stages. The "big bang" need not be a counsel of despair. The purple sea urchin *Strongylocentrotus*, for example, contains a particularly rich and diverse repertoire of innate immune receptors, which natural selection has pushed toward ever greater diversity. This is particularly true among the 220 so-called *Toll*-like receptors, rich in leucine-rich repeats, thought to modulate self-nonself recognition.[27] Some features of the sea-urchin immune system resemble the adaptive immune system of vertebrates, including a cluster of genes with similarities to *RAG1* and *RAG2*.[28]

Amphioxus and tunicates likewise possess arrays of immune-system-like genes with potential for great variety,[29] notably the Variable-region Chitin-Binding Proteins (VCBPs) that appear to form an innate immune system.[30] They are expressed in parts of the body that represent the front line of defense against possible invaders, namely the gut and the pharynx—the latter region that, in lampreys, is the site of lymphoid tissue. The amphioxus also has a gene functionally equivalent to a part of *RAG1* that, in the right laboratory circumstances, can perform certain of the DNA joining operations characteristic of the vertebrate adaptive immune system: a feat of which the sea-urchin versions of *RAG1* and *RAG2* were not capable.[31]

The roots of the vertebrate adaptive immune sytem, therefore, run deep. Many deuterostomes have complex and responsive systems of innate immunity that almost (though not quite) match the versatility of the vertebrate system. One of the keys to the vertebrate system is the *RAG* enzymes. These were present, presumably, in the common ancestor of chordates, and had some function in generating immune diversity. Lymphoid cells—the specialist white blood cells that mediate the adaptive immune response—are, however, believed to be unique to vertebrates.

13.7 THE PITUITARY GLAND

If the outside of the organism meets the inside in any particular place, it is in the pituitary gland, a small body (pea-sized in humans) located and attached to the base of the brain, and formed from the union of neurectoderm and gut. It is an important vertebrate landmark, sited at the most anterior end of the notochord, so in a sense marks a boundary between head (anterior) and trunk (posterior) as well as ectodermal (dorsal) and endodermal (ventral) tissue. The pituitary is an anatomical crossroads, important in vertebrate evolution, and a fitting capstone to this brief survey of some of the internal workings of vertebrates.

The anterior pituitary, or adenohypophysis, is the part that originates from a dorsal outpocketing of the anterior gut, an embryonic feature known as Rathke's Pouch. In development this meets and merges with a downward extension of the hypothalamus of the brain. It is this that forms the posterior pituitary. Together, this small knot of merged tissue secretes hormones that regulate virtually every aspect of life, from growth to sex drive, sleep patterns to metabolism. The pituitary, then, is not only an anatomical but a biochemical crossroads, receiving signals from the brain; translating these into hormonal messages that influence a wide variety of internal glands and other organs; and relaying the replies back again.

In most vertebrates, the pituitary is situated at the base of the brain, in the skull floor. Embryonic development, however, shows that it was once open to the outside, and might have played a role in chemosensation,[32] perhaps mediating environmental signals relevant to metabolism or sex. Earlier I reviewed molecular and anatomical evidence that Hatschek's Pit, a diverticulum of the amphioxus mouth cavity, might be a homologue of the anterior pituitary. The discovery that Hatchek's Pit secretes substances that cross-react with the vertebrate sex hormone gonadotropin—otherwise produced in the anterior pituitary—would seem to confirm that homology.[33] Homologues of the receptors for gonadotropin-releasing hormone (GnRH), an important pituitary hormone, have been discovered in the amphioxus, along with their ligands.[34]

The primary connection between the pituitary and the external environment is maintained in cyclostomes. In lampreys, the single median nostril, situated high on the head, is the external opening of the nasohypophysial duct, a blind-ended duct that runs beneath the olfactory capsule and terminates between the base of the brain and the roof of the mouth. The olfactory capsule also runs into the duct. The nasohypophyseal duct forms, in development, much further forward, and migrates backward as the young animal develops. The hagfish nasohypophysial duct is similar except that it is not blind-ended, but runs backward into the roof of the mouth.

In adult tunicates, the neural gland, connecting to the environment through a duct, might be a homologue of the pituitary.[35] This is supported by gene expression studies[36] as well as embryology, as it develops from the rudiment of the neurohypophysial duct of the larva.[37]

Taken together, the pituitary seems to have started out in the common ancestor of chordates as a relatively simple organ for detecting water-borne chemicals, perhaps hormones or other substances relevant to life history events such as mating or spawning. Eventually the pituitary was internalized and became the sophisticated interface between neural and endocrine systems we know it as today.

13.8 SUMMARY

This brief chapter has offered no more than a taste-menu of the complex and squishy interior of vertebrates. Of necessity I have treated each organ-system briefly and have left a great deal out. If there are any general themes to be drawn from this chapter, they are these: that complex internal systems are an inevitable consequence of size increase; that some very ancient systems, such as pharyngeal skeleton and the enteric nervous system, are partly or completely replaced by derivatives of the neural crest; that the distinction between the head and trunk is very hard to draw, and depends somewhat on the organ system under consideration; and that the heart plays a central role in the development of many organ systems in vertebrates, including the muscles of the head, the urogenital system, and the gut. This is perhaps appropriate given that the heart is the first organ observable as such in a developing vertebrate—it starts beating even before there is blood to pump, or vessels in which that blood might circulate.

CHAPTER 14

Some Fossil Forms

14.1 FOSSILS IN AN EVOLUTIONARY CONTEXT

Over the past twenty years a number of interesting fossil forms have come to light that could help us cross the bridge between the invertebrate and vertebrate state. I shall discuss them in this chapter, but before I get on to the fossils themselves, I hope you'll indulge me in a couple of necessary digressions.

The first concerns some terminology that seems rather arcane but, when one gets used to it, is useful when discussing the relationship of extinct forms with living ones. Because every organism is related to every other organism, whether living or extinct, any fossil you find will be related to one living organism or another. The problem is that fossil forms may not show all the features definitive of any particular extant group, and so cannot be a member of that group. What we say, then, is that such extinct forms are members of the *stem group* of an extant or *crown* group, branching off its *stem lineage*.

This concept is illustrated in fig. 14.1.

Here is a crown group A, all of whose members share a common set of

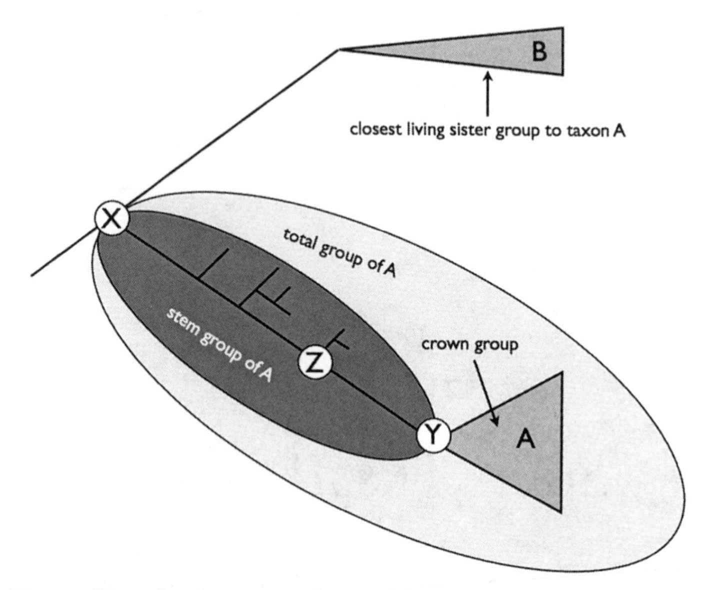

14.1 Diagram illustrating the concept of stems and crowns. A is a taxon with living and extinct members. Y is the node that defines the *crown group* of A; it represents the latest common ancestor of every member of A, whether living or extinct. That accolade cannot fall on Z, because it is the common ancestor not only of A, but also of extinct groups that fall outside it. X is the common ancestor of all organisms more closely related to A than to B, the closest extant relative (sister taxon) of A. X defines the *total group* of A. The line between X and Y defines the *stem lineage* XY. Anything branching from this stem lineage will be in the *stem group* of A.

traits. All members of crown group A have a common ancestor, Y, whose descendants contain all members of A, whether living or extinct. If, for example, crown group A is the mammals, including dogs, cats, elephants, bats, and people, then it also includes extinct forms such as woolly rhinos and saber-toothed cats, which were mammals, even though they are now extinct.

Crown group A has a more remote common ancestor, X, defining a more inclusive group, that is, all members of crown group A and crown group B, the closest living relative (sister group) of A, and all its descendants, whether living or extinct. The *total group* of A, shown as the larger of the two shaded ellipses, includes the set of organisms, living or extinct, more closely related to A than B. Between X and Y is the *stem lineage* of A. Anything branching off this stem lineage is in the *stem group* of A (the smaller, darker of the two ellipses) and will be extinct by definition. Successive stem groups can be defined as more or less crownward to crown group A depending on the number of features they share with the members of A, though of course they won't share all of them. For example, X could be the latest common ancestor of birds (crown group A) and crocodiles (crown group B): the stem group of birds includes non-avian dinosaurs, pterosaurs, and other extinct reptiles.

For present purposes, stem group A could represent chordates and B ambulacraria, with the stem lineage beween X and Y populated by various fossil groups, as I hope to show in this chapter.

My second digression concerns the difficulties faced by anyone seeking to interpret fossils proposed to be the remains of extinct deuterostomes. This constitutes a health warning to temper any optimism that follows.

The deuterostomes are a highly disparate and depauperate group. Disparate, in that the range of body form and habit is extremely varied.[1] Depauperate, in that the restricted range of extant forms and the lack of morphological intermediates means that working out the interrelationships of deuterostomes has historically been extremely difficult. The origin of the vertebrates has, of course, been a particular problem. The problem is exacerbated by the scarcity of fossil forms that might preserve some clues that might help us reconstruct the course of deuterostome evolution. A number of fossils that have been discovered relatively recently could fill in some of the gaps, but if there is a lesson to be learned about fossils, it is that one should be careful for that which one wishes. The fossils are every bit as enigmatic as extant vertebrates—far more so, because we cannot see the living animals, and the extent to which the fossilization process has destroyed many features is hard to judge. Several problems should be apparent immediately.

First: of all the deuterostomes, only vertebrates and echinoderms have extensively mineralized tissues suitable for fossilization.[2] This means that most fossils bearing on the origins and evolution of deuterostomes will be of soft-bodied organisms, which are extremely rare and peculiarly hard to interpret. Shells, spicules, bones, and teeth are resistant to decay, but the extent to which soft tissues decay before they are fossilized might influence phylogenetic interpretations. The process of decay tends to remove the more phylogenetically derived features before the more primitive ones, in which case fossils of highly decayed creatures might be interpreted as more primitive than they really were.[3]

Second: extinct creatures potentially preserve combinations of characters not seen in extant ones. For example, as we have seen, the cornute stylophoran *Cothurnocystis*, for example, has a typically echinoderm mesodermal skeleton made of calcite, but also has exhalant openings interpretable as pharyngeal slits.[4] Another stylophoran, the mitrate *Jaekelocarpus*, might even have had an atrium, very like that found in a tunicate, for all that it, like *Cothurnocystis*, was clothed in calcite.[5] These examples show that echinoderms primitively had pharyngeal slits, and even atria, but lost them. This interpretation, however, was contingent on much supporting evidence, some of it biochemical and obtained from living animals. *Cothurnocystis* and *Jaekelocarpus* are in many respects very peculiar creatures when compared with the extant fauna, and the affinities of stylophora and other extinct echinoderms are still a matter of some debate.

Third: the initial diversification of the deuterostomes took place in the Cambrian Period, between about 541 and 485 million years ago. The group might have evolved before the Cambrian, but probably no more than about 600 million years ago. By the end of the Cambrian, however, representatives of all the major extant deuterostome groups, including vertebrates, are recognizable as such. Several other forms, variously attributed to deuterostomes, are known either exclusively from the Cambrian (vetulicolians, vetulicystids), or became extinct over the next few hundred million years (cambroernids, which—if eldoniids are members—survived to the Devonian; and conodonts, which persisted to the Triassic). This means that all the major features whereby we recognize the various deuterostome groups were established in a relatively brief interval, a few tens of millions of years either side of 500 million years ago. Because all life shares a single common ancestry, any fossil organism must fall, in evolution, somewhere along the lineage of an extant group. There will be fossil organisms more closely related to tunicates than to the amphioxus, or vertebrates, or which share a common ancestor with none of these three subgroups in particular, but with chordates in general. Likewise, there will be others more closely related to echinoderms, hemichordates, or both, or whose common ancestor branched from the lineage leading to all deuterostomes. But because these creatures had not yet accumulated the half-billion-years of evolution that have since elapsed, they are likely to have looked more like one another than any member of any extant group, and this explains why some fossils have peculiar combinations of characters now seen as definitive of separate extant groups.

Fourth: were one able to travel back in time and examine Cambrian organisms as living animals, we would find that they were just like organisms today, in that they existed purely as darwinian agents involved in the struggle for life. They were not there for our reflective convenience. This means that extinct organisms might have had all sorts of unique and idiosyncratic features that might reveal nothing about their relationships with other groups. If these features have been lost in evolution, they will be very hard, if not impossible, to interpret, given that the models we have on which to base that interpretation come from extant animals, whose range of form we know is both disparate and depauperate.

With all these formidable obstacles in mind, we can take a look at some fossils interpreted as primitive deuterostomes, and see if they help or hinder our quest to cross the bridge between the worlds of vertebrates and their invertebrate relatives. I shall start by looking at problematic fossils whose affinities with extant deuterostomes is more or less debatable, and move on to discuss forms whose affinities are (relatively) more secure.

14.2 MEIOFAUNAL BEGINNINGS

We are familiar with organisms visible to the naked eye; perhaps less so with those visible under the microscope. There is, however, a shady middle ground between the macroscopic and microscopic. This is the realm of the meiofauna, organisms that are measured in fractions of a millimeter, just beneath the visibility of the naked eye.

Meiofaunal organisms live in the interstices of life, between grains of sand or mud, often in shallow marine settings. Many of the more recently discovered phyla are meiofaunal,[6] and it is likely that the meiofauna contains many as yet undiscovered creatures. If the extant meiofauna is poorly sampled, our knowledge of fossil meiofauna is likely to be fragmentary. Meiofauna are invariably soft-bodied, and even then, only very particular circumstances can preserve fossils whose entire bodies are comparable in size with the individual grains of sediment in which most fossils are entombed. To use a modern metaphor: individual meiofaunal organisms are at or below the size of a single pixel of the medium in which the fossil record is written.[7] Given that molecular clocks invariably estimate the origin of major groups at ages far greater than the earliest evidence of those groups as fossils,[8] it is possible that the earliest representatives of groups might have been meiofaunal and thus barely visible as fossils.

A small collection of minuscule, meiofaunal fossils from the lowest Cambrian of China might count as the most primitive known deuterostomes.[9] *Saccorhytus coronarius* is a potato-shaped creature known from 44 specimens, the largest of which is only 1.3 mm long. Although it seems irregular at first sight, it is bilaterally symmetrical, with a large, circular opening on the anterior, ventral surface surrounded by a concentric rings of spines or tubercles. This opening is interpreted as the mouth. On each side is a row of four much smaller openings, each looking rather like the anal pyramid of an echinoderm, and interpreted as exhalant pores.

Interpreting such a creature is extremely difficult, but it is tempting to cast it as a very simple and perhaps basal deuterostome, bearing little more to mark it out as such other than that signature feature of deuterostomes, that is, bilateral pharyngeal slits. There is nothing interpretable as an anus: if this is a deuterostome, presumably the anus debouched internally as in tunicates. The mouth, with its rings of tubercles, is, however, reminiscent of another highly controversial group of Cambrian fossils conventionally interpreted as deuterostomes, namely the vetulicolians. I shall discuss these fascinating and frustrating creatures a little later in this chapter.

14.3 CAMBROERNIDS

The cambroernids comprise an informal grouping of several fossil forms including *Herpetogaster*, from the Burgess Shale, with the goblet-shaped, stalked *Phlogites*, and the enigmatic discoidal eldoniids.[10] The former two are exclusively Cambrian: eldoniids are known mostly from the Cambrian but occur as recently as the Devonian. *Herpetogaster* was bilaterally symmetrical but curved, like a croissant. At one tip was a distinct head sprouting a pair of elaborate, arborescent tentacles. About three-fourths of the way along the segmented body was a stalk whereby the animal attached itself to the substratum. A terminal mouth led to a pharynx, a capacious gut and a terminal anus. The whole animal, including tentacles, was between three and four centimeters long. Caron et al. (2010) reconstruct the animal as a large, solitary pterobranch and argue that *Herpetogaster* was a deuterostome, branching from the main stem of ambulacraria before the divergence of the echinoderm and hemichordate lineages. This has a number of interesting implications. First, that if gill slits were once present (as our current understanding of deuterostomes implies), they were lost; yet there seems to be a clear distinction between the large, unsegmented head and the clearly segmented trunk. The presence of bilateral tentacles has implications for the earliest, possibly tentaculate feeding system of very primitive echinoderms, such as cinctans.[11]

Allowing that similarities between feeding structures might be convergent, Caron et al. (2010) unite *Herpetogaster* with *Phlogites* from the Lower Cambrian of China, a tentaculate, stalked form of hitherto debatable affinities[12]; and the equally debatable, discoidal eldoniids, sometimes compared with holothurians.[13]

14.4 VETULICYSTIDS

The vetulicystids are curious fossils from the same Chengjiang fauna of the Lower Cambrian of China that has produced vetulicolians, yunnanozoans, early vertebrates, and arthropods, many with remarkable soft-tissue preservation.

A vetulicystid consists of a blob-like theca attached to a short tail or stalk, and Shu and colleagues interpret it, essentially, as an echinoderm with no clothes.[14] The theca was presumably covered in a thick integument but there is no sign of mineralization. The theca bore three openings. Two are very like the pyramidal-shaped, plated openings seen in echinoderms. One is situated near the anterior end (that is, opposite the end bearing the tail or stalk) and is interpreted as a mouth; the other is close to the theca-stalk boundary and is interpreted as a gonopore or anus or both.

The third opening, close to the posterior pyramid-shaped opening, is a grille-like structure interpreted as a respiratory organ. Again, similar structures are seen in extinct echinoderms such as blastoids, although it might represent something similar to (or perhaps even homologous with) the structures seen in stylophorans interpreted as deuterostome-style pharyngeal slits.

As echinoderms are more or less defined by the presence of their highly distinctive skeleton of calcite stereom mesh, it's very hard to imagine what the immediate, calcite-free ancestors of echinoderms might have looked like. Whereas it is true that some holothuria have all but lost their calcite skeletons, they are immediately recognizable as echinoderms by the presence of pentameral symmetry and the water-vascular system. The most primitive echinoderms, however, were bilaterally symmetrical and probably lacked a water-vascular system, so the presence of a calcite skeleton is all we really have to go on.[15]

The phylogenetic interpretation of vetulicystids as very primitive stem-group echinoderms is reasonable but tentative, and of course debatable.[16] A phylogeny by Han et al. (2017) has vetulicystids as a sister group of vetulicolids, with *Saccorhytus* as an outgroup within a larger group of deuterostomes equally closely related to chordates and to ambulacraria.

14.5 VETULICOLIANS

The vetulicolians[17] (fig. 14.2) comprise a group of enigmatic fossils from the Cambrian. Although best known from the Chengjiang fauna of China, they have also been recovered from the Burgess Shale of Canada, the Sirius Passet Fauna of Greenland,[18] the Emu Bay Shales of southern Australia,[19] and elsewhere. Vetulicolians are a few centimeters long, bilaterally symmetrical, and divided into two very distinct halves.

One half, conventionally regarded as the anterior, is roughly egg-shaped, though in some forms it tends toward the rectangular. At the front is a large opening interpreted as the mouth. In some forms this opening is fringed by lappet-like structures that extend further forward. In others, the mouth is circular and fringed with concentric rings of spines or tubercles, as in *Saccorhytus*. Laterally and on each side is typically a groove perforated by five pores of occasionally somewhat complex internal structure. The posterior and dorsal margin may be surmounted by a triangular structure rather like a shark-like dorsal fin. The anterior half as a whole is voluminous: the fossils are filled with sediment. Sometimes the inner surfaces preserve signs of a complex arrangement or meshwork of tubercles, and a mark or groove runs along the ventral floor. The fossils consist of casts and molds made in the encasing sediment: nothing remains of the animals themselves. Because of this, the nature of the integument of the anterior half is unclear, though it seems to have been made of a number of large, stiff, cuticularized plates.

The other portion of the animal, conventionally regarded as the pos-

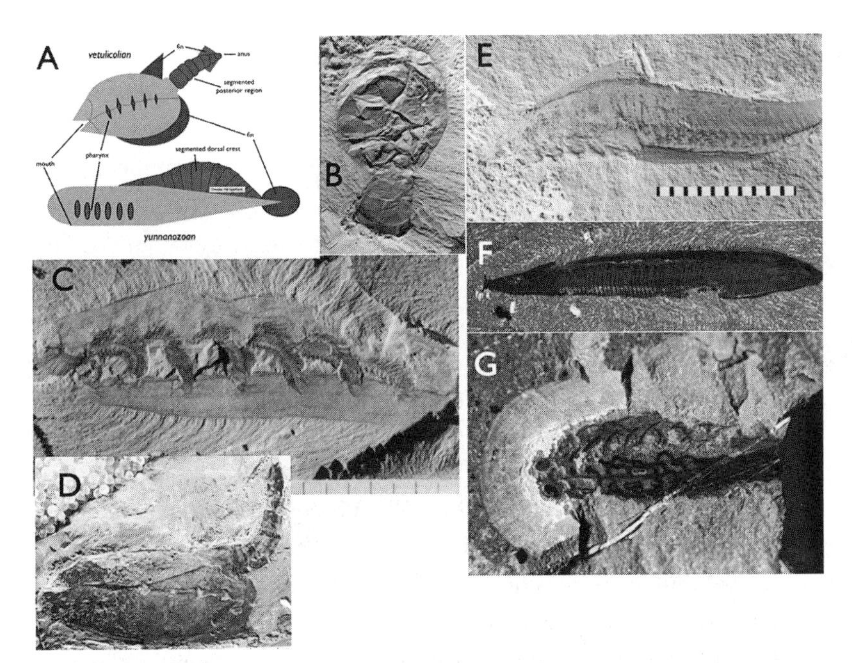

14.2 Some fossil forms. A: Cartoons of a vetulicolian and a yunnanozoon, simplified to show a consensus of only those features on which most or all researchers can agree. B: a vetulicystid; C: anterior end of the yunnanzoon *Haikouella*; D: a vetulicolian, *Vetulicola*; E: the primitive chordate *Myllokunmingia*; F: the primitive chordate *Pikaia*; G: the primitive vertebrate *Metaspriggina*. F, G courtesy of Jean-Bernard Caron/ Royal Ontario Museum; B-E courtesy of Degan Shu, Northwest University Early Life Institute, Xian.

terior, is joined to the dorsal, posterior corner of the anterior mass. The posterior is an extended structure typically consisting of seven segments, strongly reminiscent of the tail of a shrimp. In some specimens, a long, dark, and occasionally helically coiled strand runs through the posterior section, terminating at the most posterior tip. Again, the nature of the integument is unknown though the way the segments are organized is as if each were covered with fairly stiff tissue, bounded to its neighbors on either side by thinner, more flexible sections—very much like an arthropod, in fact.

There are many variations on this theme. Some vetulicolians have distinctive circular mouths ringed with tubercles or spines, whereas others do not. Some, such as *Banffia*, have many more than seven segments in the posterior section, do not seem to have lateral pores, and show a degree of torsion such that the entire body is twisted helically in a clockwise fashion when viewed from the anterior end. Others, such as *Skeemella*, have very small anterior sections and much longer, more segmented posterior tail sections,[20] whereas yet others have more bulbous anterior portions and relatively abbreviated posterior sections. Nevertheless, the vetulicolians as a whole form a morphologically distinctive and coherent group,[21] although it seems generally recognized that *Banffia* and its allies are rather different from vetulicolians proper.

When the vetulicolian *Vetulicola* was first discovered in the Chengjiang fauna of southern China it was interpreted as an arthropod.[22] This is entirely understandable. The anterior section looks very like the kind of valved carapace seen in some shrimp and many Cambrian arthropods, within which various limbs would be enclosed, and the posterior section does look very like that of an arthropod. Subsequent examination of dozens of specimens, however, has revealed no trace of limbs or other structures one would expect to find in arthropods, especially from the Chengjiang fauna, such as eyes, mouthparts, and so on.

It is possible that vetulicolians belong to the cycloneuralia, the limbless, eyeless non-arthropod branch of the ecdysozoa, which includes such creatures as nematodes and kinorhynchs.[23] The circular mouths of some vetulicolians do resemble those of some stem-arthropods such as *Anomalocaris* as well as some lobopods such as *Hallucigenia*, recent reinterpretations of which have shown strong resemblances with cycloneuralia.[24] On the other hand, vetulicolians show no sign of moulting (a key feature of all ecdysozoa); neither do they have anything like the specialized mouthparts of a nematode or a kinorhynch. The most important objection, however, is that no arthropod, or cycloneuralian, has anything like the bilateral arrangement of pores seen in vetulicolians, which look like nothing so much as deuterostome pharyngeal pores or slits.[25]

The arrangement of large, anterior mouth; the capacious pharynx; and the biserial perforation of the body wall with slits, are seen nowhere else but in deuterostomes. The interpretation of a dark, ventral strip as an endostyle[26] led some to speculate[27] that vetulicolians might be akin to chordates, or even tunicates. However, Shu and associates now regard their initial identification of an endostyle as circumstantial at best[28] and have always maintained that vetulicolians are akin to deuterostomes in general, but no extant subgroup in particular. Some researchers, however, robustly maintain an alliance between vetulicolians and tunicates, even without evidence of an endostyle: the same researchers interpret a structure in the tail region of the Australian vetulicolian *Nesonektris aldridgei* as a notochord.[29]

The narrow, dark strip seen in the posterior region of vetulicolians, terminating at the tip, is usually interpreted as a gut trace, terminating in an anus. Although tunicate tadpole larvae do have an endodermal strand in the tail, tunicates (and chordates more generally) do not have a terminal anus—a feature more associated with enteropneusts. Evidence for a notochord is much more equivocal. It could be that preservational differences between sites favor the preservation of guts over notochords, or vice-versa. Perhaps vetulicolians did have notochords that contained more muscle than fibrous connective tissue, and were therefore more prone to decay. After all, notochords in extant forms do vary in composition, and the notochord of the amphioxus contains an appreciable quantity of muscle compared with that of either tunicates or vertebrates. It could even be that some vetulicolians had notochords, whereas others did not—in which case the vetulicolians do not represent a natural group so much as an assemblage of primitive deuterostomes, some of which are more closely related to some extant groups than others.

The posterior region as a whole is as problematic as any one of its parts. It looks very much like an arthropod tail. It is segmented, but one cannot know whether the style of segmentation would have been more like that of a deuterostome or an arthropod. If a deuterostome, the lack of a notochord to act not only as a support structure, but morphologically and developmentally linked with the surrounding tissues, is puzzling. It could be that the segmentation in vetulicolians is, in terms of development and evolution, completely different from anything seen in either deuterostomes or arthropods, although one could presume that there was some "deep" homology at the molecular level.[30] Although the current consensus is that vetulicolians were probably early offshoots from the deuterostome stem, by virtue of the deuterostome-like pharynx, their phylogenetic position is still an open question.

I'd like to make the very tentative suggestion, however, that they might be the most primitive members of the chordate total group: that is, they are more closely related to chordates than to ambulacraria. My reasoning is as follows. I've described the many ways in which the vertebrate head is distinct from the trunk, especially in respect of mesodermal segmentation, and referred to Romer's insight (in his 1972 paper) that the vertebrate body is fundamentally divided into two parts: an anterior pharynx and a posterior, muscular tail, the two parts of which were once separate but which are more or less interwoven in extant forms. Although deuterostomes are distinguished by a capacious pharynx that communicates to the outside by biserial pores or slits, it is in vetulicolians that we first see the appearance of a distinct trunk or tail region which, in terms of its position relative to a bulbous, non-segmented pharynx, could be interpreted as homologous with a chordate trunk. Although appendages appear in ambulacraria that have been variously considered to be homologues of trunks or tails, these appendages are either stalks (that is, to anchor the animal to the substrate), or organs derived from stalks; whereas the appendages seen in vetulicolians betoken a fundamentally free-living existence. In addition, the appearance in evolution of a trunk after an anterior, unsegmented region could be a recapitulation of Lacalli's idea (2005) that the embryos of many deuterostomes are, essentially, heads, waiting for trunks to happen.

When vetulicolians first came to my attention[31] my initial impression was that I was looking at one of Romer's somatico-visceral animals. Although the anatomy of vetulicolians differs in many respects from Romer's ideal—notably in the terminal position of the anus and the probable absence of a notochord—it is in vetulicolians that we see the first signs of a body plan that is distinctively chordate, as opposed to ambulacrarian, or generalized deuterostome.

14.6 YUNNANOZOANS

If the vetulicolians were not mysterious enough, the Chengjiang fauna offers another peculiar and controversial group with possible deuterostome affinities. These are the so-called yunnanozoans (fig. 14.2). When sufficient fossils were collected to allow a description, the fossil *Yunnanozoon lividum* was first reconstructed as a chordate, and possibly even an early relative of the amphioxus.[32] The animal was between 2.5 and 4 cm long, fusiform, laterally flattened and with distinct anterior and posterior ends. The animal was interpreted as containing an anterior pharynx, though not nearly as capacious as that in vetulicolians; possible branchial slits; a gut, gonads, a notochord, and—most distinctively of all—muscle blocks arranged in an anterior-posterior series. These were rather unusual, however, in that they did not lie lateral to the notochord, but were positioned almost wholly dorsal to it, forming a flaring, dorsal crest.

This interpretation of *Yunnanozoon* as a chordate akin to the amphioxus was immediately challenged with a competing view, that it was, instead, a hemichordate similar to enteropneusts,[33] although perhaps one more used to swimming rather than burrowing. The proposed notochord was reinterpreted as a gut; the head was reconstructed as having a collar with a proboscis, and the segmented dorsal crest was reinterpreted as a kind of stiffened, segmented dorsal fin.

The original group raised the stakes with another fossil, *Haikouella lanceolata*, similar to *Yunnanozoon* in many ways, but which offered much better preservation, especially of the head.[34] *Haikouella* was reconstructed with a large brain, possible paired eyes, a heart and branchial circulatory system, and other features. Again, the flaring, segmented, dorsal crest was reconstructed as a set of muscle blocks surmounting a notochord. This was in turn challenged, once again, by the same school of skeptics, with the description of another species of *Haikouella* reconstructed with external gills, but none of the chordate-like features previously ascribed to the genus.[35] This time, however, comparisons with hemichordates were downplayed. Instead, yunnanozoans were regarded as deuterostomes of uncertain affinity, perhaps akin to vetulicolians, which had by then recently been redescribed as deuterostomes.

Not to be deterred, the original group presented further characters ranged in support of a vertebrate affinity for *Haikouella* in the form of a detailed cladistic analysis.[36] Reviewing the situation some years later the skeptics[37] note that whereas all researchers had access to an abundance of exquisitely preserved specimens, their interpretations about the same structures were at variance.

All agree, however, that yunnanozoans had a pharynx and gill slits, and so are deuterostomes of some sort. It could be that the segmented dorsal crest is homologous with the tail region of vetulicolians, but has shifted forward and dorsally over the pharyngeal region. This suggestion is strikingly Romerian, as an evolutionary reconciliation of the "somatic" and "visceral" components of the idealized ancestral vertebrate. In the same tentative spirit as above, for the vetulicolians, I should like to propose that yunnanozoans sit in the chordate stem group, more crownward than vetulicolians, representing a condition in which the posterior, segmented trunk has begun to integrate with the pharyngeal region.

14.7 *PIKAIA*

Perhaps no fossil from the Burgess Shale has achieved such iconic status as the small, vaguely fish-shaped *Pikaia gracilens*. This is in large part thanks to Stephen Jay Gould, who in his book *Wonderful Life* painted a picture of the modest appearance of this amphioxus-like creature compared with that of the many spectacularly spiny arthropods and other creatures of uncertain affinity found in the same strata, suggesting that had *Pikaia* and its progeny not survived, we wouldn't be here to tell the tale.

The reality is not quite so simple. Many more specimens of *Pikaia* have since been found, but more data offer a more nuanced interpretation: that whereas *Pikaia* is perhaps the most primitive known chordate, it is in many respects peculiar and bears comparison with yunnanozoans.[38]

The animal is shaped very much like an amphioxus and varies in length between 1.5 and six centimeters, with a mean of about four. It is tallest (that is, dorsoventrally) nearer its posterior end: the anterior tapers to a small head, but the pointed posterior end is more abrupt. The head is distinct, bilobed, with each lobe bearing a small, stiff, spine-like projection like an antenna or tentacle. There are no structures interpretable as eyes. Immediately behind the head are up to nine paired, feathered extensions similar to the parapodia of polychaete worms and originally identified as such. It is possible that each of these extensions is associated with a small pore. Although projecting from the body, they might have been covered by tissue in life, and in some specimens can be seen to extend inward, toward the midline, as well as outward.

The body as a whole is divided into around a hundred sinusoidal-shaped segments interpreted as muscle blocks demarcated by more decay-resistant connective tissue. These segments are relatively longer (anterior to posterior) in the midsection, becoming more closely packed toward the head and tail. They do not extend to the extreme dorsal or ventral margins. In the interior, longitudinal strands can be interpreted as a dorsal nerve cord immediately overlying a notochord; a gut, with a small ventral mouth immediately behind the head, ballooning in the pharynx but otherwise thin and terminating in an anus at the posterior end; and a ventral blood vessel. Beneath the dorsal margin is a prominent, sausage-shaped structure, once interpreted as a notochord, but referred to as the "dorsal organ." As the dorsal organ tapers toward the anterior it underlies a shield-shaped region known as the anterior dorsal unit, located on the dorsal surface immediately behind the head and most reminiscent, at least in external appearance, of the mantle of a slug.

The arrangement of segments in *Pikaia* is entirely characteristic of chordates, but many other features remain problematic, notably the bilobed head with its antennae, the bilateral series of parapodia-like structures, and the dorsal organ. Conway Morris and Caron (2012) suggest that the external appendages might be homologous with the putative gill structures of yunnanozoans, and that the dorsal organ might represent an elastic, internalized cuticle, representing, in transition, a stage in evolution between the dorsally mounted segmented structure with notochord beneath, to the more laterally equivalent arrangement seen in chordates.

Perhaps the closest analogue to *Pikaia* is the amphioxus, the animal with which it is traditionally compared, and in that light, *Pikaia* is perhaps not so unusual. After all, the amphioxus has a number of distinct and possibly unique features, particularly in the head, such as buccal cirri and the wheel organ; the anterior extension of the notochord, and the single eye spot, which are as idiosyncratic in their way as the bilobed head, peculiar extensions, and antennae of *Pikaia*, but seem less so by virtue of long familiarity. I also wonder whether the dorsal organ of amphioxus might not be homologous with the structures known as fin-ray boxes that, in the amphioxus, extend along the dorsal margin. On the other hand, *Pikaia* lacks features shared by the amphioxus as well as other chordates such as tunicates, in particular the atrium. It could, of course, represent a lineage of amphioxus-like animal that has lost this feature, but any more than that is pure speculation. In that speculative spirit I suggest that *Pikaia* might be the most primitive animal known that has a notochord and in which the fusion of pharynx and segmented trunk, only partially achieved in yunnanozoans, is essentially complete. In chordate phylogeny *Pikaia* might be the immediate outgroup of crown chordates: that is, close to the common ancestor of all extant chordates, and all its descendants, living or extinct. Crown chordates are animals that have not only a notochord, but an endostyle and an atrium.

14.8 *CATHAYMYRUS*

Cathaymyrus diadexus is the fossil of an amphioxus-like animal from the Chengjiang fauna, and is about ten million years older than *Pikaia*.[39] Known from a single specimen, *Cathaymyrus* is very much like an amphioxus in general shape, with visible muscle blocks and a longitudinal structure of the appropriate dimensions and location to be a notochord. The head is much less well-preserved but does appear to contain a pharynx with gill slits. At the time of its discovery *Pikaia* was assumed to be a cephalochordate, and *Cathaymyrus* was also ascribed to that group. Twenty years on it seems more likely that *Cathaymyrus* deserves this status more than *Pikaia*, although *Cathaymyrus* has no trace of an atrium.

Two more forms have come to light more recently.[40] *Cathaymyrus haikouensis* is similar to *C. diadexus*, with myomeres and a notochord; *Zhongxiniscus intermedius* is also similar but has a dorsal fin, and might be closer to a vertebrate than to a cephalochordate. *Pikaia* remains somewhat problematic despite a wealth of specimens; the possible amphioxus-like chordates seem much more interpretable as such, although are vanishingly rare by comparison.

14.9 THE EARLIEST FOSSIL VERTEBRATES

Because mineralized tissues make for better (or at least, more frequent) fossils than impressions of soft tissues, and because the hard tissues of vertebrates are so distinctive, the search for the earliest vertebrates has tended to concentrate on such tissues. However, the earliest mineralized remains attributable to vertebrates, from the Late Cambrian, are fragmentary and debatable,[41] so we are on surer ground with fossils of purely soft-bodied creatures from the Chengjiang fauna and the Burgess Shale. These show that vertebrates were well established by the middle of the Cambrian Period.

Myllokunmingia and *Haikouichthys* are fossils of soft-bodied, fusiform animals from the Chengjiang fauna.[42] Each described from a single fossil, they have vertebrate-like segmented muscle blocks, pharynx, notochord, possible ventral fin-folds, and pericardial cavities. *Myllokunmingia* is 28 mm long and appears to have five or six pairs of gill-pouches resembling those of cyclostomes. *Haikouichthys* is more slender and about 25 mm long. It has at least six gill arches, and possibly as many as nine, arranged in a branchial basket; structures possibly interpretable as head cartilages, and a dorsal fin with possible fin radials. The anatomy of *Haikouichthys* became clearer when more specimens came to light.[43] The animal is suggested to have had paired eyes at the extreme anterior end; possible nasal sacs and otic capsules; and vertebra-like elements (comparable with the arcualia of lampreys) associated with a clear notochord.

The hitherto enigmatic *Metaspriggina* from the Burgess Shales was recently redescribed[44] as a primitive fish, with a notochord; paired eyes at the extreme anterior end; nasal sacs; muscle blocks; post-anal tail, and branchial bars. Although clearly a jawless vertebrate of some sort, the presence of separate elements in the branchial basket (as opposed to a fused structure), and gills situated external to the gill bars (rather than internal), suggests that these features, usually seen as characteristic of gnathostomes, evolved very early, and that the condition seen in modern lampreys is peculiar to that group.

These three creatures seem to represent a group of very early vertebrates either related to the cyclostomes, or, more cautiously, as stem vertebrates, branching from before the divergence of cyclostomes and gnathostomes. In either case, their presence shows that vertebrates had not only evolved, but had become very diverse, very early in the Cambrian, implying either that the deuterostomes as a whole evolved with extraordinary speed (as part of the so-called Cambrian Explosion) or that divergence was more leisurely, Precambrian, and cryptic.

14.10 CONODONTS

Conodonts are minute tooth-like fossils found in rocks from the Late Cambrian to the Triassic (about 530 to 200 million years ago). Their distinctive and sometimes complex morphology, rapid evolution, and widespread distribution make them useful to geologists, as they can be used to establish stratigraphic succession. The identity of the creatures of which conodonts are a part, however, was a long-standing mystery until the discovery of articulated conodonts in the mouth of an otherwise soft-bodied eel-like fossil from the Carboniferous of Scotland.[45] The animal, complete with muscle blocks, a tail fin with radials, and anteriorly placed paired eyes, was clearly akin to chordates, if not vertebrates. Other conodonts associated with soft tissues indicative of eyes and muscles have since been recovered in older, Ordovician strata.[46] Although questions have been raised about the presence of eyes, and the apparent absence of gill slits or gill pouches, support for a vertebrate affinity comes from the conodonts themselves, which are phosphatic and have a microstructure very like that of enamel. It is fair to say that the interpretation of conodonts as vertebrates, and even akin to gnathostomes, has been controversial, with vigorous claims for and against.[47]

There are, in fact, three kinds of conodont, ranged in order of increasing structural and histological sophistication. The simplest are the protoconodonts, now believed to be spines from fossil chaetognaths[48] and therefore nothing to do with vertebrate evolution. The most complex are the euconodonts, and it is these that are crowned with phosphatic tissue very like enamel. In between are the paraconodonts. A recent detailed study shows that paraconodonts can be arranged into a graded series of ever more euconodont-like forms, and it seems most likely that euconodonts evolved directly from paraconodonts. This would weaken the case that conodonts are closely akin to gnathostomes, because the transition from paraconodont to euconodont is quite at variance with the evolution of vertebrate teeth, with which euconodonts had been compared and homologized.[49] This has two consequences. The first is that the enamel-like tissue in conodonts and the mineralized tissues in crown vertebrates evolved entirely convergently. This seriously weakens the case that conodonts are akin to gnathostomes, and their position among basal vertebrates remains unclear.

The second consequence is one of simplification rather than confusion. It has long been thought that what we think of as "teeth" evolved primarily in the dermis of vertebrates, migrating into the mouth only later on.[50] The skin of modern sharks is entirely clothed in denticles, each of which is a miniature tooth with enamel, dentine, and a pulp cavity, just like a regular tooth in the mouth. Many fossil fishes have scales very much like this. However, the equation of conodonts with teeth—found in the mouth but not elsewhere—confused the conventional outside-in view of tooth evolution. The relegation of conodonts as parallel structures has to some extent restored this equilibrium.

14.11 OSTRACODERMS AND PLACODERMS

Nearly all extant vertebrates are gnathostomes—that is, vertebrates with jaws. And not only jaws, but paired fins, paired nostrils, and many other refinements. As these are all vertebrates and therefore safely across the chasm that separates the world of vertebrates from those of invertebrates, I shall spend very little time with them, and those interested in the early evolution of gnathostomes may consult the several excellent sources that exist.[51] I mention the subject here to show how fossils can, in some circumstances, be illuminating as well as frustrating. The transition from jawless to jawed can, as it happens, be traced rather well using fossil forms.

The earliest jawless vertebrates, with the arguable exception of conodonts, never appear to have had any hard tissues. Mineralized exoskeletons appear in a range of fossil jawless fishes, collectively known as "ostracoderms," that lived between the latest Cambrian and the end of the Devonian. These were clothed in extensive dermal skeletons ranging from a shagreen of scales to a robust armor completely encasing the head and pharynx back as far as the heart: in other words, the ancient head region of chordates. The most primitive ostracoderms were the heterostracans. These were heavily armored but had no paired fins and little is known of their internal anatomy.

The next group, more crownward on the gnathostome stem, were the osteostracans, consisting of various groups of fishes whose head armor not only clothed the skin but also infused the underlying tissues, so much so that the internal anatomy of brain, nerves, and blood vessels can be clearly seen. Some of these fishes had paired pectoral fins, which emerged just behind the dermal armor. Although, like heterostracans and modern cyclostomes, all had a single nostril, some of the more derived forms show an incipient broadening of the face, preparatory, as it were, to the evolution of paired nostrils.[52]

The earliest jawed vertebrates, more crownward still in the gnathostome stem, are the placoderms. This large and varied group of fossil fishes lived between the Silurian and Devonian periods and was, like the ostracoderms, heavily armored, though the head armor was to some extent decoupled from that of the trunk. Fully equipped with fins, the various placoderms continue to show advancement in head anatomy that foreshadow its full gnathostome expression, such that we can say that the most derived placoderms had, for the first time, something we'd recognize as a face.[53]

14.12 SUMMARY

If this chapter illustrates anything, it is this: if to reconstruct the evolution of a group when the members have mineralized skeletons is extremely difficult, to reconstruct it when the members are exclusively soft-bodied is almost impossible. The evolutionary history of the gnathostome total group (that is, everything more closely related to gnathostomes than to cyclostomes) can be worked out because one of the first things that evolved in stem-gnathostomes was a robust exoskeleton. Even so, the emerging consensus[54] has been built on more than a century of hard work with many debates and false trails. Much the same can be said of the early history of another well-mineralized group of deuterostomes, that is, the echinoderms.

Almost everything we know about extinct, soft-bodied deuterostomes comes from vanishingly rare *Lagerstätten*—that is, sites in which highly unusual circumstances have led to exceptional preservation of soft tissues. Chief of these are the Burgess Shale of western Canada, and the somewhat older Chengjiang biota of southern China, to which can be added a small number of select sites scattered across the world from northern Greenland to southern Australia. As discussed above, interpreting the features of such fossils is fraught, given that many of the structures are difficult or impossible to homologize with structures in extant forms, even when one can account for the distorting effects of post-depositional decay and subsequent tectonic movement.

For this reason I tend to think that trying to untangle the phylogenetic history of vertebrates in any formal way, with cladistic analysis, is a Sisyphean task.[55] This is because assigning values and polarities to characters without clear extant equivalents is hard to justify without introducing circular arguments. The only way out of such an impasse is to use molecular characters, but this is problematic for extant groups that are highly disparate, and impossible for fossils of any great age. This is why the phylogenetic hypothesis I present in fig. 14.3 is not supported by any formal analysis: it is merely a scenario, a suggestion, as a way of summarizing the foregoing.

Having said that, I do not think that everything is miserable or hopeless.

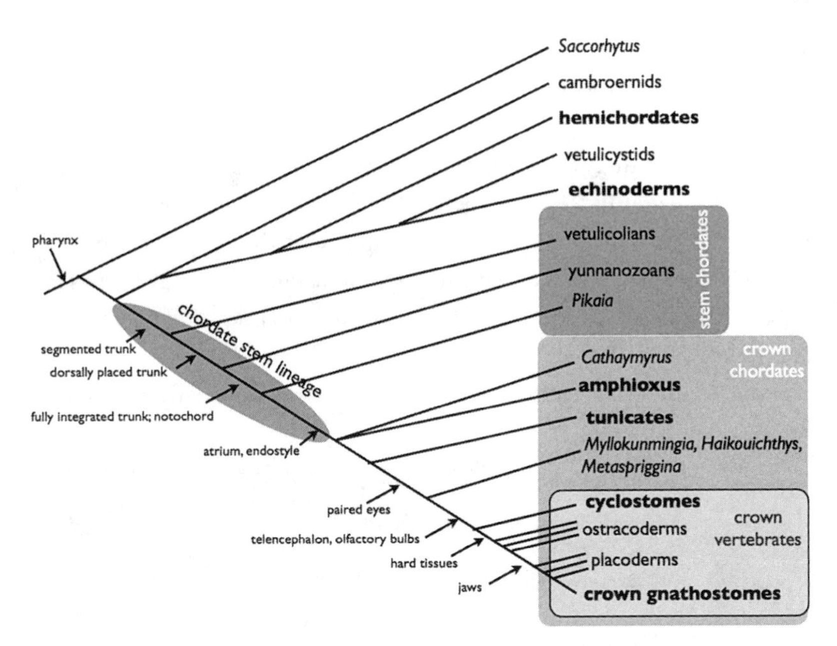

14.3 A tentative hypothesis of deuterostome relationships, summarizing the arguments about fossil forms set out in this chapter.

It is possible to construct a scenario, as I have done, in which the first innovation of crown chordates was the generation of a segmented trunk or tail to add to the primitive deuterostome body that was, in essence, a trunkless pharynx. This segmented appendage was entirely different from the stalks seen in ambulacraria. The first animals with such a structure were the vetulicolians. I think that one can place yunnanozoans in a more crownward position in the stem group of chordates, as they show signs of the integration of the appendage with the pharynx. *Pikaia* comes next, and, with the first uncontested appearance of a notochord, is the first known true chordate. We are on surer ground with *Cathaymyrus* as (possibly) a cephalochordate, and *Myllokunmingia*, *Haikouichthys*, and *Metaspriggina* as the earliest known crown vertebrates, although the relationship between these and modern cyclostomes is unclear. The position of conodont animals remains unresolved.

The astute reader will have noticed the near-total absence of tunicates from this discussion. Tunicates, of course, represent the hardest problem of all. Their close relationship with vertebrates is revealed by development and molecular phylogeny, and although they share some innovations with vertebrates not seen in the amphioxus, such as a well-developed heart, and the incipient activity of neural crest and placodes, they are so very different in structure and life habits from either vertebrates or cephalochordates that charting their evolution is almost impossible. The fossil record of tunicates is meager indeed, even when the fossils concerned might be said to be recognizably tunicate. Discovering fossils that might stand in the tunicate stem group would seem an impossible quest. The only things that have come close are the vetulicolians, but in the absence of an endostyle and (*pace* García-Bellido et al. 2014) a notochord, placing these as stem chordates rather than stem tunicates is to expose a marginally less valuable hostage to fortune.

CHAPTER 15

Breaking Branches, Building Bridges

15.1 DEFINING THE DEUTEROSTOMES

I have now outlined the various traits whereby extant vertebrates are recognized, and set these in the context of the phylogeny and development of their deuterostome relatives and animals in general. I have explored some of these traits in depth, and also investigated the degree to which various fossil forms might inform various arguments and scenarios. The time has now come to set all these things in order. Is it possible to break down the acquisition of traits into some kind of order, to minimize the enormous perceived gap between vertebrates and the non-vertebrate world?

Now, I am sufficiently aware that this problem has taxed minds as fine as Cuvier and Goethe, Aristotle and Geoffroy, Huxley and Haeckel, not to mention a host of colleagues working assiduously on various aspects of the problem to this day and whose work I have cited here. I am also aware of the long history of scenarios to "explain" vertebrate origins that have ended up in the where-are-they-now files of academic endeavor.[1] Please do not expect, therefore, a structure as imposing as Brooklyn Bridge, or even Poohsticks Bridge. It's more a Rube Goldberg arrangement of ropes and planks thrown in haste across a gorge. Some of the ropes are frayed here and there, and quite a few of the planks are rotten or have disappeared altogether. But it is a bridge nonetheless. Let's see if we can cross it without falling off.

Despite everything, the classical, developmental features of radial cleavage, deuterostomy, and enterocoely[2] remain useful as traits that define deuterostomes as a natural group. Although some protostomes develop by radial cleavage and show signs of deuterostomy, these can be said to be remote convergences. Conversely, no member of the deuterostome clade develops by anything other than radial cleavage and deuterostomy, though the commitment to enterocoely is violated here and there. Better definitions will come from analyzing the molecular underpinnings of these classical observations, and it seems that variations in blastopore fate are products of the differential activity of *Wnt* signaling related to the location and timing of mesoderm formation.[3]

A more definitive trait, however, is the presence of pharyngeal slits directed by a signature cassette of six genes found in all deuterostomes, whether they retain pharyngeal slits or have secondarily lost them.[4] Because of this we can be fairly confident that bilaterally arranged, serial perforations of the body wall immediately behind an anterior, terminal or subterminal mouth are homologues of pharyngeal slits found in deuterostomes such as the amphioxus, enteropneusts, tunicates, and ammocoetes, and so can interpret such features in fossil forms in that light with confidence, whether they are stem-echinoderms, such as stylophora; possible stem chordates such as vetulicolians and yunnanozoans; or extremely primitive deuterostomes such as *Saccorhytus* that preserve little else. The evolution of pharyngeal slits might have been accompanied by the acquisition, presumably from bacteria, of genes for sialic acid metabolism, co-opted for the formation of mucus essential for efficient pharyngeal filter-feeding.[5]

Digging deeper into metabolism, deuterostomes seem to have a tendency to sequester calcium compounds in connective tissue, perhaps forming a common basis not just for the extensive (but otherwise very different) mineralization seen in echinoderms and vertebrates, but also for the much more low-key mineralization in hemichordates and some of the exotic biochemistry of tunic formation in tunicates.[6] In addition, deuterostomes exploit various iodine-containing hormones for the direction of growth and metamorphosis.

Deuterostomes are primitively tricoelomous. That is, the body cavities are organized into three pairs of coeloms, arranged two-by-two, into the small, anterior prosomes; the larger mesosomes in the middle, and the larger metasomes posteriorly. This arrangement is seen most clearly in echinoderm larvae before the establishment of pentamery, and in hemichordates throughout life. In chordates, this pattern has been essentially overwritten by somitogenesis, but there might be traces of it in the early development of the amphioxus. As you'll recall, the most rostral somites in the amphioxus form by simultaneous enterocoely from the archenteron, whereas the rest emerge successively from the tail bud; furthermore, FGF signaling seems to be essential for the formation of these most anterior somites.[7] Given that FGF signaling seems to be a feature of the formation of the mesoderm in deuterostomes in general[8] it could be that the rostral somites represent a relic of the ancient deuterostome tricoelomous condition. Conversely, it suggests that trunk somites in chordates represent an evolutionary novelty—but that matter is something for discussion further along the bridge.

Another feature possibly common to all deuterostomes is a posterior organizer essential for the proper formation of germ layers and the anterior-posterior body axis. The ancestral organizer would have directed operations with β-catenin and *Wnt* signaling. Such an arrangement is found in enteropneusts,[9] echinoderms,[10] and vertebrates[11] and this wide distribution presumably reflects a common deuterostome heritage.

Deuterostomes also have a marked tendency toward left-right asymmetry governed by *Nodal* signaling. Although such *Nodal*-directed asymmetry is found here and there in protostomes[12] its pervasiveness in deuterostomes suggests that *Nodal* signaling plays a rather more fundamental role in axial patterning in deuterostomes than in other animal groups.[13]

In the viscera, deuterostomes often show an association between the vasculature and excretory systems as a way of draining the pericardium, that is, the coelomic compartment in which the heart resides. Enteropneusts have the heart-kidney complex, possibly homologous with the axial complex of echinoderms. It is a reasonable to ask whether the pronephros, at least, of vertebrates, might be homologous with these systems.

Finally, deuterostomes show a characteristic anterior-posterior patterning of the ectoderm, revealed by signature expression in hemichordate ectoderm of genes otherwise expressed in the Anterior Neural Ridge (ANR), Zona Limitans Intrathalamica (ZLI), and Midbrain-Hindbrain Boundary (MHB) of vertebrates[14] and, to some extent, the amphioxus. It is interesting to note that hemichordates, vertebrates, and the amphioxus represent the more slowly evolving deuterostomes, so it's a fair bet that the ectoderm of ancestral deuterostomes was patterned in a similar way.

15.2 AMBULACRARIA

The Ambulacraria was recognized as a group on the basis that its members—the echinoderms and the hemichordates—shared a distinctive larval form called the tornaria.[15] This grouping has since been validated by molecular phylogeny[16] and other genomic traits, such as the presence of three Hox genes specific to the group.[17] Other features that unite ambulacraria include the failure of the larval nervous system to persist into adulthood: in echinoderms, at least, Hox genes are not expressed in nervous tissue, a marked contrast with chordates.

But perhaps the most distinctive feature of ambulacraria is their way with mesocoels. Primitive ambulacraria tend to produce extensions of the mesocoel that are known as lophophores in hemichordates, but which might be homologies of the echinoderm water-vascular system. Initially bilateral, as in extinct ambulacraria such as the cambroernid *Herpetogaster*,[18] these extensions are modulated by the asymmetry common to deuterostomes such that the water-vascular system of echinoderms is invariably a product of the left mesocoel (hydrocoel), the right side having been suppressed.

15.3 ECHINODERMS

The mineralized skeleton of echinoderms has allowed a reasonably good fossil record for this highly distinctive group, but as we have seen, this hasn't led to a universally agreed scheme of echinoderm evolution.[19] Nevertheless, one can produce a hypothesis of various stem-echinoderm groups leading to the fully mineralized and pentameral crown echinoderms. The first and most distinctive feature of echinoderms to evolve was the calcite skeleton of stereom mesh. Because it was the first, it is extremely hard to argue that any early non-mineralized fossil, such as *Vetulicystis*, might be an echinoderm. We are on somewhat surer ground with other successive acquisitions, which were, in order, the marked asymmetry leading to suppression of right-sided structures, and, consequently, a tendency for a somewhat irregular morphology to be imposed on an ancestral bilateral symmetry; the loss of the ancestral deuterostome system of pharyngeal gill slits; the elaboration of the water-vascular system; triradiality, in forms such as helicoplacoids, and, finally, pentameral symmetry. In crown echinoderms, at least, *Nodal* has been co-opted for dorso-ventral as well as left-right patterning.[20]

15.4 HEMICHORDATES

Hemichordates are in many ways the most primitive living deuterostomes, in that they have retained a tricoelomous, bilaterally symmetrical organization with relatively little modification. Their genomes have also evolved very slowly.[21] That hemichordate symmetry is bilateral is, however, a statement that requires qualification. For such a seemingly simple body plan, the dorso-ventral and left-right body axes of hemichordates have proven hard to pin down. This could be because the ancestral deuterostome organizer has become rather weak in hemichordates, though dorso-ventral patterning is governed by BMPs and their antagonists, as seen in chordates.[22]

The problem attendant on determining the axes of hemichordates has hampered resolution of the two features of hemichordates that have, classically, attracted the most attention—the stomochord and the tubular nerve cord in the collar. There might be some mileage in the idea that the stomochord has some homology with the endostyle[23] even if it has no particular homology with the notochord. The hollow nerve cord in the collar has some similarities with the dorsal tubular nerve cord of chordates—except that there is some doubt as to whether it is really dorsal, given that enteropneusts have two main nerve cords, and that enteropneusts do not seem to have undergone the definitive dorso-ventral inversion as seen in chordates.

Hemichordates long ago adopted a tube-living habit.[24] Although this might suggest that the tube-living pterobranchs were ancestral to free-living enteropneusts, the mutual relationship between the two groups remains unresolved. It could be that the tube-living habit is not a question of absolutes, such that pterobranchs are obligate tube-dwellers and enteropneusts have always lived as free individuals. For example, the pterobranch *Cephalodiscus* can leave its tube and crawl around, and the fossil enteropneust *Spartobranchus* is associated with a tube.

15.5 CHORDATES

Chordates are recognized classically by the notochord and the dorsal, tubular nerve cord. Since the demotion of the amphioxus to the position of most basal extant chordate,[25] the possession of a segmented paraxial mesoderm can also be seen as a characteristic of chordates as a whole, even though this feature has been lost in tunicates. Another feature also seems definitively associated with chordates, and that is the inversion of the dorsoventral axis relative to that of all other animals, including other deuterostomes.

The importance of the notochord can hardly be overstated. The chordate organizer expresses *Nodal*, but because the notochord becomes, in effect, an organizer extended along the anterior posterior axis, *Nodal* becomes involved in the determination of all three body axes in chordates, integrating the formation of left-right, anterior-posterior, and dorso-ventral axes into a single process: not just left-right asymmetry (in deuterostomes in general) or left-right and dorso-ventral (as in echinoderms). The notochord also directs the formation of the nerve cord, through production of *Shh*; and is involved with the specification of various other organs and tissues such as the heart and aorta[26] as well as endoderm derivatives such as the liver and pancreas.[27]

In chapter 12, I compared and contrasted two influential papers on vertebrate origins. In one, Gans and Northcutt (1983) proposed that the vertebrate head was substantially a neomorph, a consequence of the appearance of neural crest. In the other, Romer (1972) suggested that the vertebrate body was a composite of two very different "animals"; namely the "visceral," comprising the pharynx, gut and its appendages, and the "somatic," consisting of the nervous system, and its associated musculature.

I think these papers are really about two very different things. Gans and Northcutt were concerned with the origin of the head in vertebrates, whereas Romer was really making an argument about chordates. For it does seem that chordates are distinguished by the addition of a segmented posterior region to what is, in essence, a headless trunk. To distil this argument to essentials, vertebrates are all about adding a head to a trunk, but chordates are about adding a trunk to a head. I shall discuss chordates first.

As we have seen, evidence from a number of different lines of inquiry suggests that the body plan of chordates is based on the welding, not always perfectly, of a segmented, trunk region onto a distinct anterior region consisting mainly of the pharynx. If the early Cambrian fossil *Saccorhytus* is a deuterostome, it is essentially all pharynx. One might say much the same about ambulacrarians, too. Any attachment stalks we see in echinoderms or hemichordates, while sometimes mesodermal extensions and having every appearance of segmentation, are hard to homologize with the very particular structure of a chordate trunk, which has never assumed a stalked form in any known chordate.

This distinction is recapitulated in the development of hemichordates and echinoderms: tornaria larvae are essentially floating heads (or, rather, pharynges, complete with a gut) with the hinder parts of the animal developing at metamophosis.[28] The origin of somites in the tail bud of amphioxus[29]—as opposed to the clock-and-wavefront mechanism of chordates—might reflect a primitive holdover from a very ancient feature of chordate development in which something like a pharynx started to bud mesodermal segments from its posterior end.

The origin of a segmented trunk region was accompanied by the origin of the chordate central nervous system. Aside from the remarkably detailed correspondences between gene expression in hemichordate ectoderm and in the central nervous system of chordates,[30] substantial differences remain between chordate and ambulacrarian nervous systems. The chordate CNS expresses Hox genes, and it is not likely that the chordate CNS has any particular homology with nervous systems in ambulacrarians.[31]

Work mainly on the amphioxus has revealed a remarkable degree of sophistication in the chordate central nervous system in general, an organization that must have been in place in the latest common ancestor of crown chordates, if not earlier. The neural tube has distinct regionalization and anterior-posterior patterning, into a diencephalic forebrain, and a mid- and hindbrain. The brain of the ancestral crown chordate had a distinct

ANR, ZLI, and MHB—though these need not yet have acquired organizer capability as in vertebrates—but no telencephalon.[32]

There was something like a pineal body growing upward from the diencephalon, homologous with the lamellar organ of the amphioxus[33]; an infundibulum forming a homologue of the pituitary (Hatchek's Pit in the amphioxus, the neural gland of tunicates[34]), probably developing on the left side; a single, median eyespot; and a "protoplacodal domain."[35] The protoplacodal domain is a region of non-neural ectoderm, anterior and lateral to the neural plate, in which neurosecretory cells tended to gather, expressing genes co-opted from ancient expression in the pharyngeal arches.[36]

Looking inside chordates, we find, at least primitively, a closed vascular system but a very simple heart; an endostyle; an atrium; *RAG*-like genes for a sophisticated kind of innate immunity; and perhaps a haematopoietic domain, a precursor to the cardiopharyngeal fields seen in tunicates and vertebrates.[37]

When we consider Hox expression in the chordate nervous system, this seems to be a speciality of the trunk, rather than the head, reinforcing the notion that the trunk is a rather different structure with distinct origins. As we have seen, the anterior boundary of expression of Hox paralogues in vertebrates tends to match, whereas they differ in the trunk, suggesting that Hox gene expression domains have had more time to relax, spread themselves, and generally make themselves at home in the trunk, rather than retaining their company manners, as in the head. Neural crest, too, behaves differently in the head than in the trunk.[38] There is a "newness" of neural-crest-related programming in the head, relative to that of the trunk. This could be a holdover from the earliest days of chordates, when the trunk really was new, grafted onto an older anterior region.

Throughout this book I've shown that there has been some difficulty and not a little disagreement about deciding the posterior boundary of the head in vertebrates in particular and chordates in general. Matters might be made easier if we reversed the problem and considered the anterior border of the trunk.

If the trunk started as a posterior structure (in vetulicolians) and grew over the anterior region dorsally (in yunnanozoans) before integrating (as in *Pikaia* and crown chordates), then we'd expect the anterior boundary of the trunk in chordates to show some vestige of this process. Indeed, when investigated, the anterior boundary of the trunk is further anterior on the dorsal than on the ventral side of the animal. This seems to be the case in lampreys, where the dorsal boundary seems to be more or less alongside the hindbrain, just posterior to the root of the ganglion for the vagus nerve. The ventral boundary, in contrast, is more posterior, coincident with the back wall of the pharynx, the posterior limit of neural-crest migration; the anterior limit of the pronephros and of the pericardium. This dorso-ventral difference is retained in vertebrates. As Matsuoka et al. (2005) showed, there is a distinct boundary of neural crest activity between head and trunk, dorsally, in the anterior parts of the shoulder girdle (which in the earliest vertebrates was next to the otic region) and, ventrally, in the posterior margins of the pharynx. Given that this region abuts the heart, and the increasing evidence for the involvement of cardiac mesoderm in head development, we should probably talk less about the head when what we probably mean is the cardiopharyngeal field and its derivatives.[39]

I close this section on chordates with a puzzle. In the last chapter I argued that the segmented trunk first appeared in vetulicolians, making them the most basal stem-chordates. In these controversial forms, the segmented region is entirely distinct from the pharynx. A segmented region also appears in the yunnanozoans, and although it is partially integrated with the pharyngeal region, it stands entirely dorsal to the notochord, if present.

The puzzle is that although segments (presumably mesoderm) appear in vetulicolians and yunnanozoans, the evidence for notochords and a central nervous system is equivocal at best in vetulicolians and debatable in yunnanozoans. In a way, this helps me break an important branch by suggesting that a segmented trunk appeared in the chordate stem lineage before the notochord and CNS (fig. 14.3).

On the other hand, to posit segmentation without a notochord doesn't make much biological sense, as segmented muscles would have lacked a purchase against which to move. One could speculate that the notochord in these forms, if present, was more muscular than fibrous, and therefore more prone to decay, not showing up in the fossil record. After all, the amphioxus—a primitive chordate—has a more muscular notochord than is found in tunicates or vertebrates. Perhaps these early notochords were similar to (if not necessarily homologous with) the muscular axochord of some annelids,[40] and additional biomechanical support was offered by coelomic turgor pressure (as in annelids) and the stiffened cuticle with which the segments were covered (as in arthropods). If there is no notochord, it follows that the central nervous system, if present, was not chordate-like, as the notochord is absolutely required to form it. If the segmented trunk region was used in active swimming, one is entitled to ask how it was coordinated.

This is, I stress, highly speculative, but if even a fairly cautious interpretation of the evidence suggests that a segmented trunk region was present in vetulicolians and yunnanozoans and a notochord was not, then we can break a long branch—the segmented trunk originated before the modern-style, fibrous notochord came along to support it. If so, then the segmented trunk would have grown out, segment by segment, from the posterior end of a head region, much as the posterior parts of an enteropneust grow out of a tornaria larva as it metamorphoses, or perhaps as somites develop in the amphioxus, from the tail bud forward. Only when the notochord became more established did it start to exert its role as an organizer, as it does in the chordate body today. Of course, it would be easier if vetulicolians had notochords (as argued by García-Bellido et al. 2014, for the Australian form *Nesonektris*) but one cannot simply wish features into existence to make one's evolutionary scenario more convincing.

15.6 THE AMPHIOXUS

Although the amphioxus genome has evolved very slowly, and for all that the animal looks much as one might imagine a basal chordate to look, the paucity of fossil evidence means that it's hard to disentangle genuinely primitive features from peculiarities limited to the amphioxus itself.

Such peculiarities might include a feature of neurulation in which the closing neural tube is first roofed over with non-neural ectoderm.[41] The most notable peculiarities, however, are the extension of the notochord to the extreme anterior of the animal, together with the developmental asymmetry of pharyngeal structures. It is possible that the latter is a consequence of the former, together with the suppression of the brain, and the secondary imposition of somite-like structure on what would otherwise be pre-notochordal mesoderm. The anteriormost somites are developmentally different from those in vertebrates. For example, they express T-box genes, a phenomenon not found in vertebrates.[42] The anterior extension of the notochord might be marked by a recent and amphioxus-specific duplication of the gene *brachyury*.[43]

15.7 THE COMMON ANCESTRY OF TUNICATES AND VERTEBRATES

The discovery from molecular phylogenetic evidence that tunicates are the vertebrates' closest extant sister taxon, with the relegation of the amphioxus to a more remote remove, was initially a surprise.[44] Now that we have become accustomed to the idea, we have found that it makes a great deal of sense, as tunicates and vertebrates share a number of features not seen in the amphioxus. These features bear on aspects of vertebrate anatomy once thought the exclusive preserve of vertebrates, such as the neural crest, placodes, and the development of the heart and viscera. This allows us to put down few more planks in our bridge.

The earliest stages of tunicate and vertebrate development are marked by similarities in embryonic fate maps.[45] The heart of tunicates is much more vertebrate-like than the contractile vessel found in amphioxus. The pericardium seems associated with a distinct aorta-gonads-mesonephros field; the pancreas might be a homologue of the tunicate pyloric gland; there are two main coeloms (the body cavity and the pericardium); and the potential of *SPARC*-like genes to be co-opted for mineralization.

The notochord in tunicates and vertebrates, where it exists, is more rigid and less muscular than it is in the amphioxus. This might, of course, have a bearing on the condition and preservation potential of notochords in the earliest stem-chordates, as discussed above.

The elaboration of a placodal domain anterior to the neural plate is a refinement seen only in tunicates and vertebrates, with two distinct placodal fields.[46] The anterior domain is differentiated into the primordia of the oral siphon in tunicates, and the lens, olfactory, and adenohypophysial placodes in vertebrates. The posterior subdomain, in contrast, produces atrial siphon primordia in tunicates and the otic, lateral line, and epibranchial placodes in vertebrates.

The evolution of placodes is accompanied by a limited development of what, in vertebrates, will become neural crest.[47] The statolith (gravity sensor) and ocellus (eye spot) develop from pigmented neural-plate border cells, a microcosm of the neural-crest-promoted elaboration of organs of special sense in vertebrates.

15.8 TUNICATES

The genomes of tunicates show that these remarkable animals have evolved further and faster than any other group of deuterostomes, vertebrates included. The disintegration of Hox clusters and extreme shrinkage of the genome has accompanied a mode of development that is rapid and largely determinate and that makes do with a bare minimum of cells. This shrinkage has been accompanied by the loss of characteristically chordate traits such as segmentation, and even features otherwise seen in deuterostomes generally, such as the organizer. Such radical changes might or might not be connected with the co-option of *Nodal* in the formation of the neural plate, and the promotion of notochord and neural tube formation by BMPs: neither feature is found in vertebrates and this fact could be connected with the loss of the organizer.[48]

Unlike vertebrates, tunicates have considerable regenerative capacity and are mainly hermaphrodite. Extant tunicates have a dazzling and exotic biochemistry,[49] and can synthesize cellulose[50] as well as proteins seen nowhere else in nature, as in the oikosins secreted by *Oikopleura* to generate its house.

In vertebrates, the gene *brachyury* is found in the notochord as well as posterior mesoderm. In tunicates it is restricted to the notochord. The exception is *Oikopleura*, where it is found in the posterior endoderm and hindgut.[51] Given the involvement of homologues of *brachyury* in hindgut formation in protostomes, this could represent a primitive chordate (or even bilaterian) feature and might suggest that larvacea are more primitive than other tunicates.

Stolfi and Brown (2015) suggest that tunicates have all the signature features of chordates, but have apportioned them between distinct life stages. In particular, they have exploited the essential difference between chordate anterior and trunk regions by making the best use of their functional differences. The motile, segmented trunk is ideal for a dispersive, non-feeding larval stage, whereas a sessile pharynx makes an effective feeding adult.

15.9 VERTEBRATES

Vertebrates represent a departure from the deuterostome norm at least as marked as, say, echinoderms or tunicates. Unlike all other deuterostomes, they are obligately motile predators. The ciliary movement of mucus in the pharynx is replaced by muscular ventilation. Like motile predators they have an elaborate nervous system. This is cast on the chordate neural chassis seen in other chordates but has been refined, first with the elaboration of distinct brain regions and cranial nerves; second, by an active and pervasive neural crest and highly developed derivatives of cranial placodes. Vertebrates also tend to be larger animals than other deuterostomes. This elaboration may or may not be connected with the four-fold genome duplication in the early history of vertebrates. A cost of this elaboration is a marked loss of regenerative capacity. A benefit of motility is a reduction in sexual conflict, such that sexes can afford to be separate.

The large bodies of vertebrates are supported by a cartilage endoskeleton that reinforces and eventually replaces the notochord and protects the brain and sense organs as well as supporting the pharyngeal slits.

The neural crest is totipotent and capable of mesenchymal reprogramming. It contributes to such things as pigment cells, the muscles and cellular cartilage of the pharynx, parts of the heart and major blood vessels (vertebrate blood vessels are lined with endothelium), and the nerve net that innervates the gut. Migration of Hox-coded streams of neural crest from the rhombomeres of the hindbrain remodels the cranial mesoderm and pharynx.[52] The most visible result is the creation of the dermal skeleton—that is, much of the head and face.

Vertebrate-specific elaborations of placodes include the semicircular canals and lateral lines (acoustic placode)—the precursors of paired ears—and fully developed image-forming eyes (lens placode). Unlike other chordates, which have single eyespots, vertebrates have paired eyes and a brain divided into left and right halves. The vertebrate forebrain adds the telencephalon[53] to the preexisting diencephalon, allowing the development of a single, median olfactory (nasal) capsule. Olfaction is enhanced by a single medial nostril with direct external connection to a midline pituitary, a homologue of the neural gland in tunicates and Hatschek's Pit in the amphioxus.

Some developmental peculiarities of vertebrates include the finding that notochord and neural tube formation require suppression of BMPs; that VegT, not β-catenin, patterns the vegetal pole of the blastula[54]; and that *Nodal* is important in determining endodermal and mesodermal cell fate.

Vertebrates also show some specializations in mesoderm formation and deployment. The mesoderm of the head is not segmented. Even though paraxial mesoderm is segmented by a distinctive clock-and-wavefront mechanism in which somites are formed from the front, backward, this mesoderm is further divided dorsoventrally into epaxial and hypaxial compartments. Retinoic acid buffers against lateral asymmetry, such that vertebrates appear much more symmetrical on the outside than the inside.

15.10 CYCLOSTOMES

Cyclostomes constitute a distinct if minor group of vertebrates—the lampreys and hagfishes. They are believed to have diverged from the vertebrate lineage before the acquisition of mineralized tissues such as bone, dentine, and enamel. Extant cyclostomes are distinguished by a tongue and associated musculature adapted for ectoparasitism; a pharyngeal skeleton lateral to the gills; and a peculiar adaptive immune system. As with the amphioxus, the limited diversity and specializations of extant cyclostomes make it hard to recognize stem-cyclostomes from the fossil record, especially as animals almost indistinguishable from extant forms have existed for at least 360 million years.[55] The strange fossil *Tullimonstrum* might, however, be a possible fossil cyclostome offshoot.[56]

15.11 GNATHOSTOMES

Nearly all extant vertebrates are gnathostomes—that is, vertebrates with jaws. As well as a pharyngeal skeleton modified into jaws and a cranium that has expanded to encompass the sensory capsules, they have paired limbs and other skeletal novelties such as ribs. Pectoral fins appeared first in the fossil record among the jawless stem-gnathostomes called ostracoderms; pelvic fins were added by the more crownward, jawed stem-gnathostomes called placoderms.[57]

The mineralized skeleton of gnathostomes started with expression of *SPARC*-like genes in the dermis.[58] Only later did mineralization move to encompass the internal skeleton.

Other peculiarities of gnathostomes include mesodermal head cavities; paired nostrils (and olfactory capsules); internal fertilization; an adaptive immune system, lymphoid cells, and V-D-J recombination presided over by fully functional *RAG* genes; and the neural insulation known as myelin, which is absent in cyclostomes and other chordates.[59]

15.12 THE EVOLUTION OF THE FACE

Right at the start of the book I noted that vertebrates (or, at least, gnathostomes) have faces, and this is how we instinctively recognize them. It is possible to chart the evolution of the face by breaking it down into distinct stages.

One of the innovations of the chordate lineage was a single median eyespot, along with a single gravity-sensor (a precursor of the paired ears). The median eyespot of the amphioxus is homologous with the paired eyes of vertebrates.[60] In early chordates, taste and smell were accomplished by the early homologues of the pituitary, but this happened in the roof of the pharynx: there was no separate nostril or nasohypophysial duct. The eyespot lacked a lens—an innovation of vertebrates—and there was likewise no telencephalon or olfactory capsules. The earliest chordates, therefore, didn't have much of a face. Fig. 15.1A is essentially a portrait of an amphioxus.

A number of innovations appeared later in chordate evolution, but before the appearance of crown-group vertebrates. The first was the lens placode, and this was acompanied by paired eyes. It seems to be a feature of very early vertebrates such as *Metaspriggina*, that if eyes are seen, they are at the anteriormost end and placed very close together (fig. 15.1B). This suggests that the single, median eye field in the anterior placode was as yet uninterrupted by significant developments of olfactory placodes or a telencephalon, and that the diencephalon was still the most anterior part of the forebrain.

Crown vertebrates have a telencephalon and an olfactory capsule. The brain is also divided into distinct left and right hemispheres. There is a single nostril at the end of a distinct nasohypophysial duct, communicating with the hypophysis and the olfactory capsule. Although the olfactory capsule is single, and has a single median duct, it is paired internally. The appearance of a telencephalon, the divided brain, a nostril, and an olfactory

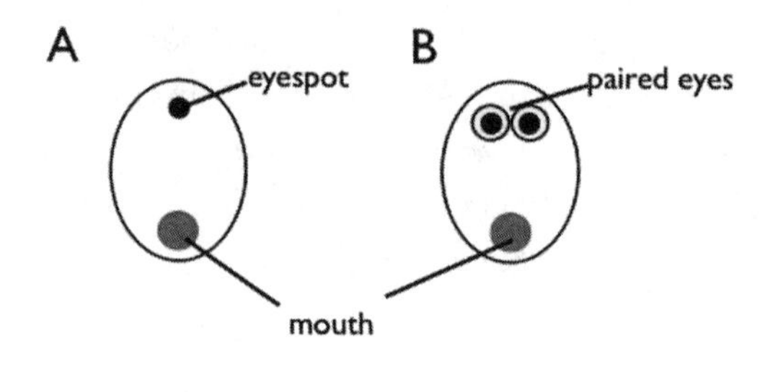

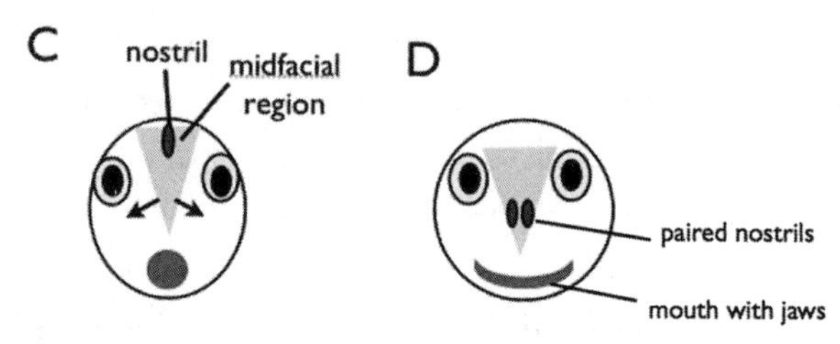

15.1 Stages in the evolution of the face, seen as cartoons in anterior view. A: Early chordates had a jawless, ventral mouth and a single, median eyespot. B: Very early vertebrates had eyes with lenses, but placed close together. C: In crown vertebrates, the appearance of the telencephalon, laterally divided brain, olfactory capsule, and median nostril spreads the eyes laterally and starts to open up the face. D: In derived placoderms and crown gnathostomes, the appearance of jaws and paired nostrils creates the face familiar in most extant vertebrates.

capsule offers an obstruction to the single eye field of the anterior cranial placode, forcing the two eyes apart laterally, and defining a new median region of the face. Fig. 15.1C is, essentially, a portrait of a lamprey, and indeed many of the stem-gnathostomes known as ostracoderms. Some of the more derived ostracoderms, however, show a pronounced broadening of the face[61] and with the evolution of jaws, the nostrils had become paired. The loss of a direct contact with the base of the brain also allowed the nostrils to move ventrally and anteriorly, so that in the most derived stem-gnathostomes such as the placoderm *Entelognathus*,[62] the modern vertebrate face had become recognizable as such (fig. 15.1D).

15.13 CROSSING THE BRIDGE

Finally I present a summary of features as they were acquired in the lineage leading from ancestral deuterostomes to gnathostome vertebrates. It's rather clumped, because one cannot yet tease apart the order in which the features appeared in any one of the numbered categories. Where I have tried to do this, I have subdivided each numbered category into lettered subsections, but where I have done this, it should be considered entirely hypothetical, and features that could have occurred in either subsection are marked as unplaced.

1. Radial cleavage; posterior organizer using β-catenin and *Wnt* signaling; enterocoely; tricoelomy; pharyngeal slits; FGF signaling in mesoderm formation; ectodermal patterning revealed by ANR, ZLI, MHB; tendency to left-right asymmetry directred by *Nodal* pathway; sialic acid metabolism; association between pericardium and excretory system; sequestration of calcium compounds in connective tissue; iodine-containing compounds for metamorphosis.

2a. Segmented trunk region, never used for attachment. Origin of segments in tail bud, as in amphioxus.

2b. Segmented tail region grows anteriorly and integrates dorsally.

2, unplaced. Notochord associated with *brachyury*; dorsal tubular nerve cord; extensive anterior-posterior regionalization in nervous system patterned by Hox genes; pineal body; infundibulum/pituitary homologue developing on left side; median eyespot; protoplacodal domain; closed vascular system; simple heart in hematopoietic domain; endostyle; atrium; *RAG*-like genes for complex but innate immune system.

3. Cardiopharyngeal field divided into first- and second- heart fields; chambered heart; pericardium associated with distinct aorta-gonads-mesonephros field; pancreas; two coeloms—the body cavity and the pericardium; SPARC-like genes involved in biomineralization; rigid, fibrous notochord; elaboration of protoplacodal field into two distinct placodal domains; limited development of neural crest; median eyespot; median gravity sensor.

4a. Genome duplication; predatory lifestyle; separate sexes; muscular ventilation; active and pervasive neural crest; cranial nerves; paired eyes with lenses; semicircular canals and lateral line; large size; extensive cartilage skeleton; circulatory system lined with endothelium; unsegmented head mesoderm; somites generated from the front, backward, by clock-and-wavefront mechanism; lateral plate mesoderm; mesonephros and Wolffian ducts; fully adaptive immune system.

4b. Telencephalon; olfactory capsule; division of brain into left and right halves; nostril communicating with hypophysis.

5. Mineralization of dermal skeleton; pectoral fins; cranium encloses brain and sensory capsules.
6. Lateral splitting of olfactory capsule, opening of mid-face, paired nostrils, jaws; mineralization of internal skeleton; ribs; cavities in head mesoderm; internal fertilization; myelinated nerves.

15.14 CONCLUSIONS

Looking back at the research amassed during the past twenty years or so, I realize how little we had to go on in the mid-1980s, when I started to think about writing the book that became *Before the Backbone*. As I wrote in the preface to this current work, fossils were few; techniques of phylogenetic reconstruction still owed much to old-fashioned adaptationist story-telling; the basic biology of many relevant organisms was still very little understood; molecular biology was generally concentrated on a few model organisms; relatively little was known of the genetics or genomics of animals. How very far we have come, in such a short time.

It is fitting that all the advances can be described most succinctly in the context of phylogenetics. Fig. 15.2 illustrates the consensus phylogeny of deuterostomes in the mid-1980s (based on Schaeffer 1987) comparing it with how it is today.

Back in the 1980s, lophophorate phyla such as brachiopods were assumed to be allied to the deuterostomes, based on shared features of embryonic development, and the presence of the feeding structure known as the lophophore, also present in pterobranch hemichordates. Molecular work has shown that the lophophore is a convergent structure; that features such as the fate of the blastopore and the style of cleavage are less fundamental than once assumed, and in any case best understood in terms of the underlying molecular biology; and that the lophophorates belong to a group of protostomes called the Lophotrochozoa.

The removal of the lophophorates has allowed a certain clarity to creep into our understanding of deuterostome evolution. We can now state with some confidence that all members of the group we have identified as deuterostomes bear the name thanks to the presence of pharyngeal slits, and the cassette of genes associated with this feature.[63] From this we can say that even if only a few protostomes have radial cleavage, have deuterostomy (the anus forming from the blastopore), and show enterocoely, these are

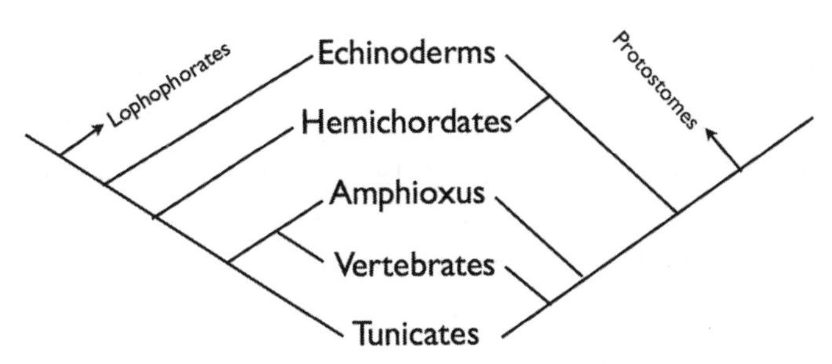

15.2 Deuterostomes then and now. With nothing more than morphology and some embryology to go on, the left-hand phylogeny tells an adaptationist story of steady acquisition of features leading to vertebrates. The right-hand phylogeny shows how molecular work has validated alternative ideas such as the grouping of echinoderms with hemichordates into the ambulacraria (Metchnikoff 1881), and the union of tunicates with vertebrates into a group sometimes called the olfactores (Jefferies 1991). The phylogenies also show how rearrangements of the protostomes, removing lophophorate phyla from the deuterostome stem, have allowed the concept of deuterostomes to become firmer.

distant convergences, because *all* bona-fide deuterostomes grouped on the basis of pharyngeal slits also share these traits, as well as all the other traits described in stage 1 above. It now seems that the deuterostomes evolved directly from early bilaterian animals, most plausibly reconstructed as motile, wormlike, coelomate creatures not too dissimilar from enteropneusts. The earliest deuterostomes did not go through a sessile stage. The massive revisions of the animal kingdom that started with the creation of the new groups Lophotrochozoa and Ecdysozoa, with older groups such as Coelomata falling by the wayside, were properly all about invertebrates in general and protostomes in particular, but a consequence has been to affirm the status of deuterostomes and garland them with definable features. The first steps of the bridge between invertebrates and vertebrates look much more secure than they once did.

The old view of vertebrate origins placed echinoderms as an outgroup of all other deuterostomes, mainly because they seemed to share little with other deuterostomes beyond traits of early embryonic life. The most conspicuous of these was a bilateral, planktonic larva very similar to that of enteropneusts. This led to a view of vertebrate origins largely based on the theoretical transformation of the motile larval forms of sessile adults[64] into sexually mature, motile adults, a concept that became known as paedomorphosis.[65]

Much less attention was devoted to the adults, but with the molecular revival of the old concept of Ambulacraria, zoologists were forced to find shared similarities between the otherwise highly dissimilar echinoderms and hemichordates. Perhaps the most important was the realization that the identification of pharyngeal slits in some fossil echinoderms[66] really were homologues of pharyngeal slits in hemichordates, and therefore of chordates.[67] Echinoderms once had pharyngeal slits, but lost them. This new idea has allowed some progress to be made in the vexed issue of the interrelationships of the earliest echinoderms.

Progress has also been made in the phylogeny of enteropneusts, though the relationships between enteropneusts and pterobranchs are still debated. At issue is whether hemichordates were primitively sessile. Given our emerging understanding of deuterostome evolution in general, it is unlikely that either echinoderms or hemichordates had sessile ancestors. In any case the issue hardly matters, as we know that echinoderms and hemichordates can alternate between stalked and motile forms during the same life history; and some even nominally stalked ambulacrarians can move around.

In general, though, ambulacrarians have adopted the stalked habit at various times, but an important consequence of the new phylogeny is that we can say with some confidence that the stalked habit is peculiar to that group. In my view, the common ancestor of chordates was motile, and among chordates the sessile habit was only secondarily adopted by tunicates. We can therefore rule out all ideas that posit a sessile chordate ancestor.

Moving up the old-style phylogeny, we find hemichordates as intermediates between echinoderms and chordates, as their name suggests. They share larval forms with echinoderms—but gill slits with chordates. Schaeffer (1987) unites hemichordates with chordates under the name Pharyngotremata, on the basis of the shared presence of pharyngeal slits. The position of the hemichordates, closer to chordates than to echinoderms, tended to be overwelcoming of the interpretation of particular enteropneust structures such as the collar cord and stomochord as homologous with various chordate features, when such homologies might not in fact be justifiable. Enteropneusts, however, have emerged as model organisms in their own right for understanding the condition of the earliest deuterostomes, rather than that of the earliest chordates.

So far, so good. The reorganization of the animal kingdom has allowed us to posit a much firmer basis for deuterostomes as a natural group than before, and the revival of the Ambulacraria as a group has allowed us to get a better grasp on the early history of echinoderms; the status of enteropneusts as credible model organisms for early deuterostome evolution; and the idea that the sessile, stalked habit is a peculiarity of Ambulacraria. We have reached the center span of the bridge more or less unscathed. The next few steps are probably the shakiest of the whole traverse.

The first thing to recognize about chordates is that they aren't stalked. With the exception of the secondary sessility of extant tunicates, no chordate shows the sessile habit. The second thing to learn is that the first major chordate innovation is a segmented trunk, as in stage 2a of the previous section. All other deuterostomes essentially begin life as mobile heads.[68] The earliest chordates underwent the dorsoventral inversion first discussed by Geoffroy, and then engendered a segmented posterior section, perhaps initially reinforced by a cuticular exoskeleton if the notochord was muscular, otherwise non-fibrous, or nonexistent.

This posterior region extended dorsally and anteriorly, integrating with the older, pharyngeal region, a process suggested by Gislén (1930), taken up by Jefferies (e.g., Jefferies 1986), and captured, possibly, and successively, by fossil forms such as vetulicolians and yunnanozoans. The other chordate traits, such as the notochord and central nervous system, originated either together with, or just after the origin of the segmented trunk. The earliest chordates had a distinct, regionated brain, but it was not divided into hemispheres, and sense organs were limited to an anterior eyespot.

If we have managed to leap across this somewhat unsteady section of the bridge without falling into the swirling waters far beneath, we find ourselves on a much steadier structure.

The 1980s chordate phylogeny has tunicates standing outside a group made of the amphioxus and vertebrates, what Schaeffer (1987) called "somitic" chordates. The assumption was that tunicates were an offshoot from the main line that, having acquired chordate features such as a notochord and nerve cord, veered off before the evolution of somites. The discovery in the first decade of the present century that tunicates, not the amphioxus, are the closest invertebrate relatives of vertebrates[69] was revolutionary.

The ancestors of tunicates once had somites, but lost them, much as the ancestors of echinoderms had had pharyngeal slits. The amphioxus, once held as being a close relative of vertebrates, was seen in a new light, as a model representative of the earliest phases of chordate evolution, in the same way that enteropneusts now stand for early deuterostomes. The discoveries that these creatures have very slowly evolving, conservatively structured genomes[70] reinforces this view. At the same time, we must be sensitive to traits peculiar to the amphioxus lineage alone, such as the duplication of *brachyury* and the rostral extension of the notochord, with the consequent distortion of anterior structures.

Once the shock had passed the new status of tunicates made a great deal of sense. In a pre-Delsuc paper, Simões-Costa et al. (2005) lamented the then-current phylogeny on the grounds that the tunicate heart was so much more like a vertebrate heart than the contractile vessel seen in the amphioxus. Since Delsuc et al. the similarities between tunicates and vertebrates have accumulated, so we can now cut a long branch in two—stage 3 shows what features we'd expect to see in a common ancestor of tunicates and vertebrates that excludes the amphioxus. These include simple versions of the neural crest and placodes once thought the exclusive preserve of vertebrates. Our tread on the bridge is now sure and steady.

In the old-style phylogeny, tunicates were seen as cast-offs, irrelevant to the main business of vertebrate evolution (although, it has to be said, some early ideas had tunicates as degenerate vertebrates). The new phylogeny forces us to confront tunicates in their wonder and diversity and appreciate that they have achieved great success by evolving in ways entirely at variance with the vertebrate course.

From stages 4a onward we are on surer ground, partly because of the mineralization of the vertebrate skeleton. The earliest vertebrates underwent a genome duplication, not once but twice. I suggest, however, that the evolution of a telencephalon and olfactory capsules was a distinct event, as fossil evidence from the earliest fishes such as *Metaspriggina* suggest that the paired eyes were at the extreme anterior of the animal, uninterrupted by any kind of midface, and so did not have a telencephalon or any means of chemosensation aside from the hypophysis.

What remains to be discovered? As someone once said, prediction is very hard—especially about the future. But thanks to the monumental achievements made since I last surveyed this field we can sketch out a number of topics that might benefit from further investigation. One might be the unpicking of the earliest stages of chordate evolution. This might have to await further fossil discoveries, but there might be some mileage in phylostratigraphic work for estimating the relative ages of the genetic programs that, for example, govern somitogenesis and notochord formation.

Lastly, much work remains to be done on tunicates. How did these creatures come to have evolved so quickly? Genome shrinkage, gene loss, determinate development, and short generation times have undoubtedly contributed to the enormous speed of tunicate evolution. I think Stolfi and Brown (2015) have identified at least part of the answer. That is, tunicates have taken the two halves of the chordate body and assigned each to the life stage to which it is best suited. The "somatic" half is best optimized as a motile, dispersal stage; the "visceral" as a dedicated feeding and reproductive structure. After all, burdening a motile phase with a pharynx and its appendages would only be a handicap, whereas a feeding structure doesn't really need an expensive brain or muscles to get around—especially if it can solve the problems of sessility with hermaphroditism, to get round problems of gamete dispersal; and exotic biochemistry to deter predators.

Splitting the somatic and visceral animals, then, not only optimizes life history strategy but saves a great deal of energy that can be devoted to other things. It amazes me that *Oikopleura dioica*, the size of a rice grain, can whip up a complex walnut-sized house of mucus and specialized proteins, only to discard the whole thing and replace it with another within hours; and that the discarded houses of larvaceans form a substantial part of the oceanic carbon budget. These creatures really do seem to have energy to burn.

Vertebrates, on the other hand, have pursued a strategy of integration whose energy requirements are supported by an obligately predatory habit, which in turn imposes further energetic requirements such as the development of a sophisticated brain and sense organs. It could be that the duplication of the pre-vertebrate genome and the expansion of the neural crest effectively imposed this course on our distant ancestors, forcing them to be mobile to make a hard living, banishing them from the effectively mindless Eden the tunicates still inhabit. In which case, one must wonder which, after all, has made the better bargain.

Notes

PREFACE

1. Contrast Garstang (1894), in which the assumption was that chordates evolved from free-swimming ancestors, with Garstang (1928), in which the ancestor was presumed to have been sessile.

2. Holland et al. 2015.

3. With a respectful nod to Romer 1972.

4. Gee et al. 2015.

5. I think that this might be an example of bathos. I'm not sure.

CHAPTER ONE

1. I recommend Peter Holland's book *The Animal Kingdom: A Very Short Introduction* (Oxford, 2011), for a wide overview in a small space.

2. I wanted to include pentastomids, but it's now generally thought that these parasitic worms are highly derived crustaceans (Lavrov et al. 2004).

3. Members of the phylum Platyhelminthes (another *p*).

4. Tarazona et al. 2016.

5. There are exceptions. The otoliths or ear-stones of some fishes are made of calcium carbonate, and the shells of some clam-like creatures called brachiopods (lamp shells) contain calcium phosphate—but these are minor.

6. Some vertebrates such as baleen whales and very large sharks have adopted a filter-feeding mode of life, but this is a secondary adaptation. Their immediate cousins have jaws full of teeth.

7. Gillis et al. 2012.

8. This structure is not the same as the superficially similar feeding structure

found in brachiopods and some other invertebrates, collectively the lophophorates, but I am getting ahead of myself.

9. Smith 2005.

10. Corallo et al. 2015.

11. Green et al. 2015.

12. Abitua et al. 2012, 2015.

13. L. Z. Holland 2009; Pani et al. 2012.

14. P. W. H. Holland et al. 1994; Dehal and Boore 2005; Kuraku et al. 2008; L. Z. Holland 2013; Smith et al. 2013.

15. Smith and Keinath 2015.

16. Amores et al. 1998; Christoffels et al. 2004; Jaillon et al. 2004; Meyer and Van de Peer 2005; Brunet et al. 2006; Kassahn et al. 2009; Berthelot et al. 2014; Glasauer and Neuhauss 2014.

17. I offer a critique of the idea of "missing links" in my book *The Accidental Species*.

18. This is the problem that besets interpretation of the strange creatures known as fossils from the Ediacaran Period, the interval of time immediately before the Cambrian. Although the Cambrian Period saw the emergence of most of the major phyla that are with us today—allowing the fossils to be interpreted, in the main, according to models from the present-day fauna, such as arthropods or chordates—nothing like an Ediacaran organism survives today. This makes them hard if not impossible to understand. Ediacaran fossils are plainly the remains of organized living forms, but they are so different from anything from more recent periods, or alive today, that there is much debate about what kinds of creatures they were (Narbonne 2005; Xiao and Laflamme 2009). This hasn't, however, stopped people from making ingenious deductions about their biology (Mitchell et al. 2015).

19. Delsuc et al. 2006, 2008.

20. Schaeffer 1987.

CHAPTER TWO

1. See Tobey Appel's altogether splendid book *The Cuvier-Geoffroy Debate* for an account of the relationship between Cuvier and Geoffroy and its profound influence on nineteenth-century biology. It sounds stuffy but it isn't. I have been saying this for years: someone should make it into a movie. But nobody cares what I think.

2. I discuss some of these ideas in my book *Jacob's Ladder* (2004).

3. Dobzhansky 1973.

4. Gee 2002.

5. This stems from the assumption that new species form from existing ones, creating either two new ones, or a new one alongside the old. However, speciation might also occur by hybridization, when two species merge (Mallet 2007). Even

when speciation can be regarded as a bifurcation, after the fact, close examination shows that the two incipient species might hybridize with each other and continue to do so to an extent even as the two new species start to emerge and separate. This leads to a phenomenon known as "genealogical discordance" or "incomplete lineage sorting" in which estimating evolutionary relationships from single genes can give differing results for the same two species, depending on the history of the genes concerned during the speciation event. See Degnan and Rosenberg (2009) for a review.

6. Litopterns, which numbered many browsing and grazing animals in their ranks, only some of which looked like horses, turn out to have been distant relations of the odd-toed ungulates, the group to which true horses, tapirs, and rhinos belong (Welker et al. 2015).

7. Bartelt 2000.

8. Gould and Lewontin 1979.

9. I discuss cladistics in more detail in my book *In Search of Deep Time* (1999). Historians of science will enjoy the detailed account of the birth and growth of cladistics in *Science as a Process* (1988), by David L. Hull.

10. That when choosing between competing hypotheses, one should select the one that makes the fewest prior assumptions. Attributed to the Franciscan friar William of Ockham (c. 1287–1347).

11. Theobold 2010.

12. The astute reader will notice the cluster of traits in the lineage segment leading to echinoderms, but as I shall explain later, the rich fossil record of echinoderms has helped resolve this into a series of shorter segments in a way that has not been possible with the generally soft-bodied chordates or vertebrates.

13. Linnaeus, whose system of formal nomenclature we use to this day, was—of course—a pre-evolutionary scholar, and his classifications included minerals as well as plants and animals.

CHAPTER THREE

1. Miller 1988.

2. Gee 2002, 2013.

3. At this stage you might be tempted to ask which layer of tissue is evolutionarily the older—ectoderm or endoderm? You might legitimately ask whether it's possible to ask such a question as it seems moot, at first, whether one germ layer can be regarded as distinct without the presence of the other, in opposition, as it were. But sponges (Srivastava et al. 2010) and *Trichoplax* (Srivastava et al. 2008) have more complicated genomes than one might assume from their structure, so you could ask whether, if they have one type of cell, these cells form more than just a primordial mass but might instead behave more like the ectoderm or endoderm

in animals that have both. In other words, do ectoderm and endoderm have characteristics that allow them to be identified in isolation? The answer seems to be "yes." Work on the evolutionary antiquity of genes expressed in the embryonic development of a range of animals supports the idea that the endoderm was the primordial germ layer, the ectoderm evolving as a second germ layer freed from ancestral feeding functions (Hashimshony et al. 2015).

4. There are, however, animals with mesoderm and body cavities, but closer inspection suggests that these body cavities are not lined with mesoderm and therefore not true coeloms, but persistent remnants of the blastocoel. These are the so-called pseudocoelomates and include nematodes (roundworms), rotifers (wheel-animalcules), and some other animals.

5. As ever, biology treasures its exceptions. Ctenophores, or comb-jellies, are diploblasts, with what at first appears to be a single opening serving as mouth and anus. Closer inspection, however, reveals what appears to be a functionally tripartite through gut, with a number of anal openings (Presnell et al. 2016).

6. See Gee 1996, and Holland et al. 2015, for reviews.

7. See Hillier et al. 2005, for a review.

8. Lüter 2000.

9. Martín-Durán et al. 2016.

10. This may be related to secondary evolution of very small size and a lack, in some tunicates, of retinoic-acid-based signaling, which in other chordates helps lay down the body plan. See Cañestro and Postlethwait 2007; L. Z. Holland 2007.

11. Kaul-Strehlow and Stach 2011; Dilly 2014.

12. Urata et al. 2014.

13. Willmer 1990.

14. Field et al. 1988.

15. The identity of the gene does not matter—the important thing is that it was known to evolve very slowly, so cognates could be recognized in a very wide range of phyla that started to evolve away from one another more than 600 million years ago. Although mutations in genes generate the data from which phylogenies are constructed, one must take care not to select genes evolving so fast that all signals of evolutionary relationship have been overwritten by mutations.

16. Wainright et al. 1993; Müller 1995; Lake 1990; Nosenko et al. 2013.

17. Paps et al. 2013.

18. See for example King et al. 2008; Suga et al. 2013; Fairclough et al. 2013; Richter and King 2013; Williams et al. 2014.

19. Edgecombe et al. 2011.

20. Ryan et al. 2013; Moroz et al. 2014.

21. Pisani et al. 2015.

22. Dunn et al. 2015.

23. Putnam et al. 2007; Miller and Ball 2008.

24. See Baguñà et al. 2008 for a discussion on what the earliest bilaterians might have been like.

25. Achatz et al. 2013.

26. Halanych et al. 1995; Gee 1995; Passamaneck and Halanych 2006.

27. Telford and Holland 1993; Papillon et al. 2004.

28. Ivanova-Kazas 2007.

29. McHugh 1997.

30. Bourlat et al. 2006; Philippe et al. 2011.

31. Rouse et al. 2016; Cannon et al. 2016.

32. Ruppert 2005.

33. If deuterostomy is such a poor guide to phylogenetic placement compared with the presence of pharyngeal slits, one might legitimately ask why deuterostomes are still named deuterostomes, rather than something like, say, "pharyngophores." The reason, presumably, is academic inertia, and I shall continue to refer to deuterostomes as such in this book. If, however, you choose to take up the cause of pharyngophores, remember, you read it here first.

34. Halanych et al. 1995; Halanych 2004; Helmkampf et al. 2008a.

35. Aguinaldo et al. 1997.

36. Philippe et al. 2005.

37. It was first proposed to me by Jim Lake, in a hotel bar in Los Angeles, but that's enough about my private life.

38. For a recent review see Liu and Dunlop 2014.

39. Conway Morris 1977.

40. The Jarts in Greg Bear's novel *Eternity*. Greg Bear confirmed this to me personally when I discussed the matter with him in 2007.

41. Smith and Caron 2015.

42. I sometimes wonder whether, had embryology not sprung from nature philosophy, evolution would have had as close a link with development as it does today; and if that link had not been forged, whether our understanding of evolution would have been retarded or advanced.

CHAPTER FOUR

1. Lewis 1978.

2. McGinnis et al. 1984.

3. Riddihough 1992.

4. Nüsslein-Volhard and Wieschaus 1980.

5. Crow et al. 2006.

6. Garcia-Fernàndez and Holland 1994.

7. Carroll 1995; Bailey et al. 1997; Fried et al. 2003.

8. Wagner et al. 2003.

9. Larroux et al. 2007.

10. Brooke et al. 1998; Garstang and Ferrier 2013.

11. Degnan et al. 1995; Ramos et al. 2012.

12. Slack et al. 1993.

13. Ferrier and Holland 2001; Balavoine et al. 2002; Ryan et al. 2007; Ferrier 2012.

14. Chourrout et al. 2006.

15. Lemons and McGinnis 2006.

16. Seo et al. 2004.

17. Awgulewitsch and Jacobs 1992.

18. Malicki et al. 1992.

19. Nübler-Jung and Arendt 1994; De Robertis and Sasai 1996.

20. Nübler-Jung and Arendt 1999; N. D. Holland 2003; Lowe et al. 2015; L. Z. Holland 2015.

21. Irie and Kuratani 2011.

22. Irie and Kuratani 2014.

23. Bininda-Emonds et al. 2003.

24. Richardson and Keuck 2002; Richards 2009.

25. Kalinka et al. 2010; Domazet-Lošo and Tautz 2010; Prud'homme and Gompel 2010.

26. Owen 2007.

27. Shubin et al. 2009.

28. Quiring et al. 1994; Gehring and Ikeo 1999.

CHAPTER FIVE

1. Grobben 1908.

2. Chea et al. 2005.

3. Peterson and Eernisse 2001.

4. Wada and Satoh 1994; Halanych et al. 1995; Cameron et al. 2000; Blair and Hedges 2005; Philippe et al. 2005; Gerhart 2006.

5. Martín-Durán et al. 2012.

6. Wada and Satoh 1994; Peterson and Eernisse 2001.

7. Ruppert 2005; Gillis et al. 2012; Simakov et al. 2015.

8. Metchnikoff 1881; Turbeville et al. 1994; Halanych 1995; Bromham and Degnan 1999; Peterson and Eernisse 2001; Furlong and Holland 2002; Peterson et al. 2004; Blair and Hedges 2005.

9. Schaeffer 1987.

10. Ruppert 2005; Blair and Hedges 2005; Delsuc et al. 2006, 2008.

11. Gee 2001b.

12. Delsuc et al. 2006; Gee 2006.

13. Turbeville et al. 1994.

14. Wada and Satoh 1994; Winchell et al. 2002; Zeng and Swalla 2005.

15. Swalla and Smith 2008.

16. The journal *Genesis* devoted a special issue to left-right asymmetry in 2014, containing reviews by Halpern et al. Warner et al. Su, and Manigai et al. . See also Boorman and Shimeld 2002a; Raya and Izpisúa-Belmonte 2005; Tabin 2006.

17. Those interested may consult Yoshiba and Hamadi 2014; Blum et al. 2014.

18. Yasui et al. 2000; Yu et al. 2002a; Boorman and Shimeld 2002b; Soukup et al. 2015.

19. Duboc et al. 2005; Molina et al. 2013; Warner et al. 2014; Su 2014.

20. Duboc and Lepage 2008.

21. Manigai et al. 2014.

22. Grande and Patel 2009.

CHAPTER SIX

1. Clausen and Smith 2005; Bottjer et al. 2006; Swalla and Smith 2008.

2. Sodergren et al. 2006; Livingston et al. 2006; Bottjer et al. 2006.

3. Livingston et al. 2006.

4. Veis et al. 2002.

5. Cameron and Bishop 2012.

6. Smith 2005; Zamora and Rahman 2014.

7. Rahman et al. 2015.

8. Daley 1995.

9. Jefferies 1973; Dominguez et al. 2002

10. Durham and Caster 1963; Sprinkle and Wilbur 2005.

11. Shu et al. 2004; Smith 2004; Swalla and Smith 2008; Clausen et al. 2010.

12. Livingston et al. 2006; Cameron and Bishop 2012.

13. Matsumoto 1929; Gislén 1930; Gregory 1936.

14. Jefferies 1990.

15. Peterson 1995.

16. Jefferies 1997.

17. Dominguez et al. 2002.

18. Clausen and Smith 2005.

19. Sodergren et al. 2006.

20. Those interested in the minutiae of the calcichordate theory should consult the extensive treatment in *Before the Backbone*.

21. Martinez et al. 1999; Mito and Endo 2000; Méndez et al. 2000; Long and Byrne 2001.

22. Arenas-Mena et al. 2000.

CHAPTER SEVEN

1. See Röttinger and Lowe (2012) and Kaul-Strehlow and Röttinger (2015) for excellent overviews.

2. Dunn 2015; Kaul-Strehlow and Röttinger 2015.

3. Holland et al. 2005; Gage 2005; Osborn et al. 2012.

4. Worsaae et al. 2012.

5. Kowalevsky 1866a; Bateson 1886.

6. Lowe et al. 2015.

7. Röttinger and Lowe 2012.

8. Garstang 1922, 1928.

9. Lacalli 2005.

10. Gonzalez and Cameron 2009.

11. Kaul-Strehlow and Stach 2011.

12. Balser and Ruppert 1990.

13. Biology treasures its exceptions. Torquaratorid enteropneusts lack a stomochord as adults.

14. Luttrell et al. 2012.

15. Peterson et al. 1999.

16. Satoh et al. 2014.

17. Rychel et al. 2006; Rychel and Swalla 2007; Gillis et al. 2012.

18. Simakov et al. 2015.

19. Miyamoto and Wada 2013, but see Luttrell et al. 2012.

20. Kaul and Stach 2010.

21. Nomaksteinsky et al. 2009. This raises the question of whether traces of a centralized nervous system might be found in echinoderms. I have mentioned in passing the basal ganglion of crinoids and it would be interesting to know whether the development of this structure bears comparison with any part of the enteropneust nervous system. Jefferies (1986) reconstructed elaborate central nervous systems in stylophora, and although Clausen and Smith (2005) refuted some of these claims, Smith (2005) acknowledges that at least some of Jefferies' reconstructions of prominent sensory nerve trunks in mitrates might have been substantially correct.

22. Cunningham and Casey 2014.

23. Aronowicz and Lowe 2006; Lowe et al. 2006; Gee 2007.

24. Pani et al. 2012.

25. Lemons et al. 2010.

26. Holland 2003.

27. Lowe et al. 2003; Gerhart et al. 2005; Pani et al. 2012; L. Z. Holland et al. 2013; L. Z. Holland 2015a; Albuixech-Crespo et al. 2017.

28. Lowe et al. 2006; Röttinger and Martindale 2011.

29. Duboc et al. 2005; Duboc and Lepage 2008.
30. Kaul-Strehlow and Röttinger 2015.
31. Röttinger et al. 2015.
32. Darras et al. 2011.
33. Angerer et al. 2011.
34. Niehrs 2010.
35. Green et al. 2013.
36. Simakov et al. 2015.
37. Röttinger and Lowe 2012.
38. Lacalli 2005.
39. Gonzalez et al. 2017.
40. Freeman et al. 2012.
41. Putnam et al. 2008.
42. Ogasawara et al. 1999.
43. Simakov et al. 2015.
44. Caron et al. 2013; Nanglu et al. 2016.
45. Stach et al. 2012.
46. Lester 1985.
47. Sato et al. 2008.
48. Dilly 1993; Mitchell et al. 2013.
49. Palmer and Rickards 1991.
50. See for example Hou et al. 2011.
51. Cannon et al. 2014.
52. Cameron et al. 2000; Stach 2013; Cannon et al. 2013.
53. Cannon et al. 2009.

CHAPTER EIGHT

1. They are grouped into three genera—*Branchiostoma*, *Epigonichthys* (a single species), and *Asymmetron* (two known species). The latter two genera are similar to *Branchiostoma* except that they have gonads on the right side only. I recommend Linda Holland's review (2015b) as a comprehensive, accessible treatment of the biology and development of the amphioxus, and much of what follows uses this review as a starting point.

2. I feel qualified to be so familiar, given that at a conference many years ago I was invited by the late Professor Carl Gans to a party held for "Friends of the Amphioxus." QED.

3. The evolutionary roots of the notochord are contentious, as Annona et al. explore in an amusing and informative historical review (2015). The structure and development of the notochord have elicited comparisons with a structure in some annelids called the axochord (Lauri et al. 2014), as well as with the stomochord in

enteropneusts (see above), although until more is learned about the genetic underpinnings of notochord development—especially the role of the gene *Brachyury*—the default option is that the notochord is a genuine chordate novelty.

4. Yasuo and Satoh 1994.

5. P. W. H. Holland et al. 1995.

6. Levin et al. 1995.

7. Shimeld 1999.

8. Putnam et al. 2008.

9. Venkatesh et al. 2014; Yue et al. 2014.

10. Holland et al. 2008; Putnam et al. 2008.

11. Garcia-Fernàndez and Holland 1994.

12. See for example Huang et al. 2008; Yuan et al. 2015.

13. Gans and Northcutt 1983.

14. Lacalli 1996a; Šestak and Domazet-Lošo 2015; L. Z. Holland 2015c.

15. Toresson et al. 1998.

16. Olsson et al. 1994; Lacalli and Kelly 2000.

17. Lacalli 2008b.

18. Holland et al. 1992.

19. See for example Holland et al. 1996.

20. As you'll recall from the section on hemichordates, Lowe and colleagues (Pani et al. 2012) have described similar gene expression patterns in the ectoderm of enteropneusts, suggesting that this feature of chordate organization has very deep deuterostome roots. Kozmik et al. (1999), however, find no expression of *AmphiPax2/5/8* in the region of the amphioxus neural tube that might correspond with the vertebrate MHB, suggesting that a vertebrate-style MHB is not present. More recent work by Albuixech-Crespo et al. (2017) has shown how the vertebrate midbrain and the cognate region in the amphioxus differ in various ways, such that there is not always a clear correspondence between one and the other.

21. Candiani et al. 2012.

22. Suzuki et al. 2015.

23. Lacalli et al. 1994; Wicht and Lacalli 2005.

24. Lacalli et al. 1994; Lacalli 1996b; Lacalli 2008b; Wicht and Lacalli 2005.

25. Flood 1966; Lacalli 2002a.

26. Azevedo et al. 2009; Candiani et al. 2012.

27. L. Z. Holland (2015b).

28. De Robertis 2006.

29. Yu et al. 2007. It is possible that the roots of the Spemann organizer run even more deeply, since the discovery that the ventral ectoderm of sea-urchin embryos has similar properties. See Lapraz et al. 2015.

30. Escrivà et al. 2002a and b; Onai et al. 2009 2010.

31. Beaster-Jones et al. 2008.

32. Onai et al. 2015.

33. N. D. Holland et al. 1996. I am well aware that this is fighting talk. I discuss the issue of head segmentation in more detail in chapter 12.

34. Yu et al. 2002b; Ono et al. 2014.

35. At this point one might mention *Amphioxides*, a creature discovered in the early twentieth century that looked like a giant amphioxus larva. It is now suspected to have been a larva of *Asymmetron* that had failed to metamorphose.

36. Well, sort of. It has also been interpreted as a mesodermal coelomic duct that opens to the exterior, similar to (and perhaps a homologue of) the left hydropore in echinoderms (Kaji et al. 2016).

37. Ruppert 1996.

38. Glardon et al. 1998.

39. Boorman and Shimeld 2002b

40. Kozmik et al. 1999.

41. Lacalli 2008a; N. D. Holland et al. 2009.

42. Holland et al. 2003. The gene in *Drosophila* that, when mutated, results in the absence of a heart, is called *tinman*. And, yes, if you were wondering, as I was, there is a related gene in *Drosophila* involved in brain development called *scarecrow* (Zaffran et al. 2000). We're not in Kansas any more, Toto.

43. Yasui et al. 2014.

44. Delsuc et al. 2006, 2008.

45. Conway Morris and Caron 2012.

CHAPTER NINE

1. See Lemaire (2011) and L. Z. Holland (2016) for a brief and accessible introduction, and Stolfi and Brown (2015) for a more detailed treatment. My account draws heavily on these, as well as Lemaire et al. (2008) and, for adult ascidian anatomy, the lucid account in Jefferies (1986).

2. Huxley 1851.

3. Kowalevsky 1866b, 1871.

4. Garstang 1922, 1928; Berrill 1955.

5. Delsuc et al. 2006, 2008.

6. Simakov et al. 2015.

7. Schaeffer 1987.

8. Kurabayashi et al. 2003.

9. Monahan-Earley et al. 2013.

10. Veeman et al. 2010.

11. Stolfi et al. 2010; Tzahor and Evans 2011; Tolkin and Christiaen 2012; Diogo et al. 2015.

12. See for example Lebar et al. 2010.

13. Nakashima et al. 2004; Matthysse et al. 2004.

14. Sasakura et al. 2005.

15. I am indebted to Mark Potter for alerting me to this.

16. Imai et al. 2000; Imai 2003.

17. Logan et al. 1999; Miyawaki et al. 2003.

18. Zhang et al. 1998.

19. See Lemaire et al. (2008) for more discussion and references on the roles of FGF, Nodal, and BMP in ascidian development.

20. See Kumano and Nishida 2007, for a review.

21. Lemaire 2009.

22. Oda-Ishii et al. 2005; Stolfi et al. 2014.

23. Although the "typical" ascidian larva is well known as a developmental system (reviewed at book length by Satoh 1994), one should note that there are many variations on the theme. Tailless larvae, in which the development of the notochord, nerve cord, and muscles is aborted or absent, have evolved several times in tunicates. In some cases there may be little distinction between larva and adult and such tunicates are said to have direct development. In other species, the tail is beefed up with extra muscles, and the larva might take on features of the adult before losing its tail. See Jeffery and Swalla (1992) for a review. Of course, some colonial tunicates, as well as thaliaceans, may reproduce asexually, by budding, eschewing the tadpole stage completely, and single zooids of the colonial ascidian *Botryllus*—or even pieces of blood vessels—can reconstitute an entire colony. See Lemaire (2011) for discussion and references.

24. "For all these reasons I think I am fully justified in saying that the axial cylinder of the tunicate tail is functionally, developmentally and genetically equivalent to the notochord in the amphioxus." (My translation.)

25. Jiang and Smith 2007.

26. Crowther and Whittaker 1994.

27. Yasuo and Satoh 1998.

28. Cole and Meinertzhagen 2004.

29. Horie et al. 2011.

30. Acampora et al. 2001.

31. Dufour et al. 2006; Ikuta and Saiga 2007.

32. Davidson and Swalla 2002.

33. Detailed in Stolfi and Brown 2015, and references therein.

34. D'Agati and Cammarata 2006.

35. Heyland and Hodin 2004.

36. Paris et al. 2008b.

37. Hiruta et al. 2005.

38. Wada 2001.

39. Jeffery et al. 2004; Mazet and Shimeld 2005; Jeffery 2007; Abitua et al. 2012; Stolfi et al. 2015.

40. Abitua et al. 2015.

41. Garstang 1922, 1928; Berrill 1955.

42. Alldredge 1977.

43. Hosp et al. 2012.

44. Kugler et al. 2011; Ferrier 2011.

45. Wada 1998; Swalla et al. 2000; Zeng et al. 2006; Delsuc et al. 2006, 2008; Tsagkogeorga et al. 2009; Govindajaran et al. 2010.

46. Fujii et al. 2008; Stach et al. 2008.

47. Dehal et al. 2002; Holland and Gibson-Brown 2003.

48. The smallness of tunicate genomes has made them ideal platforms for teasing out networks of gene expression and transcription. Imai et al. (2006) have exploited this to tease out, from studies on ascidian embryos, a regulatory "blueprint" for chordates in general.

49. The human genome contains more than 3,000Mb of DNA.

50. Vinson et al. 2005, Small et al. 2007.

51. Voskoboynik et al. 2013.

52. Denoeud et al. 2010.

53. See Stolfi and Brown 2015.

54. Linda Holland (2016) reports that the genome of *Salpa thompsoni* has been sequenced, but at the time of writing it is not publicly available.

55. Putnam et al. 2008; Simakov et al. 2015.

56. Seo et al. 2004.

57. Ikuta et al. 2010.

58. Ferrier and Holland 2002; Spagnuolo et al. 2003; Ikuta et al. 2004.

59. Lemaire et al. 2008. The fate map of the amphioxus appears different from those of tunicates and vertebrates, too (Holland and Holland 2007), suggesting that the fate map we see in tunicates might reflect traits held in common between tunicates and vertebrates to the exclusion of the amphioxus.

60. Hill et al. 2008.

61. Kourakis and Smith 2005. The presence of an organizer in vertebrates and possibly the amphioxus (Yu et al. 2007), and its apparent absence in tunicates, might in part explain the very different roles of transcription factors such as BMP and *Nodal* in ascidians with respect to vertebrates (reviewed by Lemaire et al. 2008).

62. Seo et al. 2004.

63. Sobral et al. 2009.

64. Oda-Ishii et al. 2005.

65. Imai et al. 2006.
66. Imai et al. 2009.
67. Moret et al. 2005.
68. Mazet and Shimeld 2005.
69. Dufour et al. 2006.
70. Davidson 2007; Stolfi et al. 2010.
71. Swalla and Jeffery 1990.
72. Swalla and Jeffery 1996.
73. Shu et al. 2001b.
74. Chen et al. 2003.
75. Hou et al. 2006.
76. Caron et al. 2010.

CHAPTER TEN

1. My account of the basic anatomy and development of vertebrates draws heavily on Jefferies' book *The Ancestry of the Vertebrates* (1986). Although written to preface his own idiosyncratic views on vertebrate origins (discussed briefly in chapter 6) this account is in my view hard to beat for clarity and scholarship. I have also consulted Janvier's book *Early Vertebrates* (1996), another masterpiece of lucidity—and, for cyclostomes, Shimeld and Donoghue (2012).
2. Müller 1873.
3. See for example Scofield et al. 1982; Khalturin and Bosch 2007.
4. Pancer et al. 2004, 2005.
5. Janvier 2003.
6. Zintzen et al. 2011.
7. Bullock et al. 1984.
8. Janvier 1981; Forey and Janvier 1993.
9. Yalden 1985.
10. Oisi et al. 2013.
11. Ota et al. 2011.
12. Gabbott et al. 2016.
13. Kuraku et al. 1999; Delarbre et al. 2002; Furlong and Holland 2002; Mallatt and Sullivan 1998; Heimberg et al. 2010.
14. Gorbman 1997.
15. Ota et al. 2007.
16. Kuratani et al. 2016.
17. Bardack 1991; Janvier 1996; Gess et al. 2006; Janvier 2006.
18. Chang et al. 2006.
19. Clements et al. 2016; McCoy et al. 2016; Kuratani and Hirasawa 2016.
20. Bardack and Richardson 1977; Shu et al. 1999a.

21. Smith et al. 2013.

22. Ohno 1999.

23. Kuraku et al. 2008.

24. Escrivà et al. 2002a and b.

25. Mehta et al. 2013.

26. Brazeau and Friedman 2015.

27. Long et al. 2008, 2009, 2015; Ahlberg et al. 2009.

28. As I shall show in the next chapter, the choice of sharks is in some respects unfortunate. Although their embryology was examined by great biologists of the past such as Balfour and Goodrich, these studies have colored our understanding of many vertebrate features, in particular the question of whether the head of vertebrates is primitively segmented.

29. I describe the course of human embryology in my book *Jacob's Ladder*. My source for that account was Larsen 1998.

30. Venkatesh et al. 2007, 2014.

CHAPTER ELEVEN

1. As discussed earlier, acoels are very simple acoelomate bilaterians with a mouth and no anus, formerly allied with turbellarian flatworms such as planaria but now generally regarded as a distinct group, branching close to the origin of Bilateria. They are united with another group of simple worms, the nemertodermatida, in a group called Acoelomorpha. See Ruiz-Trillo et al. 1999; Jondelius et al. 2002; Philippe et al. 2007.

2. Nakano et al. 2013.

3. Nielsen 2010.

4. Franzén and Afzelius 1987; Ehlers and Sopott-Ehlers 1997; Lundin 1998; Raikova et al. 2000.

5. Telford et al. 2003, Ruiz-Trillo et al. 1999.

6. Stach et al. 2005.

7. Pedersen and Pedersen 1986.

8. Pedersen and Pedersen 1988.

9. Norén and Jondelius 1997; Israelsson 1997.

10. Israelsson 1999.

11. Israelsson and Budd 2005.

12. Bourlat et al. 2003; Gee 2003; Perseke et al. 2007; Fritzsch et al. 2008.

13. Bourlat et al. 2006, 2009; Telford 2008.

14. Philippe et al. 2011; Lowe and Pani 2011; Maxmen 2011.

15. Dunn et al. 2008.

16. Paps et al. 2009.

17. Hejnol et al. 2009.

18. Achatz et al. 2013.
19. Ruiz-Trillo and Paps 2015.
20. References in Hejnol and Martindale 2008b.
21. Hejnol and Martindale 2008a and b.
22. Simakov et al. 2015.
23. Jondelius et al. 2002.
24. Rouse et al. 2016.
25. Cannon et al. 2016.
26. Arroyo et al. 2016.
27. Telford and Copley 2016.
28. Ball and Miller 2006.
29. Shinn 1994; Shinn and Roberts 1994.
30. Telford and Holland 1993.
31. Papillon et al. 2004; Shen et al. 2015.
32. Faure and Casanova 2006; Matus et al. 2006.
33. Halanych 1996.
34. Helmkampf et al. 2008b.
35. Marlétaz et al. 2006, 2008.
36. Matus et al. 2007.
37. Papillon et al. 2003.
38. Szaniawski 1982; Doguzhaeva et al. 2002.
39. Chen and Huang 2002; Vannier et al. 2007.
40. Conway Morris 1977b, 2009.
41. Nanglu et al. 2016.

CHAPTER TWELVE

1. Spemann and Mangold 1924. Other vertebrates have structures homologous to the dorsal lip organizer of amphibians, but they labor under a variety of names such as the Primitive Knot or Hensen's Node.

2. For this section I have drawn on the reviews by Gerhart (2001), Niehrs (2004), and De Robertis (2006, 2009). If I have made Mangold's work sound simple in a with-one-bound-Jack-was-free style, this is simply a result of brevity. The work was arduous and took many years. Of the hundreds of transplants Mangold performed as part of her dissertation, almost all were ruined by infectious agents in the pondwater in which the embryos were bathed (sterile, standard media being a later innovation), and only five yielded meaningful results. Life outside the lab was also somewhat fraught. In 1924, mere months after completing her dissertation, Mangold died from serious burns received when her kitchen stove exploded as she was warming milk for her infant son. She was just short of her twenty-sixth birthday, and did not live to see her research published. Her husband, Herr Mangold,

an assistant to her supervisor Hans Spemann, was active in the Nazi party. Her son grew up only to be killed in World War II. Spemann coined the term "organizer" for the kind of special embryonic tissue represented by the dorsal lip of the blastopore. He received the Nobel Prize for the work in 1935. He died of heart failure in 1941, aged 72.

3. Stewart and Gerhart 1990.

4. These factors help determine left-right asymmetry in chordate embryos, as I described in chapters 3 and 4, but that generally happens later.

5. Dosch and Niehrs 2000.

6. Reversade and De Robertis 2005.

7. Bouwmeester et al. 1996.

8. Shawlot and Behringer 1994.

9. Tung et al. 1962, 1965.

10. Yu et al. 2007; Garcia-Fernàndez et al. 2007.

11. Kourakis and Smith 2005.

12. Imai et al. 2004.

13. Lowe et al. 2006.

14. Lapraz et al. 2015.

15. For this section I have drawn on Technau 2001; Stemple 2005; Nibu et al. 2013; Hejnol and Lowe 2014 and references therein.

16. Stemple et al. 1996; Amacher and Kimmel 1998; Fekany et al. 1999.

17. Fouquet et al. 1997; Sumoy et al. 1997.

18. Cleaver and Krieg 2001.

19. Chesley 1935; Herrmann 1991; Herrmann et al. 1990; Wilkinson et al. 1990; Beddington et al. 1992; Wilson et al. 1995.

20. Yasuo and Satoh 1993; Corbo et al. 1997; P. W. H. Holland et al. 1995.

21. Yasuo and Satoh 1994, 1998.

22. Bassham and Postlethwait 2000.

23. Papaioannou and Silver 1998.

24. See references in Nibu et al. 2013.

25. Scholz and Technau 2003; Yamada et al. 2010.

26. Singer et al. 1996; Kispert et al. 1994.

27. Gross and McClay 2001.

28. Tagawa et al. 1998.

29. Peterson et al. 1999a, b.

30. Lauri et al. 2014; Hejnol and Lowe 2014.

31. Technau 2001; Arendt et al. 2001.

32. For this section I have drawn extensively on Maroto et al. 2012; Hubaud and Pourquié 2014; Dequéant and Pourquié 2008 and references therein.

33. Cooke and Zeeman 1976.

34. Gomez et al. 2008.

35. References in Hubaud and Pourquié 2014.

36. Hornstein and Tabin 2005, and references therein.

37. Wellik 2007.

38. Onai et al. 2015b.

39. Bertrand et al. 2011.

40. Schubert et al. 2001.

41. Mazet and Shimeld 2003; Beaster-Jones et al. 2008.

42. Bertrand et al. 2015.

43. Passamaneck et al. 2007.

44. Graham et al. 2014.

45. Noden and Trainor 2005.

46. The two views persist to this day, and just to show I'm not biased, I drew heavily on segmentationists (L. Z. Holland et al. 2008) and antisegmentationists (Kuratani and Schilling 2008; Kuratani 2008) as I researched this section.

47. Balfour was a keen mountaineer as well as zoologist. In the fall of 1882 he was due to take up a professorship created especially for him by the University of Cambridge in recognition of his prodigious talent. In the event he never delivered a single lecture: that July he died while scaling a hitherto unclimbed route up Mont Blanc in the Alps. He was not yet thirty-one years old.

48. Goodrich 1930.

49. Romer 1972.

50. Graham et al. 2014.

51. Wicht and Lacalli 2005; Castro et al. 2006; references in L. Z. Holland et al. 2008.

52. Kuratani 2008; Kuratani and Schilling 2008; Adachi and Kuratani 2012; Adachi et al. 2012; Kuratani and Adachi 2016; Kuratani et al. 2016; Onai et al. 2014, 2015a.

53. Ultrastructural studies suggested the existence of embryonic divisions in head mesoderm known as "somitomeres." Subsequent studies have failed to identify these. But as Antone Jacobson (1988) joked to me in an elevator once, anyone who doesn't believe in their existence is "antisomitic."

54. Adachi et al. 2012.

55. Beaster-Jones et al. 2006.

56. See for example Herr et al. 2003.

57. Pani et al. 2012.

58. The anterior extension of the notochord might also explain the very peculiar and highly asymmetric development of the pharynx, with the larval mouth on the left side—speculation that goes back more than a hundred years (Lankester and Willey 1890: Willey 1893).

59. Putnam et al. 2008.

60. See Ryan and Grant 2009, for an in-depth review of the origin of synapses.

61. The nervous system of ctenophores appears differently organized from those in cnidarians and bilaterians, leading to claims that ctenophores represent an entirely distinct branch of multicellular animals in which the nervous system evolved independently (Ryan et al. 2013; Moroz et al. 2014). This relies on the idea that sponges are primitively nerveless. It remains a possibility that sponges (and placozoa) might once have had nervous systems but lost them (Ryan and Chiodin 2015).

62. Mizutani and Bier 2008.

63. Hejnol and Rentzch 2015.

64. Nübler-Jung and Arendt 1994, 1999. By convention, the position of the mouth in any bilaterian, if not precisely terminal, defines the ventral surface. The inversion of the dorso-ventral axis in the ancestral chordate would have left the mouth stranded on the dorsal side, such that it would have to have migrated ventrally again.

65. See the review by Copp et al. (2003) for a succinct account of mammalian neurulation and its problems.

66. Reviewed by L. Z. Holland 2015b.

67. This is described in detail by Stolfi and Brown 2015.

68. Lacalli et al. 1994; Lacalli 1996b, 2008b; Wicht and Lacalli 2005; L. Z. Holland 2015c, but see Albuixech-Crespo et al. 2017.

69. Toresson et al. 1998.

70. Candiani et al. 2012; Vopalensky et al. 2012; Suzuki et al. 2015.

71. Lacalli et al. 1994; Wicht and Lacalli 2005.

72. Kozmik et al. 1999; reviewed in L. Z. Holland 2015c.

73. Albuixech-Crespo et al. 2017.

74. Cole and Meinterzhagen 1994; Horie et al. 2011.

75. Ryan et al. 2016.

76. Acampora et al. 2001; Dufour et al. 2006; Ikuta and Saiga 2007.

77. L. Z. Holland 2015c; P. W. H. Holland et al. 1994.

78. Šestak and Domazet-Lošo 2015.

79. Irimia et al. 2010.

80. Conway Morris and Caron 2014.

81. Gai et al. 2011.

82. Chiang et al. 1996; Roessler et al. 1996. See also Gilbert 2000.

83. Nakano et al. 2009.

84. Garstang and Garstang 1926; Garstang 1928.

85. Arenas-Morena et al. 2000.

86. Kaul and Stach 2010.

87. N. D. Holland 2003; Lowe et al. 2003; reviewed by L. Z. Holland 2015a.

88. Mizutani and Bier 2008.

89. Denes et al. 2007; Arendt et al. 2008; Nomaksteinsky et al. 2009; L. Z. Holland 2015a.

90. Cunningham and Casey 2014.

91. Miyamoto et al. 2010.

92. Luttrell et al. 2012; Miyamoto and Wada 2013.

93. Green et al. 2015.

94. Tarazona et al. 2016.

95. Jandzik et al. 2015.

96. Jeffery 2007; Jeffery et al. 2004, 2008.

97. Abitua et al. 2012.

98. Ivashkin and Adameyko 2013.

99. Gostling and Shimeld 2003; Yu et al. 2008.

100. Durston et al. 2011.

101. This doesn't mean that Hox genes play no part in defining the structure of the anterior part of the head—only those genes corresponding to the canonical *antennapedia-bithorax* cluster. A gene called *orthodenticle* in *Drosophila* contains a homeodomain and specifies anterior structures. Cognates of this gene do much the same in bilateria. In vertebrates it is known as *Otx* where it specifies anterior brain structures (Simeone et al. 1993; Leuzinger et al. 1998; Acampora et al. 2001, 2005) and the anteriormost part of the neural tube in the amphioxus (Williams and Holland 1998) and tunicates (Wada et al. 1998). Schilling and Knight (2001) suggest that genes in the *Otx* family are to the anterior brain as Hox-cluster genes are to the hindbrain and spinal cord—that is, near-universal patterning agents in the anterior-posterior axis of bilaterians.

102. Wada et al. 1999.

103. Seo et al. 2004; Ikuta et al. 2004.

104. See Meier and Packard (1984) for an excellent example.

105. Much of our knowledge about the role of the neural crest in skull formation comes from the work of Nicole Le Douarin and her associates, who use the clever technique of hybrid embryos made of chicks and quails, the cells of which can be told apart. This method is reminiscent of Mangold's use of differentially pigmented newts to unpick the role of the Organizer, as I described at the beginning of this chapter. For a brief review, see Le Douarin (2012).

106. In the amphioxus, Hox expression does not extend further anterior in the neural tube than the level of the second somite: fig. 7 in Wada et al. 1999.

107. Hunt et al. 1991; Kessel and Gruss 1991.

108. Matsuoka et al. 2005.

109. For this section I have drawn extensively on two connected articles by Schlosser et al. (2014) and Patthey et al. (2014).

110. Schlosser 2008.

111. Graham and Shimeld 2013; Abitua et al. 2015.

112. Schlosser 2010.

113. Schlosser 2005.

114. Mazet et al. 2003.

115. L. Z. Holland 2005; Gillis et al. 2012.

116. Šestak et al. 2013.

117. Veis et al. 2002; Livingston et al. 2006.

118. Tarazona et al. 2016.

119. Jandzik et al. 2015.

120. Donoghue and Sansom 2002; Brazeau and Friedman 2015.

121. Kawasaki and Weiss 2003; Kawasaki et al. 2004.

122. Wilt et al. 2003.

123. Lambert et al. 1989.

124. Donoghue and Sansom 2002; Donoghue et al. 2006.

125. As Per Erik Ahlberg has pointed out in a recent lecture: see https://www.youtube.com/watch?v=So6lknLVWjY

126. Kuratani and Schilling 2008.

CHAPTER THIRTEEN

1. Green et al. 2015.

2. This duality might have very ancient origins indeed. Nomaksteinsky et al. (2013) show that a distinction between somatic and visceral neurons might have been characteristic of the earliest Bilateria.

3. A recent finding shows that the striated muscle of the esophagus originates from cranial, not somitic, mesoderm (Gopalakrishnan et al. 2015).

4. Garstang 1922, 1928.

5. Delsuc et al. 2006, 2008.

6. Grenier et al. 2009.

7. For this section I have drawn on Burns 2005; Kuo and Erickson 2011; Furness 2012; Bondurand and Sham 2013; Musser and Southard-Smith 2013; Newgreen et al. 2013, and references therein.

8. Interestingly, the ENS in the sea lamprey derives entirely from trunk neural crest, with no vagal contribution (Green et al. 2017).

9. Kuo and Erickson 2011, and references therein.

10. Diogo et al. 2015.

11. Lescroart et al. 2015.

12. Tzahor and Evans 2011.

13. Tzahor 2009; Grifone and Kelly 2007.

14. Sambasivan et al. 2011.

15. Simões-Costa et al. 2005.

16. Simões-Costa et al. 2005; Davidson 2007; Kaplan et al. 2015; Anderson and Christiaen 2016.

17. Cano et al. 2013.

18. Ichimura and Sakai 2015.

19. For summaries of the the morphology and development of the kidney, see Faa et al. 2014, and Chang et al. 2016.

20. Cano et al. 2013.

21. Summarized by Grapin-Botton (2005), who notes the seminal early work of Nicole Le Douarin on the patterning of the gut, often less well-remembered than her pioneering work on neural crest derivatives using chick-quail chimeras.

22. Kim et al. 1997.

23. For this section I have drawn on Flajnik 2004, Litman et al. 2010, and references therein.

24. Although the study of immunity is notoriously difficult for anyone other than immunologists to understand, I recommend Davis (2013) as a primer. And even though Tonegawa's 1983 review is long in years, it remains highly accessible as a guide to the basics of V-D-J recombination.

25. Flajnik 2004; Pancer et al. 2004, 2005.

26. Witness the recently discovered and in many ways remarkably vertebrate-like system in insects, as described by Dong et al. (2006).

27. Hibino et al. 2006.

28. Fugmann et al. 2006.

29. Cannon et al. 2002; Dishaw et al. 2008; Huang et al. 2008; Liberti, et al. 2015.

30. Why bind chitin? Chitin, whose full name is beta-(1-4)-poly-*N*-acetyl D-glucosamine, is the second most common polysaccharide in nature after cellulose. It is found in virtually every organism. As well as a structural substance (the exoskeletons of insects are made of it), it is important in a variety of biochemical and cellular processes. An ability to bind and neutralize varieties of chitin seems like a good facility to have for a filter-feeding organism welcoming large numbers of potential pathogens in through its pharynx.

31. Zhang et al. 2014.

32. Gorbman 1995.

33. Gorbman et al. 1999.

34. Roch et al. 2014.

35. Ruppert 1990.

36. Christiaen et al. 2002.

37. Manni et al. 2005.

CHAPTER FOURTEEN

1. Disparity is easy to understand intuitively but much harder to measure. For a guide, with some applications to Cambrian fossils, see Wills et al. 1994.

2. Graptolites, which are now recognized as hemichordates, also have an extensive fossil record, in that their decay-resistant stolons and thecae are frequently preserved as carbonaceous films.

3. Sansom et al. 2010; Briggs et al. 2010; Sansom 2014.

4. Jefferies 1967, 1986.

5. Dominguez et al. 2002.

6. Kristensen 2002.

7. This doesn't mean that fossil meiofauna are absent. They are sometimes found in fossil deposits with the so-called Orsten style of preservation in which minute creatures are preserved in phosphate or silica in exquisite detail and in three dimensions. See, for example, Eriksson and Waloszek 2016.

8. Wray 2015.

9. Han et al. 2017.

10. Caron et al. 2010.

11. Zamora and Rahman 2014.

12. Chen et al. 2003; Hou et al. 2006.

13. Durham 1974.

14. Shu et al. 2004.

15. This exemplifies a notorious problem in phylogeny: that is, recognizing the earliest members of a given group, given that the group is defined on the basis of features acquired since it first evolved from its common ancestor with an extant sister-taxon. You can trace the lineage back through evolution, losing first this character and then that, until you are left with an indefinable blob, characterizable—if at all—solely on any peculiar features it might have which, while interesting, need say nothing about wider evolutionary relationship. At this point I shall introduce another playground joke, and it goes like this. Q: Why are elephants large, gray, and wrinkled? A: Because if they were small, white, and round, they'd be table-tennis balls. Discuss.

16. Swalla and Smith 2008; Clausen et al. 2010.

17. Shu et al. 2001a; Gee 2001a.

18. Vinther et al. 2011.

19. García-Bellido et al. 2014.

20. Briggs et al. 2005.

21. Aldridge et al. 2007; Shu et al. 2010.

22. Hou 1987.

23. Briggs et al. 2005.

24. Smith and Caron 2015.

25. Ou et al. 2012; Smith 2012.

26. Shu et al. 2001a.

27. Gee 2001a; Lacalli 2002b.

28. Ou et al. 2012.

29. García-Bellido et al. 2014.

30. Shubin et al. 2009.

31. Shu et al. 2001a.

32. Chen et al. 1995.

33. Shu et al. 1996a.

34. Chen et al. 1999.

35. Shu et al. 2003a.

36. Mallatt and Chen 2003.

37. Shu et al. 2010.

38. Conway Morris and Caron 2012.

39. Shu et al. 1996b.

40. Luo et al. 2001.

41. Young et al. 1996; and see Janvier 2015, for a review.

42. Shu et al. 1999b.

43. Shu et al. 2003b.

44. Conway Morris and Caron 2014.

45. Briggs et al. 1983.

46. Gabbott et al. 1995.

47. See Donoghue et al. (2000) in the vertebrates corner; Turner et al. (2010) in the non-vertebrate corner, with Janvier (2013) acting as a referee.

48. Szaniawski 1982.

49. Murdock et al. 2013.

50. Donoghue and Sansom 2002; Donoghue et al. 2006.

51. I recommend the concise review by Brazeau and Friedman (2015) as a very good place to start, with the magisterial Janvier (1996) adding strength in depth.

52. Gai et al. 2011.

53. Dupret et al. 2014; Zhu et al. 2013; Friedman and Brazeau 2013.

54. Brazeau and Friedman 2015.

55. Those who know me will understand that I do not say such things lightly. Having fallen into bad company in my youth (as I explain in my book *In Search of Deep Time*), I remain the last pattern-cladist in town.

CHAPTER FIFTEEN

1. Holland et al. 2015.

2. Grobben 1908.

3. Martín-Durán et al. 2016.
4. Ogasawara et al. 1999; Simakov et al. 2015.
5. Simakov et al. 2015.
6. Veis et al. 2002; Livingston et al. 2006.
7. Bertrand et al. 2011.
8. Green et al. 2013.
9. Darras et al. 2011.
10. Angerer et al. 2011; Lapraz et al. 2015.
11. Niehrs 2010.
12. See, for example, Grande and Patel 2009.
13. Duboc and Lepage 2008.
14. Pani et al. 2012.
15. Metchnikoff, 1881.
16. Halanych et al. 1995.
17. Freeman et al. 2012.
18. Caron et al. 2010.
19. Smith 2005.
20. Duboc et al. 2005; Duboc and Lepage 2008.
21. Simakov et al. 2015.
22. Lowe et al. 2006.
23. Satoh et al. 2014.
24. Caron et al. 2013.
25. Delsuc et al. 2006.
26. Fouquet et al. 1997; Sumoy et al. 1997.
27. Cleaver and Krieg 2001.
28. Lacalli 2005; Röttinger and Lowe 2012.
29. Beaster-Jones et al. 2008.
30. Pani et al. 2012.
31. Arenas-Morena et al. 2000.
32. Lacalli 1996a; Šestak and Domazet-Lošo 2015; L. Z. Holland 2015c; Toresson et al. 1998.
33. Lacalli et al. 1994; Wicht and Lacalli 2005.
34. Ruppert 1990; Gorbman et al. 1999; Christiaen et al. 2002.
35. Schlosser et al. 2014.
36. Schlosser 2005; L. Z. Holland 2005; Gillis et al. 2012.
37. Pascual-Anaya et al. 2013.
38. Green et al. 2015.
39. Diogo et al. 2015.
40. Lauri et al. 2014.
41. L. Z. Holland 2015b.

42. Beaster-Jones et al. 2006.
43. P. W. H. Holland et al. 1995.
44. Delsuc et al. 2006, 2008.
45. Lemaire et al. 2008.
46. Mazet and Shimeld 2005; Šestak et al. 2013.
47. Abitua et al. 2012, 2015.
48. Lemaire et al. 2008.
49. Lebar et al. 2010.
50. Nakashima et al. 2004; Matthysse et al. 2004.
51. Bassham and Postlethwait 2000.
52. Noden and Trainor 2005.
53. Šestak and Domazet-Lošo 2015.
54. Zhang et al. 1998.
55. Bardack 1991; Janvier 1996, 2006; Gess et al. 2006.
56. Clements et al. 2016; McCoy et al. 2016; Kuratani and Hirasawa 2016.
57. Brazeau and Friedman 2015.
58. Donoghue and Sansom 2002; Donoghue et al. 2006.
59. Bullock et al. 1984; Li and Richardson 2016.
60. Lacalli et al. 1994; Wicht and Lacalli 2005; Candiani et al. 2012; Suzuki et al. 2015.
61. Gai et al. 2011.
62. Dupret et al. 2014; Zhu et al. 2013; Friedman and Brazeau 2013.
63. Simakov et al. 2015.
64. Garstang 1928; Romer 1972.
65. Garstang 1922.
66. Gislén 1930; Jefferies 1986.
67. Halanych et al. 1995.
68. Lacalli 2005.
69. Delsuc et al. 2006, 2008.
70. Putnam et al. 2008; Simakov et al. 2015.

References

Abitua, P. B., Wagner, E., Navarrete, I. A., and Levine, M. 2012. "Identification of a rudimentary neural crest in a non-vertebrate chordate." *Nature* 492: 104–107.

Abitua, P. B., et al. 2015. "The pre-vertebrate origins of neurogenic placodes." *Nature* 524: 462–465.

Acampora, Dario, Gulisano, Massimo, Broccoli, Vania, and Simeone, Antonio. 2001. "Otx genes in brain morphogenesis." *Progress in Neurobiology* 64: 69–95.

Acampora, Dario, et al. 2005. "*Otx* genes in the evolution of the vertebrate brain." *Brain Research Bulletin* 66: 410–420.

Achatz, Johannes G., et al. 2013. "The Acoela: On their kind and kinships, especially with nemertodermatids and xenoturbellids (Bilateria *incertae sedis*)." *Organisms Diversity and Evolution* 13: 267–286.

Adachi, Noritaka, and Kuratani, Shigeru. 2012. "Development of the head and trunk mesoderm in the dogfish, *Scyliorhinus torazame*, I: Embryology and morphology of the head cavities and related structures." *Evolution and Development* 14: 234–256.

Adachi, Noritaka, Takechi, Masaki, Hirai, Tamami, and Kuratani, Shigeru. 2012. "Development of the head and trunk mesoderm in the dogfish, *Scyliorhinus torazame*, II: Comparison of gene expression between the head mesoderm and somites with reference to the origin of the vertebrate head." *Evolution and Development* 14: 257–276.

Aguinaldo, A. M. A., et al. 1997. "Evidence for a clade of nematodes, arthropods and other moulting animals." *Nature* 387: 489–493.

Ahlberg, Per, Trinajstic, Kate, Johanson, Zerina, and Long, John A. 2009. "Pelvic claspers confirm chondrichthyan-like internal fertilization in arthrodires." *Nature* 460: 888–889.

Albuixech-Crespo, Beatriz, et al. 2017. "Molecular regionalization of the developing amphioxus neural tube challenges major partitions of the vertebrate brain." *PLOS Biology*. https://doi.org/10.1371/journal.pbio.2001573

Aldridge, Richard J., et al. 2007. "The systematics and phylogenetic position of vetulicolians." *Palaeontology* 50: 131–168.

Alldredge, Alice L. 1977. "House morphology and mechanisms of feeding in the Oikopleuridae (Tunicata, Appendicularia)." *Journal of Zoology* 181: 175–188.

Amacher, S. L., and Kimmel, C. B. 1998. "Promoting notochord fate and repressing muscle development in zebrafish axial mesoderm." *Development* 125: 1397–1406.

Amores, Angel, et al. 1998. "Zebrafish hox clusters and vertebrate genome evolution." *Science* 282: 1711–1714.

Anderson, Heather Evans, and Christiaen, Lionel. 2016. "*Ciona* as a simple chordate model for heart development and regeneration." *Journal of Cardiovascular Development and Disease* 3. doi:10.3390/jcdd3030025.

Angerer, Lynne M., Yaguchi, Shinsuke, Angerer, Robert C., and Burke, Robert D. 2011. "The evolution of nervous system patterning: insights from sea urchin development." *Development* 138: 3613–3623.

Annona, Giovanni, Holland, Nicholas D., and D'Aniello, Salvatore. 2015. "Evolution of the notochord." *EvoDevo* 6. doi:10.1186/s13227-015-0025-3.

Appel, Tobey. 1987. *The Cuvier-Geoffroy Debate: French Biology in the Decades before Darwin*. New York: Oxford University Press.

Arenas-Mena, C., Cameron, A. R., and Davidson, E. H. 2000. "Spatial expression of Hox cluster genes in the ontogeny of a sea urchin." *Development* 127: 4631–4643.

Arendt, Detlev, Technau, Ulrich, and Wittbrodt, Joachim. 2001. "Evolution of the bilaterian larval foregut." *Nature* 409: 81–85.

Arendt, Detlev, Denes, Alexandru S., Jékely, Gáspár, and Tessmar-Raible, Kristin. 2008. "The evolution of nervous system centralization." *Philosophical Transactions of the Royal Society of London B* 363: 1523–1528.

Aronowicz, J., and Lowe, C. J. 2006. "Hox gene expression in the hemichordate *Saccoglossus kowalevskii* and the evolution of deuterostome nervous systems." *Integrative and Comparative Biology* 46: 890–901.

Arroyo, Alicia S., López-Escardó, David, de Vargas, Colomban, and Ruiz-Trillo, Iñaki, 2016. "Hidden diversity of Acoelomorpha revealed through metabarcoding." *Biology Letters* 12: 10.1098/rsbl.2016.0674

Awgulewitsch, Alexander, and Jacobs, Donna. 1992. "Deformed autoregulatory element from *Drosophila* functions in a conserved manner in transgenic mice." *Nature* 358: 341–345.

Azevedo, F. A., et al. 2009. "Equal numbers of neuronal and nonneuronal cells make the human brain an isometrically scaled-up primate brain." *Journal of Comparative Neurology* 513: 532–541.

Baguñà, Jaume, Martinez, Pere, Paps, Jordi, and Riutort, Marta. 2008. "Back in time: A new systematic proposal for the Bilateria." *Philosophical Transactions of the Royal Society B* 363: 1481–1491.

Bailey, W. J., Kim, J., Wagner, G. P., and Ruddle, F. H. 1997. "Phylogenetic reconstruction of vertebrate Hox cluster duplications." *Molecular Biology and Evolution* 14: 843–853.

Balavoine, G., de Rosa, R., and Adoutte, A. 2002. "Hox clusters and bilaterian phylogeny." *Molecular Phylogenetics and Evolution* 24: 366–373.

Ball, Eldon, and Miller, David J. 2006. "The continuing classificatory conundrum of chaetognaths." *Current Biology* 16: R593–R596.

Balser, Elizabeth J., and Ruppert, Edward E. 1990. "Structure, ultrastructure, and function of the preoral heart-kidney in *Saccoglossus kowalevskii* (Hemichordata, Enteropneusta) including new data on the stomochord." *Acta Zoologica* 71: 235–249.

Bardack, D. 1991. "First fossil hagfish (Myxinoidea): A record from the Pennsylvanian of Illinois." *Science* 254: 701–703.

Bardack, D., and Richardson, E. S. 1977. "New agnathous fishes from the Pennyslvanian of Illinois." *Fieldiana Geology* 33: 489–510.

Bartelt, Karen. 2000. "Review: Evolution." *Reports of the National Center for Science Education* 20: 28–39.

Bassham, Susan, and Postlethwait, John. 2000. "*Brachyury* (*T*) expression in embryos of a larvacean urochordate, *Oikopleura dioica*, and the ancestral role of *T*." *Developmental Biology* 220: 322–332.

Bateson, William. 1886. "The ancestry of the chordata." *Quarterly Journal of Microscopical Science* 26: 535–571.

———. 1894. *Materials for the Study of Variation, treated with especial regard to Discontinuity in the Origin of Species*. London: MacMillan.

Beaster-Jones, Laura, Horton, A. C., Gibson-Brown, J. J., Holland, N. D., and Holland, L. Z. 2006. "The amphioxus T-box gene, *AmphiTbx15/18/22*, illuminates the origins of chordate segmentation." *Evolution and Development* 8: 119–129.

Beaster-Jones, Laura, Kaltenbach, S., Koop, D., Yuan, S. C., Chastain, R., Holland, L. Z. 2008. "Expression of somite segmentation genes in amphioxus: A clock without a wavefront?" *Development Genes and Evolution* 218: 599–611.

Beddington, R. S. P., Rashbass, P., and Wilson, V. 1992. "*Brachyury*—a gene affecting mouse gastrulation and early organogenesis." *Development* 116: 157–165.

Berrill, N. J. 1955. *The Origin of Vertebrates*. Oxford: Clarendon.

Berthelot, Camille, et al. 2014. "The rainbow trout genome provides novel insights into evolution after whole-genome duplication in vertebrates." *Nature Communications* 5. doi:10.1038/ncomms4657.

Bertrand, Stéphanie, et al. 2011. "Amphioxus FGF signaling predicts the acquisition of vertebrate morphological traits." *Proceedings of the National Academy of Sciences of the United States of America* 108: 9180–9165.

Bertrand, Stéphanie, et al. 2015. "Evolution of the role of RA and FGF signals in the control of somitogenesis in chordates." *PLOS ONE* 10: e136587. doi:10.1371/journal.pone.0136587.

Binida-Emonds, Olaf R. P., Jeffery, Jonathan E., and Richardson, Michael K. 2003. "Inverting the hourglass: quantitative evidence against the phylotypic stage in vertebrate development." *Proceedings of the Royal Society of London B* 270. doi:10.1098/rspb.2002.2242.

Blair, Jaime E., and Hedges, S. Blair 2005. "Molecular phylogeny and divergence times of deuterostome animals." *Molecular Biology and Evolution* 22: 2275–2284.

Blum, Martin, Feistel, Kerstin, Thumberger, Thomas, and Schweickert, Axel. 2014. "The evolution of left-right patterning mechanisms." *Development* 141: 1603–1613.

Bondurand, Nadege, and Sham, Mai Har. 2013. "The role of *SOX10* during enteric nervous system development." *Developmental Biology* 382: 330–343.

Boorman, Clive J., and Shimeld, Sebastian M. 2002a. "The evolution of left-right asymmetry in chordates." *BioEssays* 24: 1004–1011.

———. 2002b. "*Pitx* homeobox genes in *Ciona* and amphioxus show left-right asymmetry in a conserved chordate character and define the ascidian adenohypophysis." *Evolution and Development* 4: 354–365.

Bottjer, David J., Davidson, Eric H., Peterson, Kevin J., and Cameron, R. Andrew. 2006. "Paleogenomics of echinoderms." *Science* 314: 956–960.

Bourlat, Sarah J., et al. 2003. "*Xenoturbella* is a deuterostome that eats mollusks." *Nature* 424: 925–928.

Bourlat, Sarah J., et al. 2006. "Deuterostome phylogeny reveals monophyletic chordates and the new phylum *Xenoturbellida*." *Nature* 444: 85–88.

Bourlat, Sarah J., Rota-Stabelli, Omar, Lanfear, Robert, and Telford, Maximilian J. 2009. "The mitochondrial genome structure of *Xenoturbella bocki* (phylum Xenoturbellida) is ancestral within the deuterostomes." *BMC Evolutionary Biology* 9. doi:10.1186/1471-2148-9-107.

Bouwmeester, Tewis, et al. 1996. "*Cerberus* is a head-inducing secreted factor expressed in the anterior endoderm of Spemann's organizer." *Nature* 382: 595–601.

Brazeau, Martin D., and Friedman, Matt. 2015. "The origin and early phylogenetic history of jawed vertebrates." *Nature* 520: 490–497.

Briggs, Derek E. G., et al. 2010. "Decay distorts ancestry." *Nature* 463: 741–743.

Briggs, Derek E. G., Clarkson, Euan N. K., and Aldridge, Richard J. 1983. "The conodont animal." *Lethaia* 16: 1–14.

Briggs, Derek E. G., Lieberman, Bruce S., Halgedahl, Susan L., and Jarrard, Richard D. 2005. "A new metazoan from the Middle Cambrian of Utah and the nature of the Vetulicolia." *Palaeontology* 48: 681–686.

Bromham, Lindell D., and Degnan, Bernard M. 1999. "Hemichordates and deuterostome evolution: robust molecular phylogenetic support for a hemichordate + echinoderm clade." *Evolution and Development* 1: 166–171.

Brooke, Nina M., Garcia-Fernàndez, Jordi, and Holland, P. W. H. 1998. "The *ParaHox* gene cluster is an evolutionary sister of the Hox gene cluster." *Nature* 392: 920–922.

Brunet, Frédéric G., et al. 2006. "Gene loss and evolutionary rates following whole genome duplication in teleost fishes." *Molecular Biology and Evolution* 23: 1808–1816.

Bullock, Theodore H., Moore, Jean K., and Fields, R. Douglas. 1984. "Evolution of myelin sheaths: Both lamprey and hagfish lack myelin." *Neuroscience Letters* 48: 145–148.

Burns, Alan J. 2005. "Migration of neural crest-derived enteric nervous system precursor cells to and within the gastrointestinal tract." *International Journal of Developmental Biology* 49: 143–150.

Cameron, C. B., and Bishop, C. D. 2012. "Biomineral ultrastructure, elemental constitution and genomic analysis of biomineral-related proteins in hemichordates." *Proceedings of the Royal Society of London B* 279: 3041–3048.

Cameron, Chris B., Garey, James R., and Swalla, Billie J. 2000. "Evolution of the chordate body plan: New insights from phylogenetic analyses of deuterostome phyla." *Proceedings of the National Academy of Sciences of the United States of America* 97: 4469–4474.

Candiani, Simona, et al. 2012. "A neurochemical map of the developing amphioxus nervous system." *BMC Neuroscience*. doi:10.1186/1471-2202-13-59.

Cañestro, Cristian, and Postlethwait, John H. 2007. "Development of a chordate anterior-posterior axis without classical retinoic acid signaling." *Developmental Biology* 305: 522–538.

Cannon, Johanna, et al. 2009. "Molecular phylogeny of hemichordata, with updated status of deep-sea enteropneusts." *Molecular Phylogenetics and Evolution* 52: 17–24.

Cannon, Johanna T., Swalla, Billie J., and Halanych, Kenneth M. 2013. "Hemichordate molecular phylogeny reveals a novel cold-water clade of Harrimaniid acorn worms." *Biological Bulletin* 225: 194–204.

Cannon, Johanna T., et al. 2014. "Phylogenomic resolution of the hemichordate and echinoderm clade." *Current Biology* 24: 2827–2832.

Cannon, Johanna T., et al. 2016. "Xenacoelomorpha is the sister group to Nephrozoa." *Nature* 530: 89–93.

Cannon, John P., Haire, Robert N., and Litman, Gary W. 2002. "Identification of diversified genes that contain immunoglobulin-like variable regions in a protochordate." *Nature Immunology* 3: 1200–1207.

Cano, Elena, Carmona, Rita, and Muñoz-Chápuli, Ramón. 2013. "Evolutionary origin of the proepicardium." *Journal of Developmental Biology* 1: 3–19.

Caron, Jean-Bernard, Conway Morris, Simon, and Shut, Degan. 2010. "Tentaculate fossils from the Cambrian of Canada (British Columbia) and China (Yunnan) interpreted as primitive deuterostomes." *PLOS ONE* 5: e9586. doi:10.1371/journal.pone.0009586.

Caron, Jean-Bernard, Conway Morris, Simon, and Cameron, Christopher B. 2013. "Tubiculous enteropneusts from the Cambrian period." *Nature* 495: 503–506.

Carroll, Sean B. 1995. "Homeotic genes and the evolution of arthropods and chordates." *Nature* 376: 479–485.

Castro, L. Filipe C., et al. 2006. "A *Gbx* homeobox gene in amphioxus: Insights into ancestry of the ANTP class and evolution of the midbrain/hindbrain boundary." *Developmental Biology* 295: 40–51.

Chang, Hao-Han, Naylor, Richard W., and Davidson, Alan J. 2016. "Organogenesis of the zebrafish kidney." In *Organogenetic Gene Networks*, edited by James Castelli-Gair Hombría and Paola Bovolenta, 213–233. Switzerland: Springer International.

Chang, Mee-Mann, Zhang, Jiangyong, and Miao, Desui. 2006. "A lamprey from the Cretaceous Jehol biota of China." *Nature* 111: 972–974.

Chea, Helen K., Wright, Christopher V., and Swalla, Billie J. 2005. "Nodal signaling and the evolution of deuterostome gastrulation." *Developmental Dynamics* 234: 269–278.

Chen, Jun-Yuan, and Huang, Di-Ying. 2002. "A possible Lower Cambrian chaetognath." *Science* 298: 187.

Chen, Jun-Yuan, et al. 2003. "The first tunicate from the Early Cambrian of South China." *Proceedings of the National Academy of Sciences of the United States of America* 100: 8314–8318.

Chen, J. Y., Dzik, J., Ramsköld, L., and Zhou, G.-Q. 1995. "A possible Early Cambrian chordate." *Nature* 377: 720–722.

Chen, Jun-Yuan, Huang, Di-Ying, and Li, Chia-Wei. 1999. "An early Cambrian craniate-like chordate." *Nature* 402: 518–522.

Chesley, Paul. 1935. "Development of the short-tailed mutant in the house mouse." *Journal of Experimental Zoology* 70: 429–459.

Chiang, C., et al. 1996. "Cyclopia and defective axial patterning in mice lacking Sonic hedgehog gene function." *Nature* 383: 407–413.

Chourrout, D., et al. 2006. "Minimal *ProtoHox* cluster inferred from bilaterian and cnidarian Hox complements." *Nature* 442: 684–687.

Christiaen, Lionel, et al. 2002. "*Pitx* genes in Tunicates provide new molecular insight into the evolutionary origin of pituitary." *Gene* 287: 107–113.

Christoffels, Alan, et al. 2004. "Fugu genome analysis provides evidence for a whole-genome duplication early during the evolution of ray-finned fishes." *Molecular Biology and Evolution* 21: 1146–1151.

Clausen, Sébastien, and Smith, Andrew B. 2005. "Palaeoanatomy and biological affinities of a Cambrian deuterostome (Stylophora)." *Nature* 438: 351–354.

Clausen, Sébastien, Hou, Xian-Guang, Bergström, Jan, and Franzén, Christina. 2010. "The absence of echinoderms from the Lower Cambrian Chengjiang fauna of China: Palaeoecological and palaeogeographical implications." *Palaeogeography, Palaeoclimatology, Palaeoecology* 294: 133–141.

Cleaver, Ondine, and Krieg, Paul A. 2001. "Notochord patterning of the endoderm." *Development* 234: 1–12.

Clements, Thomas, et al. 2016. "The eyes of *Tullimonstrum* reveal a vertebrate affinity." *Nature* 532: 500–503.

Cole, Alison G., and Meinertzhagen, Ian A. 2004. "The central nervous system of the ascidian larva: Mitotic history of cells forming the neural tube in late embryonic *Ciona intestinalis*." *Developmental Biology* 271: 239–262.

Cole, W. C., and Youson, J. H. 1982. "Morphology of the pineal complex of the anadromous sea lamprey, *Petromyzon marinus* L." *American Journal of Anatomy* 165: 131–163.

Conway Morris, S. 1977a. "A new metazoan from the Cambrian Burgess Shales of British Columbia." *Palaeontology* 20: 623–640.

———. 1977b. "A redescription of the Middle Cambrian worm *Amiskwia sagittiformis* Walcott from the Burgess Shale of British Columbia." *Paläontologische Zeitschrift* 51: 271–287.

———. 2009. "The Burgess Shale animal *Oesia* is not a chaetognath: A reply to Szaniawski (2005)." *Acta Palaeontologica Polonica* 54: 175–179.

Conway Morris, Simon, and Caron, Jean-Bernard. 2012. "*Pikaia gracilens* Walcott, a stem-group chordate from the Middle Cambrian of British Columbia." *Biological Reviews* 87: 480–512.

———. 2014. "A primitive fish from the Cambrian of North America." *Nature* 512: 419–422.

Cooke, J., and Zeeman, E. C. 1976. "A clock and wavefront model for control of the number of repeated structures during animal morphogenesis." *Journal of Theoretical Biology* 58: 455–476.

Copp, Andrew J., Greene, Nicholas D. E., and Murdoch, Jennifer N. 2003. "The genetic basis of mammalian neurulation." *Nature Reviews Genetics* 4: 784–793.

Corallo, Diana, Trapani, Valeria, and Bonaldo, Paolo. 2015. "The notochord: Structure and functions." *Cell and Molecular Life Sciences* 72: 2989–3008.

Corbo, J. C., Levine, M., and Zeller, R. W. 1997. "Characterization of notochord-specific enhancer from the Brachyury promoter region of the ascidian, *Ciona intestinalis*." *Development* 124: 589–602.

Crow, Karen D., et al. 2006. "The 'fish-specific' Hox cluster duplication is coincident with the origin of teleosts." *Molecular Biology and Evolution* 23: 121–136.

Crowther, R. J., and Whittaker, J. R. 1994. "Serial repetition of cilia pairs along the tail surface of an ascidian larva." *Journal of Experimental Zoology* 268: 9–16.

Cunningham, Doreen, and Casey, Elena Silva. 2014. "Spatiotemporal development of the embryonic nervous system of *Saccoglossus kowalevskii*." *Developmental Biology* 386: 252–263.

D'Agati, Paolo, and Cammarata, Matteo. 2006. "Comparative analysis of thyroxine distribution in ascidian larvae." *Cell and Tissue Research* 323: 529–535.

Daley, Paul Edward John. 1995. "Anatomy, locomotion and ontogeny of the solute *Castericystis vali* from the Middle Cambrian of Utah." *Geobios* 28: 585–614.

Darras, Sébastien, et al. 2011. "β-Catenin specifies the endomesoderm and defines the posterior organizer of the hemichordate *Saccoglossus kowalevskii*." *Development* 138: 959–970.

Davidson, Brad. 2007. "*Ciona intestinalis* as a model for cardiac development." *Seminars in Cell and Developmental Biology* 18: 16–26.

Davidson, Brad, and Swalla, Billie J. 2002. "A molecular analysis of ascidian metamorphosis reveals activation of an innate immune response." *Development* 129: 4739–4751.

Davis, Daniel M. 2013. *The Compatibility Gene*. London: Allen Lane.

Degnan, Bernard M., Degnan, Sandie M., Giusti, Andrew, and Morse, Daniel E. 1995. "A hox/hom homeobox gene in sponges." *Gene* 155: 175–177.

Degnan, James H., and Rosenberg, Noah A. 2009. "Gene tree discordance, phylogenetic inference and the multispecies coalescent." *Trends in Ecology and Evolution* 24: 332–340.

Dehal, Paramvir, et al. 2002. "The draft genome of *Ciona intestinalis*: Insights into chordate and vertebrate origins." *Science* 298: 2157–2167.

Dehal, Paramvir, and Boore, Jeffrey L. 2005. "Two rounds of whole genome duplication in the ancestral vertebrate." *PLOS Biology*. doi:10.1371/journal.pbio.0030314.

Delarbre, Christiane, et al. 2002. "Complete mitochondrial DNA of the hagfish, *Eptatretus burgeri*: The comparative analysis of mitochondrial DNA sequences strongly supports the cyclostome monophyly." *Molecular Phylogenetics and Evolution* 22: 184–192.

Delsuc, Frédéric, Brinkmann, Henner, Chourrout, Daniel, and Philippe, Hervé.

2006. "Tunicates and not cephalochordates are the closest living relatives of vertebrates." *Nature* 439: 965–968.

Delsuc, Frédéric, Tsagkogeorga, Georgia, Lartillot, Nicolas, and Philippe, Hervé. 2008. "Additional molecular support for the new chordate phylogeny." *Genesis* 46: 592–604.

Denes, Alexandru S., et al. 2007. "Molecular architecture of annelid nerve cord supports common origin of nervous system centralization in bilateria." *Cell* 129: 277–288.

Denoeud, France, et al. 2010. "Plasticity of animal genome architecture unmasked by rapid evolution of a pelagic tunicate." *Science* 330: 1381–1385.

Dequéant, Mary-Lee, and Pourquié, Olivier, 2008. "Segmental patterning of the vertebrate embryonic axis." *Nature Reviews Genetics* 9: 370–382.

De Robertis, E. M. 2006. "Spemann's organizer and self-regulation in amphibian embryos." *Nature Reviews Molecular Cell Biology* 7: 296–301.

———. 2009. "Spemann's organizer and the self-regulation of embryonic fields." *Mechanisms of Development* 126: 925–941.

De Robertis, E. M., and Sasai, Y. 1996. "A common plan for dorsoventral patterning in bilateria." *Nature* 380: 37–40.

Dilly, P. N. 1993. "*Cephalodiscus graptoloides* sp. nov. a probable extant graptolite." *Journal of Zoology* 229: 69–78.

Dilly, Peter N. 2014. "*Cephalodiscus* reproductive biology (Pterobranchia, Hemichordata)." *Acta Zoologica* 95: 111–124.

Diogo, Rui, et al. 2015. "A new heart for a new head in vertebrate cardiopharyngeal evolution." *Nature* 520: 466–473.

Dishaw, Larry J., et al. 2008. "Genomic complexity of the variable region-containing chitin-binding proteins in amphioxus." *BMC Genetics* 9:78. doi 10.1186/1471-2156-9-78.

Dobzhansky, Theodosius. 1973. "Nothing in biology makes sense except in the light of evolution." *American Biology Teacher* 35: 125–129.

Doguzhaeva, L. A., Mutvei, H., and Mapes, R. H. 2002. "Chaetognath grasping spines from the Upper Mississippian of Arkansas (USA)." *Acta Palaeontologica Polonica* 47: 421–430.

Domazet-Lošo, Tomislav, and Tautz, Diethard. 2010. "A phylogenetically based transcriptome age index mirrors ontogenetic divergence patterns." *Nature* 468: 815–818.

Dominguez, Patrício, Jacobson, Antone G., and Jefferies, Richard P. S. 2002. "Paired gill slits in a fossil with a calcite skeleton." *Nature* 417: 841–844.

Dong, Yuemei, Taylor, Harry E., and Dimopoulos, George. 2006. "AgDscam, a hypervariable immunoglobulin domain-containing receptor of the *Anopheles gambiae* innate immune system." *PLOS Biology*. doi:10.1371/journal.pbio.0040229.

Donoghue, Philip C. J., Forey, Peter L., and Aldridge, Richard J. 2000. "Conodont affinity and chordate phylogeny." *Biological Reviews* 75: 191–251.

Donoghue, Philip C. J., and Sansom, Ivan J. 2002. "Origin and early evolution of vertebrate skeletonization." *Microscopy Research and Technique* 59: 352–372.

Donoghue, Philip C. J., Sansom, Ivan J., and Downs, Jason P. 2006. "Early evolution of vertebrate skeletal tissues and cellular interactions, and the canalization of skeletal development." *Journal of Experimental Zoology B* 206B: 278–294.

Dosch, Roland, and Niehrs, Christof. 2000. "Requirement for anti-dorsalizing morphogenetic protein in organizer patterning." *Mechanisms of Development* 90: 195–203.

Duboc, Véronique, and Lepage, Thierry. 2008. "A conserved role for the nodal signalling pathway in the establishment of dorso-ventral and left-right axes in deuterostomes." *Journal of Experimental Zoology* 310B: 41–53.

Duboc, Véronique, Röttinger, Eric, Besnardeau, Lydia, and Lepage, Thierry. 2004. "Nodal and BMP2/4 signaling organizes the oral-aboral axis of the sea urchin embryo." *Developmental Cell* 6: 397–410.

Duboc, Véronique, et al. 2005. "Left-right asymmetry in the sea urchin embryo is regulated by nodal signaling on the right side." *Developmental Cell* 9: 147–158.

Dufour, Héloïse D., et al. 2006. "Precraniate origin of cranial motoneurons." *Proceedings of the National Academy of Sciences of the United States of America* 103: 8727–8732.

Dunn, Casey W. 2015. "Acorn worms in a nutshell." *Nature* 527: 448–449.

Dunn, Casey W., et al. 2008. "Broad phylogenomic sampling improves resolution of the animal tree of life." *Nature* 452: 745–749.

Dunn, Casey W., Leys, Sally P., and Haddock, Steven H. D. 2015. "The hidden biology of sponges and ctenophores." *Trends in Ecology and Evolution* 30: 282–291.

Dupret, Vincent, et al. 2014. "A primitive placoderm sheds light on the origin of the jawed vertebrate face." *Nature* 507: 500–503.

Durham, J. W. 1974. "Systematic position of *Eldonia ludwigi* Walcott." *Journal of Palaeontology* 48: 750–755.

Durham, J. W., and Caster, K. E. 1963. "Helicoplacoidea: A new class of echinoderms." *Science* 140: 820–822.

Durston, Antony J., Jansen, Hans J., In der Rieden, Paul, and Hooiveld, Michiel H. W. 2011. "Hox collinearity—a new perspective." *International Journal of Developmental Biology* 55: 899–908.

Edgecombe, Gregory D., et al. 2011. "Higher-level metazoan relationships: recent progress and remaining questions." *Organisms Diversity and Evolution* 11: 151–172.

Ehlers, Ulrich, and Sopott-Ehlers, Beate. 1997. "Ultrastructure of the subepidermal musculature of *Xenoturbella bocki*, the adelphotaxon of the Bilateria." *Zoomorphology* 117: 71–79.

Eriksson, Mats E., and Waloszek, Dieter. 2016. "Half-a-billion-year-old microscopic treasures—the Cambrian 'Orsten' fossils of Sweden." *Geology Today* 32: 115–120.

Escrivà, Hector, Manzon, Lori, Youson, John, and Laudet, Vincent. 2002a. "Analysis of lamprey and hagfish genes reveals a complex history of gene duplications during early vertebrate evolution." *Molecular Biology and Evolution* 19: 1440–1450.

Escrivà, Hector, et al. 2002b. "The retinoic acid signaling pathway regulates anterior/posterior patterning in the nerve cord and pharynx of amphioxus, a chordate lacking neural crest." *Development* 129: 2905–2916.

Faa, Gavino, et al. 2014. "Development of the human kidney: Morphological events." In *Kidney Development in Renal Pathology*, edited by G. Faa and V. Fanos, 1–12. New York: Springer Science and Business Media.

Fairclough, Stephen R., et al. 2013. "Premetazoan genome evolution and the regulation of cell differentiation in the choanoflagellate *Salpingoeca rosetta*." *Genome Biology* 14. http://genomebiology.com/2013/14/2/R15

Faure, E., and Casanova, J.-P. 2006. "Comparison of chaetognath mitochondrial genomes and phylogenetical implications." *Mitochondrion* 6: 258–262.

Fekany, Kimberly, et al. 1999. "The zebrafish bozozok locus encodes Dharma, a homeodomain protein essential for induction of gastrula organizer and dorsoanterior embryonic structures." *Development* 128: 1427–1438.

Ferrier, David E. K. 2011. "Tunicates push the limits of animal evo-devo." *BMC Biology* 9:3.

———. 2012. "Evolution of the Hox gene cluster." *Wiley Online Library eLS*. doi:10.1002/9780470015902.a0023989.

Ferrier, David E. K., and Holland, P. W. H. 2001. "Ancient origin of the Hox gene cluster." *Nature Reviews Genetics* 2: 33–38.

———. 2002. "*Ciona intestinalis* ParaHox genes: evolution of Hox/ParaHox cluster integrity, developmental mode, and temporal colinearity." *Molecular Phylogenetics and Evolution* 24: 412–417.

Field, K. G., et al. 1988. "Molecular phylogeny of the animal kingdom." *Science* 239: 748–753.

Flajnik, Martin F. 2004. "Another manifestation of GOD." *Nature* 430: 157–158.

Flood, Per R. 1966. "A peculiar mode of muscular innervation in amphioxus: Light and electron microscopic studies of the so-called ventral roots." *Journal of Comparative Neurology* 126: 181–217.

Forey, Peter, and Janvier, Philippe. 1993. "Agnathans and the origin of jawed vertebrates." *Nature* 361: 129–134.

Fouquet, Bernadette, Weinstein, Brant M., Serluca, Fabrizio C., and Fishman, Mark C. 1997. "Vessel patterning in the embryo of the zebrafish: Guidance by notochord." *Developmental Biology* 183: 37–48.

Franzén, Åke, and Afzelius, Björn A. 1987. "The ciliated epidermis of *Xenoturbella bocki* (Platyhelminthes, Xenoturbellida) with some phylogenetic considerations." *Zoologica Scripta* 16: 9–17.

Freeman, Robert, et al. 2012. "Identical genomic organization of two hemichordate Hox clusters." *Current Biology* 22: 2053–2058.

Fried, Claudia, Prohaska, Sonja J., and Stadler, Peter F. 2003. "Independent Hox cluster duplications in lampreys." *Journal of Experimental Zoology* 299B: 18–25.

Friedman, Matt, and Brazeau, Martin. 2013. "A jaw-dropping fossil fish." *Nature* 502: 175–177.

Fritzsch, Guido, et al. 2008. "PCR survey of *Xenoturbella bocki* Hox genes." *Journal of Experimental Zoology* 310B: 278–284.

Fugmann, Sebastian D., et al. 2006. "An ancient evolutionary origin of the Rag1/2 gene locus." *Proceedings of the National Academy of Sciences of the United States of America* 103: 3728–3733.

Fujii, Setsuko, Nishio, Takaya, and Nishida, Hiroki. 2008. "Cleavage pattern, gastrulation, and neurulation in the appendicularian, *Oikopleura dioica*." *Development Genes and Evolution* 281: 69–79.

Furlong, Rebecca F., and Holland, P. W. H. 2002. "Bayesian phylogenetic analysis supports monophyly of ambulacraria and cyclostomes." *Zoological Science* 19: 593–599.

Furness, John B. 2012. "The enteric nervous system and neurogastroenterology." *Nature Reviews Gastroenterology and Hepatology* 9: 286–294.

Gabbott, S. E., Aldridge, R. J., and Theron, J. N. 1995. "A giant conodont with preserved muscle tissue from the Upper Ordovician of South Africa." *Nature* 374: 800–803.

Gabbott, S. E., et al. 2016. "Pigmented anatomy in Carboniferous cyclostomes and the evolution of the vertebrate eye." *Proceedings of the Royal Society of London B* 283: 20161151. doi:10.1098/rspb.2016.1151.

Gage, John. 2005. "Deep-sea spiral fantasies." *Nature* 434: 283–284.

Gai, Zhikun, et al. 2011. "Fossil jawless fish from China foreshadows early jawed vertebrate anatomy." *Nature* 476: 324–327.

Gans, C., and Northcutt, R. G. 1983. "Neural crest and the origin of vertebrates: A new head." *Science* 220: 268–274.

García-Bellido, Diego C., et al. 2014. "A new vetulicolian from Australia and its bearing on the chordate affinities of an enigmatic Cambrian group." *BMC Evolutionary Biology* 14: 214.

Garcia-Fernàndez, Jordi, and Holland, Peter W. H. 1994. "Archetypal organization of the amphioxus Hox gene cluster." *Nature* 370: 563–566.

Garcia-Fernàndez, Jordi, D'Aniello, Salvatore, and Escrivà, Hector. 2007. "Organizing chordates with an organizer." *BioEssays* 29: 619–624.

Garstang, Myles, and Ferrier, David E. K. 2013. "Time is of the essence for ParaHox homeobox gene clustering." *BMC Biology* 11: 72.

Garstang, S. L., and Garstang, W. 1926. "On the development of Botrylloides and the ancestry of vertebrates." *Proceedings of the Leeds Philosophical Society* 1: 81–86.

Garstang, Walter. 1894. "Preliminary note on a new theory of the phylogeny of the Chordata." *Zoologischer Anzeiger* 17: 122–125.

———. 1922. "The theory of recapitulation: a critical restatement of the Biogenetic Law." *Journal of the Linnean Society (Zoology)* 35: 81–101.

———. 1928. "The morphology of the Tunicata, and its bearings on the phylogeny of the Chordata." *Quarterly Journal of Microscopical Science* 72: 51–187.

Gee, Henry. 1995. "Lophophorates prove likewise variable." *Nature* 374: 493.

———. 1996. *Before the Backbone; Views on the Origin of the Vertebrates*. London: Chapman and Hall.

———. 1999. *In Search of Deep Time: Beyond the Fossil Record to a New History of Life*. New York: Free Press.

———. 2001a. "On being Vetulicolian." *Nature* 414: 407–409.

———. 2001b. "Deuterostome phylogeny: The context for the origin and evolution of vertebrates." In *Major Events in Early Vertebrate Evolution*, edited by P. E. Ahlberg, 1–14. London: Taylor & Francis.

———. 2002. "Aspirational thinking." *Nature* 420: 611.

———. 2003. "You aren't what you eat." *Nature* 424: 885–886.

———. 2004. *Jacob's Ladder: The History of the Human Genome*. New York: Norton.

———. 2006. "Careful with that amphioxus." *Nature* 439: 923–924.

———. 2007. "This worm is not for turning." *Nature* 445: 33–34.

———. 2013. *The Accidental Species: Misunderstandings of Human Evolution*. Chicago: University of Chicago Press.

Gee, Henry, et al. 2015. "Origin and evolution of vertebrates." *Nature* 520: 449–497.

Gehring, Walter J., and Ikeo, Kazuho. 1999. "Pax 6: Mastering eye morphogenesis and eye evolution." *Trends in Genetics* 15: 371–377.

Gerhart, John. 2001. "Evolution of the organizer and the chordate body plan." *International Journal of Developmental Biology* 45: 133–153.

———. 2006. "The deuterostome ancestor." *Cellular Physiology* 209: 677–685.

Gerhart, John, Lowe, Christopher, and Kirschner, Marc. 2005. "Hemichordates and the origin of chordates." *Current Opinion in Genetics and Development* 15: 461–467.

Gess, Robert W., Coates, Michael I., and Rubidge, Bruce S. 2006. "A lamprey from the Devonian period of South Africa." *Nature* 443: 981–984.

Gilbert, Scott F. 2000. *Developmental Biology*. 6th ed. Sunderland, MA: Sinauer.

Gillis, J. Andrew, Fritzenwanker, Jens H., and Lowe, Christopher J. 2012. "A stem-deuterostome origin of the vertebrate pharyngeal transcriptional network." *Proceedings of the Royal Society of London B* 279: 237–246.

Gislén, T. 1930. "Affinities between the Echinodermata, Enteropneusta and Chordonia." *Zoologiska Bidrag frän Uppsala* 12: 199–304.

Glardon, S., Holland, L. Z., Gehring, W. J., and Holland, N. D. 1998. "Isolation and developmental expression of the amphioxus Pax-6 gene (AmphiPax-6): insights into eye and photoreceptor evolution." *Development* 125: 2701–2710.

Glasauer, Stella M. K., and Neuhauss, Stephan C. F. 2014. "Whole-genome duplication in teleost fishes and its evolutionary consequences." *Molecular Genetics and Genomics* 289: 1045–1060.

Gomez, C., et al. 2008. "Control of segment number in vertebrate embryos." *Nature* 454: 335–339.

Gonzalez, Paul, and Cameron, Christopher B. 2009. "The gill slits and pre-oral ciliary organ of *Protoglossus* (Hemichordata: Enteropneusta) are filter-feeding structures." *Biological Journal of the Linnean Society* 98: 898–906.

Gonzalez, Paul, Uhlinger, Kevin R., and Lowe, Christopher J. 2017. "The adult body plan of indirect developing hemichordates develops by adding a Hox-patterned trunk to an anterior larval territory." *Current Biology* 27: R21–R24.

Goodrich, E. S. 1930. *Studies on the Structure and Development of Vertebrates*. London: MacMillan.

Gopalakrishnan, Swetha, et al. 2015. "A cranial mesoderm origin for esophagus striated muscles." *Developmental Cell* 34: 694–704.

Gorbman, Aubrey. 1995. "Olfactory origins and evolution of the brain-pituitary endocrine system." *General and Comparative Endocrinology* 97: 171–178.

———. 1997. "Hagfish development." *Zoological Science* 14: 375–390.

Gorbman, Aubrey, Nozaki, Masumi, and Kubokawa, Kaoru. 1999. "A Brain-Hatschek's Pit connection in amphioxus." *General and Comparative Endocrinology* 113: 251–254.

Gostling, Neil J., and Shimeld, Sebastian M. 2003. "Protochordate Zic genes define primitive somite compartments and highlight molecular changes underlying neural crest evolution." *Evolution and Development* 5: 136–144.

Gould, S. J. 1990. *Wonderful Life: The Burgess Shale and the Nature of History*. New York: Norton.

Gould, S. J., and Lewontin, R. C. 1979. "The spandrels of San Marco and the Panglossian Paradigm: A critique of the adaptationist programme." *Proceedings of the Royal Society of London B* 205: 581–598.

Govindajaran, Annette F., Bucklin, Ann, and Madin, Laurence P. 2010. "A molecular phylogeny of the Thaliacea." *Journal of Plankton Research*. doi:10.1093/plankt/fbq197.

Graham, Anthony, and Shimeld, Sebastian M. 2013. "The origin and evolution of ectodermal placodes." *Journal of Anatomy* 222: 32–40.

Graham, Anthony, Butts, Thomas, Lumsden, Andrew, and Kiecker, Clemens. 2014. "What can vertebrates tell us about segmentation?" *EvoDevo* 5. doi:10.1186/2041-9139-5-24.

Grande, Cristina, and Patel, Nipam H. 2009. "Nodal signalling is involved in left-right asymmetry in snails." *Nature* 457: 1007–1011.

Grapin-Botton, Anne. 2005. "Antero-posterior patterning of the vertebrate digestive tract: 40 years after Nicole Le Douarin's PhD thesis." *International Journal of Developmental Biology* 49: 335–347.

Green, Stephen A., Norris, Rachael P., Terasaki, M., and Lowe, Christopher J. 2013. "FGF signaling induces mesoderm in the hemichordate *Saccoglossus kowalevskii*." *Development* 140: 1024–1033.

Green, Stephen A., Simoes-Costa, Marcos, and Bronner, Marianne E. 2015. "Evolution of vertebrates as viewed from the crest." *Nature* 520: 474–482.

Green, Stephen A., Uy, Benjamin R., and Bronner, Marianne E. 2017. "Ancient evolutionary origin of vertebrate enteric neurons from trunk-derived neural crest." *Nature* 544: 88–91.

Gregory, W. K. 1936. "The transformation of organic designs: a review of the origin and deployment of the earlier vertebrates." *Biological Reviews* 11: 311–344.

Grenier, Julien, et al. 2009. "Relationship between neural crest cells and cranial mesoderm during head muscle development." *PLOS ONE* 4: e4381. doi:10.1371/journal.pone.0004381.

Grifone, Raphaëlle, and Kelly, Robert G. 2007. "Heartening news for head muscle development." *Trends in Genetics* 23: 365–369.

Grobben, Karl. 1908. "Die Systematische Einteilung des Tierreiches." *Verhandlungen der Zoologisch-Botanischen Gesellschaft in Wien* 5: 491–511.

Gross, Jeffrey M., and McClay, David R. 2001. "The role of Brachyury (T) during gastrulation movements in the sea urchin *Lytechinus variegatus*." *Developmental Biology* 239: 132–147.

Halanych, Kenneth M. 1995. "The phylogenetic position of the pterobranch hemichordates based on 18S rDNA sequence data." *Molecular Phylogenetics and Evolution* 4: 72–76.

———. 1996. "Testing hypotheses of chaetognath origins: long branches revealed by 18S ribosomal DNA." *Systematic Biology* 45: 223–246.

———. 2004. "The new view of animal phylogeny." *Annual Review of Ecology, Evolution and Systematics* 35: 229–256.

Halanych, Kenneth M., et al. 1995. "Evidence from 18S ribosomal DNA that the lophophorates are protostome animals." *Science* 267: 1641–1643.

Halpern, Marnie E., Hobert, Oliver, and Wright, Christopher V. E. 2014. "Left-right asymmetry: Advances and enigmas." *Genesis* 52: 451–454.

Han, Jian, et al. 2017. "Meiofaunal deuterostomes from the basal Cambrian of Shaanxi." *Nature* 542: 228–231.

Hashimshony, Tamar, et al. 2015. "Spatiotemporal transcriptomics reveals the evolutionary history of the endoderm germ layer." *Nature* 519: 219–222.

Heimberg, Alysha M., et al. 2010. "microRNAs reveal the interrelationships of hagfish, lampreys, and gnathostomes and the nature of the ancestral vertebrate." *Proceedings of the National Academy of Sciences of the United States of America* 107: 19379–19383.

Hejnol, Andreas, and Lowe, Christopher J. 2014. "Stiff or squishy notochord origins?" *Current Biology* 24: R1131–R1133.

Hejnol, Andreas, and Martindale, Mark Q. 2008a. "Acoel development indicates the independent evolution of the bilaterian moth and anus." *Nature* 456: 382–386.

———. 2008b. "Aceol development supports a simple planula-like urbilaterian." *Philosophical Transactions of the Royal Society of London B* 363: 1493–1501.

Hejnol, Andreas, and Rentzch, Fabian. 2015. "Neural nets." *Current Biology* 25: R782–R786.

Hejnol, Andreas, et al. 2009. "Assessing the root of bilaterian animals with scalable phylogenomic methods." *Proceedings of the Royal Society of London B* 276: 4261–4270.

Helmkampf, Martin, Bruchhaus, Iris, and Hausdorf, Bernhard. 2008a. "Phylogenomic analyses of lophophorates (brachiopods, phoronids and bryozoans) confirm the Lophotrochozoa concept." *Proceedings of the Royal Society of London B*. doi:10.1098/rspb.2008.0372.

———. 2008b. "Multigene analysis of lophophorate and chaetognath phylogenetic relationships." *Molecular Phylogenetics and Evolution* 46: 206–214.

Herr, Alexander, et al. 2003. "Expression of mouse Tbx22 supports its role in palatogenesis and glossogenesis." *Developmental Dynamics* 226: 579–586.

Herrmann, Bernhard G. 1991. "Expression pattern of the Brachyury gene in whole-mount TWIs/TWis mutant embryos." *Development* 113: 913–917.

Herrmann, Bernhard G., et al. 1990. "Cloning of the T gene required in mesoderm formation in the mouse." *Nature* 343: 617–622.

Heyland, Andreas, and Hodin, Jason. 2004. "Heterochronic developmental shift caused by thyroid hormone in larval sand dollars and its implications for phenotypic plasticity and the evolution of nonfeeding development." *Evolution* 58: 524–538.

Hibino, Taku, et al. 2006. "The immune gene repertoire encoded in the purple sea urchin genome." *Developmental Biology* 300: 349–365.

Hill, Matthew M., et al. 2008. "The *C. savignyi* genetic map and its integration with the reference sequence facilitates insights into chordate genome evolution." *Genome Research* 18: 1369–1379.

Hillier, LaDeana W., et al. 2005. "Genomics in *C. elegans*: So many genes, such a little worm." *Genome Research* 15: 1651–1660.

Hiruta, Jin, et al. 2005. "Comparative expression analysis of transcription factor genes in the endostyle of invertebrate chordates." *Developmental Dynamics* 233: 1031–1037.

Holland, Linda Z. 2005. "Non-neural ectoderm is really neural: Evolution of developmental patterning mechanisms in the non-neural ectoderm of chordates and the problem of sensory cell homologies." *Journal of Experimental Zoology B* 304B: 304–323.

———. 2007. "A chordate with a difference." *Nature* 447: 153–155.

———. 2008. "The amphioxus genome illuminates vertebrate origins and cephalochordate biology." *Genome Research* 18: 1100–1111.

———. 2009. "Chordate roots of the vertebrate nervous system: Expanding the molecular toolkit." *Nature Reviews Neuroscience* 10: 736–746.

———. 2013. "Evolution of new characters after whole genome duplications: Insight from amphioxus." *Seminars in Cell and Developmental Biology* 24: 101–109.

———. 2015a. "Evolution of basal deuterostome nervous systems." *Journal of Experimental Biology* 218: 637–645.

———. 2015b. "Cephalochordata." In *Evolutionary Developmental Biology of Invertebrates 6: Deuterostomia*, edited by A. Wanniger, 92–133. Vienna: Springer-Verlag.

———. 2015c. "The origin and evolution of chordate nervous systems." *Philosophical Transactions of the Royal Society of London B* 370. doi:10.1098/rstb.2015.0048.

———. 2016. "Tunicates." *Current Biology* 26: R146–R152.

Holland, Linda Z., and Gibson-Brown, Jeremy J. 2003. "The *Ciona intestinalis* genome: When the constraints are off." *BioEssays* 25: 529–532.

Holland, Linda Z., and Holland, N. D. 2007. "A revised fate map for amphioxus and the evolution of axial patterning in chordates." *Integrative and Comparative Biology* 47: 360–372.

Holland, Linda Z., Holland, Nicholas D., and Gilland, Edwin. 2008. "Amphioxus and the evolution of head segmentation." *Integrative and Comparative Biology* 48: 630–646.

Holland, Linda Z., et al. 2013. "Evolution of bilaterian central nervous systems: A single origin?" *EvoDevo* 4: 27.

Holland, Nicholas D. 2003. "Early central nervous system evolution: An era of skin brains?" *Nature Reviews Neuroscience* 4: 617–627.

Holland, Nicholas D., Holland, L. Z., and Holland, P. W. H. 2015. "Scenarios for the making of vertebrates." *Nature* 520: 450–455.

Holland, Nicholas D., Holland, Linda Z., and Kozmik, Z. 1995. "An amphioxus Pax gene, *AmphiPax1*, expressed in embryonic endoderm, but not mesoderm: Implications for the evolution of class I paired box genes." *Molecular Marine Biology and Biotechnology* 4: 206–214.

Holland, Nicholas D., Panganiban, G., Henyey, E. L., and Holland, L. Z. 1996. "Sequence and developmental expression of *AmphiDll*, an amphioxus Distal-less gene transcribed in the ectoderm, epidermis and nervous system: Insights into evolution of craniate forebrain and neural crest." *Development* 122: 2911–2920.

Holland, N. D., et al. 2003. "*AmphiNK2-tin*, an amphioxus homeobox gene expressed in myocardial progenitors: Insights into evolution of the vertebrate heart." *Developmental Biology* 255: 128–137.

Holland, Nicholas D., et al. 2005. "'Lophenteropneust' hypothesis refuted by collection and photos of new deep-sea hemichordates." *Nature* 434: 374–376.

Holland, Nicholas D., Paris, Mathilde, and Koop, Demian. 2009. "The club-shaped gland of amphioxus: Export of secretion to the pharynx in pre-metamorphic larvae and apoptosis during metamorphosis." *Acta Zoologica* 90: 372–379.

Holland, P. W. 2011. *The Animal Kingdom—A Very Short Introduction*. Oxford: Oxford University Press.

Holland, P. W. H., Garcia-Fernàndez, J., Williams, N. A., and Sidow, A. 1994. "Gene duplications and the origins of vertebrate development." *Development Supplement* 125–133.

Holland, P. W. H., Holland, L. Z., Williams, N. A., and Holland, N. D. 1992. "An amphioxus homeobox gene: Sequence conservation, spatial expression during development and insights into vertebrate evolution." *Development* 116: 653–661.

Holland, P. W. H., Koschorz, B., Holland, L. Z., and Herrmann, B. G. 1995. "Conservation of *Brachyury* (T) genes in amphioxus and vertebrates: Developmental and evolutionary implications." *Development* 121: 4283–4291.

Horie, Takeo, et al. 2011. "Ependymal cells of chordate larvae are stem-like cells that form the adult nervous system." *Nature* 469: 525–528.

Hornstein, Eran, and Tabin, Clifford J. 2005. "Asymmetrical threat averted." *Nature* 435: 155–156.

Hosp, Julia, Sagane, Yoshimasa, Danks, Gemma, and Thompsson, Eric M. 2012. "The evolving proteome of a complex extracellular matrix, the *Oikopleura* house." *PLOS ONE* 7: e40172. doi:10.1371/journal.pone.0040172.

Hou, Xian-Guang. 1987. "Early Cambrian large bivalved arthropods from Chengjiang, eastern Yunnan." *Acta Palaeontologica Sinica* 26: 286–298.

Hou, Xian-Guang, Bergström, J., Ma, Xiao-Ya, and Zhao, Je. 2006. "The Lower Cambrian *Phlogites* Luo & Hu re-considered." *GFF* 128: 47–51.

Hou, Xian-Guang, et al. 2011. "An early Cambrian hemichordate zooid." *Current Biology* 21: 612–616.

Huang, Shengfeng, et al. 2008. "Genomic analysis of the immune gene repertoire of amphioxus reveals extraordinary innate complexity and diversity." *Genome Research* 18: 1112–1126.

Hubaud, Alexis, and Pourquié, Olivier. 2014. "Signalling dynamics in vertebrate segmentation." *Nature Reviews Molecular Cell Biology* 15: 709–721.

Hull, David L. 1988. *Science as a Process: An Evolutionary Account of the Social and Conceptual Development of Science*. Chicago: University of Chicago Press.

Hunt, Paul, et al. 1991. "A distinct Hox code for the branchial region of the vertebrate head." *Nature* 353: 861–864.

Huxley, T. H. 1851. "Observations upon the anatomy and physiology of *Salpa* and *Pyrosoma*." *Philosophical Transactions of the Royal Society of London* 141: 567–593.

Ichimura, Koichiro, and Sakai, Tatsuo. 2015. "Evolutionary morphology of podocytes and primary urine-producing apparatus." *Anatomical Science International* doi:10.1007/s12565-015-0317-7.

Ikuta, Tetsuro, and Saiga, Hidetoshi. 2007. "Dynamic change in the expression of developmental genes in the ascidian central nervous system: Revisit to the tripartite model and the origin of the midbrain-hindbrain boundary region." *Developmental Biology* 312: 631–643.

Ikuta, Tetsuro, Satoh, Nori, and Saiga, Hidestoshi. 2010. "Limited functions of Hox genes in the larval development of the ascidian *Ciona intestinalis*." *Development* 137: 1505–1513.

Ikuta, Tetsuro, Yoshida, Natsue, Satoh, Nori, and Saiga, Hidetoshi. 2004. "*Ciona intestinalis* Hox gene cluster: Its dispersed structure and residual colinear expression in development." *Proceedings of the National Academy of Science of the United States of America* 101: 15188–15123.

Imai, Kaoru S. 2003. "Isolation and characterization of β-catenin downstream genes in early embryos of the ascidian *Ciona savignyi*." *Differentiation* 71: 346–360.

Imai, Kaoru, et al. 2004. "Gene expression profiles of transcription factors and signaling molecules in the ascidian embryo: Towards a comprehensive understanding of gene networks." *Development* 131: 4047–4058.

Imai, Kaoru S., Levine, Michael, Satoh, Nori, and Satou, Yutaka. 2006. "Regulatory blueprint for a chordate embryo." *Science* 312: 1183–1187.

Imai, Kaoru S., Stolfi, Alberto, Levine, Michael, and Satou, Yutaka. 2009. "Gene regulatory networks underlying the compartmentalization of the *Ciona* central nervous system." *Development* 136: 285–293.

Imai, K., Takada, N., Satoh, N., and Satou, Y. 2000. "β-catenin mediates the specification of endoderm cells in ascidian embryos." *Development* 127: 3009–3020.

Irie, Naoki, and Kuratani, Shigeru. 2011. "Comparative transcriptome analysis reveals vertebrate phylotypic period during organogenesis." *Nature Communications* 2: 248. doi:10.1038/ncomms1238.

———. 2014. "The developmental hourglass model: A predictor of the basic body plan?" *Development* 141: 4649–4655.

Irimia, Manuel, et al. 2010. "Conserved developmental expression of FezF in chordates and *Drosophila* and the origin of the Zona Limitans Intrathalamica (ZLI) brain organizer." *EvoDevo* 1:7.

Israelsson, Olle. 1997. ". . . and molluskan embryogenesis." *Nature* 390: 32.

———. 1999. "New light on the enigmatic *Xenoturbella* (phylum uncertain): Ontogeny and phylogeny." *Proceedings of the Royal Society of London B* 266. doi:10.1098/rspb.1999.0713.

Israelsson, O., and Budd, G. E. 2005. "Eggs and embryos of *Xenoturbella* (phylum uncertain) are not ingested prey." *Development Genes and Evolution* 215: 358–363.

Ivanova-Kazas, O. M. 2007. "On the problem of the origin of Pogonophora." *Russian Journal of Marine Biology* 33: 228–342.

Ivashkin, Evgeniy, and Adameyko, Igor. 2013. "Progenitors of the protochordate ocellus as an evolutionary origin of the neural crest." *EvoDevo* 4:12. doi:10.1186/2041-9139-4-12.

Jacobson, Antone G. 1988. "Somitomeres: Mesodermal segments of vertebrate embryos." *Development* 104: 209–220.

Jaillon, Olivier, et al. 2004. "Genome duplication in the teleost fish *Tetraodon nigroviridis* reveals the early vertebrate proto-karyotype." *Nature* 431: 946–957.

Jandzik, David, et al. 2015. "Evolution of the new vertebrate head by co-option of an ancient chordate skeletal tissue." *Nature* 518: 534–537.

Janvier, Philippe. 1981. "The phylogeny of the Craniata, with particular reference to the significance of the fossil 'agnathans.'" *Journal of Vertebrate Paleontology* 1: 121–159.

———. 1996. *Early Vertebrates*. Oxford: Oxford University Press.

———. 2003. "Skeleton keys." *Nature* 424: 493.

———. 2006. "Modern look for ancient lamprey." *Nature* 443: 921–924.

———. 2013. "Inside-out turned upside-down." *Nature* 502: 457–458.

———. 2015. "Facts and fancies about early fossil chordates and vertebrates." *Nature* 520: 483–489.

Jefferies, R. P. S. 1967. "Some fossil chordates with echinoderm affinities." *Symposia of the Zoological Society of London* 20: 163–208.

———. 1973. "The Ordovician fossil *Lagynocystis pyramidalis* (Barrande) and the ancestry of amphioxus." *Philosophical Transactions of the Royal Society of London B* 265: 409–469.

———. 1986. *The Ancestry of the Vertebrates*. London: British Museum (Natural History).

———. 1990. "The solute *Dendrocystoides scoticus* from the Upper Ordovician of Scotland and the ancestry of chordates and echinoderms." *Palaeontology* 33: 631–679.

———. 1991. "Two types of bilateral symmetry in the Metzoa: Chordate and bilaterian." In *Biological Asymmetry and Handedness*, edited by G. R. Bock and J. Marsh, 94–127. Wiley: Chichester.

———. 1997. "A defence of the calcichordates." *Lethaia* 30: 1–10.

Jeffery, William R. 2007. "Chordate ancestry of the neural crest: New insights from ascidians." *Seminars in Cell and Developmental Biology* 18: 481–491.

Jeffery, William R., Strickler, Allen G., and Yamamoto, Yoshiyuki. 2004. "Migratory neural crest-like cells form body pigmentation in a urochordate embryo." *Nature* 431: 696–699.

Jeffery, William R., and Swalla, Billie J. 1992. "Evolution of alternate modes of development in ascidians." *Bioessays* 14: 219–226.

Jeffery, William R., et al. 2008. "Trunk lateral cells are neural crest-like cells in the ascidian *Ciona intestinalis*: Insights into the ancestry and evolution of the neural crest." *Developmental Biology* 324: 152–160.

Jiang, Di, and Smith, William C. 2007. "Ascidian notochord morphogenesis." *Developmental Dynamics* 236: 1748–1757.

Jondelius, U., Ruiz-Trillo, I., Baguñà, J., and Riutort, M. 2002. "The Nemertodermatida are basal bilaterians and not members of the Platyhelminthes." *Zoologica Scripta* 31: 201–215.

Kaji, Takao, et al. 2016. "Amphioxus mouth after dorso-ventral inversion." *Zoological Letters* 2. doi:10.1186/s40851-016-0038-3.

Kalinka, Alex T., et al. 2010. "Gene expression divergence recapitulates the developmental hourglass model." *Nature* 468: 811–814.

Kaplan, Nicole, Razy-Krajka, Florian, and Christiaen, Lionel. 2015. "Regulation and evolution of cardiopharyngeal cell identity and behavior: Insights from simple chordates." *Current Opinion in Genetics and Development* 32: 119–128.

Kassahn, Karin S., et al. 2009. "Evolution of gene function and regulatory control after whole-genome duplication: Comparative analyses in vertebrates." *Genome Research* 19: 1404–1418.

Kaul, S., and Stach, T. 2010. "Ontogeny of the collar cord: Neurulation in the hemichordate *Saccoglossus kowalevskii*." *Journal of Morphology* 271: 1240–1259.

Kaul-Strehlow, S., and Röttinger, E. 2015. "Hemichordata." In *Evolutionary Developmental Biology of Invertebrates*, vol. 6: *Deuterostomia*, edited by A. Wanninger, 59–89. Vienna: Springer.

Kaul-Strehlow, Sabrina, and Stach, Thomas. 2011. "The pericardium in the deuterostome *Saccoglossus kowalevskii* (Enteropneusta) develops from the ectoderm by schizocoely." *Zoomorphology* 130: 107–120.

Kawasaki, Kazuhiko, Suzuki, Tohru, and Weiss, Kenneth M. 2004. "Genetic basis for the evolution of vertebrate mineralized tissue." *Proceedings of the National Academy of Sciences of the United States of America* 101: 11356–11361.

Kawasaki, Kazuhiko, and Weiss, Kenneth M. 2003. "Mineralized tissue and vertebrate evolution: The secretory calcium-binding phosphoprotein gene cluster." *Proceedings of the National Academy of Sciences of the United States of America* 100: 4060–4065.

Kessel, Michael, and Gruss, Peter. 1991. "Homeotic transformations of murine vertebrae and concomitant alteration of Hox codes induced by retinoic acid." *Cell* 67: 89–104.

Khalturin, Konstantin, and Bosch, Thomas G. 2007. "Self/nonself discrimination at the basis of chordate evolution: Limits on molecular conservation." *Current Opinion in Immunology* 19: 4–9.

Kim, S. K., Hebrok, M., and Melton, D. A. 1997. "Notochord to endoderm signaling is required for pancreas development." *Development* 124: 4243–4252.

King, Nicole, et al. 2008. "The genome of the choianoflagellate *Monosiga brevicollis* and the origin of metazoans." *Nature* 451: 783–788.

Kispert, A., Hermann, B. G., Leptin, M., and Reuter, R. 1994. "Homologs of the mouse *Brachyury* gene are involved in the specification of posterior terminal structures in *Drosophila*, *Tribolium*, and *Locusta*." *Genes and Development* 8: 2137–2150.

Kourakis, Matthew J., and Smith, William C. 2005. "Did the first chordates organize without the organizer?" *Trends in Genetics* 21: 506–510.

Kowalevsky, A. 1866a. "*Anatomie des Balanoglossus*." *Mémoires de l'Académie Impériale de St. Pétersbourg* 7th series 10 (2): 1–18.

———. 1866b. "*Entwicklungsgeschichte der einfachen Ascidien*." *Mémoires de l'Académie des Sciences de St Pétersbourg* 7th series 10 (15): 1–19.

———. 1871. "*Weitere Studien über die Entwickelung der einfachen Ascidien*." *Archiv fur mikroskopische Anatomie (Bonn)* 7: 101–130.

Kozmik, Zbynek, et al. 1999. "Characterization of an amphioxus paired box gene, AmphiPax2/5/8: Developmental expression patterns in optic support cells, nephridium, thyroid-like structures and pharyngeal gill slits, but not in the midbrain-hindbrain boundary region." *Development* 126: 1295–1304.

Kristensen, Reinhardt M. 2002. "An introduction to Loricifera, Cycliophora and Micrognathozoa." *Integrative and Comparative Biology* 42: 641–651.

Kugler, Jamie E., et al. 2011. "Evolutionary changes in the notochord genetic toolkit: A comparative analysis of notochord genes in the ascidian *Ciona* and the larvacean *Oikopleura*." *BMC Evolutionary Biology* 11: 21. doi:10.1186/1471-2148-11-21.

Kumano, Gaku, and Nishida, Hiroki. 2007. "Ascidian embryonic development: An emerging model system for the study of cell fate specification in chordates." *Developmental Dynamics* 236: 1732–1747.

Kuo, Bryan R., and Erickson, Carol A. 2011. "Vagal neural crest cell migratory behaviour: a transition between the cranial and trunk crest." *Developmental Dynamics* 240: 2084–2100.

Kurabayashi, Atsushi, et al. 2003. "Phylogenetic position of a deep-sea ascidian, *Megalodicopia hians*, inferred from the molecular data." *Zoological Science* 20: 1243–1247.

Kuraku, Shigehiro, Meyer, Axel, and Kuratani, Shigeru. 2008. "Timing of genome duplications relative to the origin of the vertebrates: Did cyclostomes diverge before or after?" *Molecular Biology and Evolution* 26: 47–59.

Kuraku, Shigehiro, et al. 1999. "Monophyly of lampreys and hagfishes supported by nuclear DNA-coded genes." *Journal of Molecular Evolution* 49: 729–735.

Kuratani, Shigeru. 2008. "Is the vertebrate head segmented?—Evolutionary and developmental considerations." *Integrative and Comparative Biology* 48: 647–657.

Kuratani, Shigeru, and Adachi, Noritaka. 2016. "What are head cavities? History of studies on the vertebrate head segmentation." *Zoological Science* 33: 213–228.

Kuratani, Shigeru, and Hirasawa, Tatsuya. 2016. "Getting the measure of a monster." *Nature* 532: 447–448.

Kuratani, Shigeru, Horigome, Naoto, and Horano, Shigeki. 1999. "Developmental morphology of the head mesoderm and reevaluation of segmental theories of the vertebrate head: Evidence from embryos of an agnathan vertebrate, *Lampetra japonica*." *Developmental Biology* 210: 381–400.

Kuratani, Shigeru, Oisi, Yasuhiro, and Ota, Kinya. 2016. "Evolution of the vertebrate cranium." *Zoological Science* 33: 229–238.

Kuratani, Shigeru, and Schilling, Thomas. 2008. "Head segmentation in vertebrates." *Integrative and Comparative Biology* 48: 604–610.

Lacalli, Thurston C. 1996a. "Landmarks and subdomains in the larval brain of Branchiostoma: Vertebrate homologs and invertebrate antecedents." *Israel Journal of Zoology* 42 (supplement 1): S131–S146.

———. 1996b. "Frontal eye circuitry, rostral sensory pathways and brain organization in amphioxus larvae: Evidence from 3D reconstructions." *Philosophical Transactions of the Royal Society B* 351: 243–263.

———. 2002a. "The dorsal compartment locomotory control system in amphioxus larvae." *Journal of Morphology* 252: 227–237.

———. 2002b. "Vetulicolians—are they deuterostomes? chordates?" *BioEssays* 24: 208–211.

———. 2005. "Protochordate body plan and the evolutionary role of larvae: Old controversies resolved?" *Canadian Journal of Zoology* 83: 216–224.

———. 2008a. "Mucus secretion and transport in amphioxus larvae: Organization and ultrastructure of the food trapping system, and implications for head evolution." *Acta Zoologica* 89: 219–230.

———. 2008b. "Basic features of the ancestral chordate brain: A protochordate perspective." *Brain Research Bulletin* 75: 319–323.

Lacalli, T. C., Holland, N. D., and West, J. E. 1994. "Landmarks in the anterior central nervous system of amphioxus larvae." *Philosophical Transactions of the Royal Society of London B* 344. doi:10.1098/rstb.1994.0059.

Lacalli, Thurston C., and Kelly, Samantha. 2000. "The infundibular balance organ in amphioxus larvae and related aspects of cerebral vesicle organization." *Acta Zoologica* 81: 37–47.

Lake, J. A. 1990. "Origin of the metazoa." *Proceedings of the National Academy of Sciences of the United States of America* 87: 763–766.

Lambert, G., Lambert, C. C., and Lowenstam, H. A. 1989. "Protochordate biomineralization." In *Skeletal Biomineralization: Patterns, Processes and Evolutionary Trends*, edited by J. G. Carter, 165–173. Washington, DC: American Geophysical Union. doi:10.1029/SC005p0165.

Lankester, E. R., and Willey, A. 1890. "The development of the atrial chamber of amphioxus." *Quarterly Journal of Microscopical Science* 31: 445–466.

Lapraz, François, Haillot, Emmaneul, and Lepage, Thierry. 2015. "A deuterostome origin of the Spemann organiser suggested by Nodal and ADMPs functions in echinoderms." *Nature Communications* 6. doi:10.1038/ncomms9434.

Larroux, Claire, et al. 2007. "The KH homeobox gene cluster predates the origin of hox genes." *Current Biology* 17: 706–710.

Larsen, William J. 1998. *Essentials of Human Embryology*. New York: Churchill Livingstone.

Lauri, A., et al. 2014. "Development of the annelid axochord: Insights into notochord evolution." *Science* 345: 1365–1368.

Lavrov, Dennis V., Brown, Wesley M., and Boore, Jeffrey L. 2004. "Phylogenetic position of the Pentastomida and (pan)crustacean relationships." *Proceedings of the Royal Society of London B* 271. doi:10.1098/rspb.2003.2631.

Lebar, Matthew D., et al. 2010. "Accumulation of vanadium, maganese, and nickel in Antarctic tunicates." *Polar Biology* 34: 587–590.

Le Douarin, Nicole M. 2012. "Piecing together the vertebrate skull." *Development* 139: 4293–4296.

Lemaire, Patrick. 2009. "Unfolding a chordate developmental program, one cell at a time: Invariant cell lineages, short-range inductions and evolutionary plasticity in ascidians." *Developmental Biology* 332: 48–60.

———. 2011. "Evolutionary crossroads in developmental biology: The tunicates." *Development* 138: 2143–2152.

Lemaire, Patrick, Smith, William C., and Nishida, Hiroki. 2008. "Ascidians and the plasticity of the chordate developmental program." *Current Biology* 18: R620–R631.

Lemons, D., and McGinnis, William. 2006. "Genomic evolution of Hox gene clusters." *Science* 313: 1918–1922.

Lemons, Derek, et al. 2010. "Co-option of an anteroposterior head axis patterning system for proximodistal patterning of appendages in early bilaterian evolution." *Developmental Biology* 344: 358–362.

Lescroart, Fabienne, et al. 2015. "Clonal analysis reveals a common origin between nonsomite-derived neck muscles and heart myocardium." *Proceedings of the National Academy of Sciences of the United States of America* 112: 1446–1451.

Lester, S. M. 1985. "*Cephalodiscus* sp. (Hemichordata: Pterobranchia): Observations of functional morphology, behavior and occurrence in shallow water around Bermuda." *Marine Biology* 85: 263–268.

Leuzinger, S., et al. 1998. "Equivalence of the fly orthodenticle gene and the human OTX genes in embryonic brain development of *Drosophila*." *Development* 125: 1703–1710.

Levin, Michael, Johnson, Randy L., Sterna, Claudio D., Kuehn, Michael, and Tabin, Cliff. 1995. "A molecular pathway determining left-right asymmetry in chick embryogenesis." *Cell* 82: 803–814.

Lewis, E. B. 1978. "A gene complex controlling segmentation in *Drosophila*." *Nature* 276: 565–570.

Li, Huiliang, and Richardson, William D. 2016. "Evolution of the CNS myelin gene regulatory program." *Brain Research* 1641A: 111–121.

Liberti, Assunta, et al. 2015. "An immune effector system in the protochordate gut sheds light on fundamental aspects of vertebrate immunity." In *Pathogen-Host Interactions: Antigenic Variation v. Somatic Adaptations*. Vol. 57, *Results and Problems in Cell Differentiation*, edited by E. Hsu and L. Du Pasquire, 159–173. Springer.

Litman, Gary W., Rast, Jonathan P., and Fugmann, Sebastian D. 2010. "The origins of vertebrate adaptive immunity." *Nature Reviews Immunology* 10: 543–553.

Liu, Jianni, and Dunlop, Jason. 2014. "Cambrian lobopodians: A review of recent progress in our understanding of their morphology and evolution." *Palaeogeography, Palaeoclimatology, Palaeoecology* 398: 4–15.

Livingston, B. T., et al. 2006. "A genome-wide analysis of biomineralization-related proteins in the sea urchin *Strongylocentrotus purpuratus*." *Developmental Biology* 300:335–348.

Logan, C. Y., Miller, J. R., Ferkowicz, M. J., and McClay, D. R. 1999. "Nuclear beta-catenin is required to specify vegetal cell fates in the sea urchin embryo." *Development* 126: 345–357.

Long, John A., Trinajstic, Kate, and Johanson, Zerina. 2009. "Devonian arthrodire embryos and the origin of internal fertilization in vertebrates." *Nature* 457: 1124–1127.

Long, John A., Trinajstic, Kate, Young, Gavin C., and Senden, Tim. 2008. "Live birth in the Devonian period." *Nature* 453: 650–652.

Long, John A., et al. 2015. "Copulation in antiarch placoderms and the origin of gnathostome internal fertilization." *Nature* 517: 196–199.

Long, Suzanne, and Byrne, Maria. 2001. "Evolution of the echinoderm Hox gene cluster." *Evolution and Development* 3: 302–311.

Lowe, Christopher J., and Pani, Ariel M. 2011. "A soap opera of unremarkable worms." *Current Biology* 21: R151–R153.

Lowe, Christopher J., et al. 2003. "Anteroposterior patterning in hemichordates and the origins of the chordate nervous system." *Cell* 113: 853–865.

Lowe, Christopher J., et al. 2006. "Dorsoventral patterning in hemichordates: Insights into early chordate evolution." *PLOS Biology*. doi:10.1371/journal.pbio.0040291.

Lowe, Christopher J., et al. 2015. "The deuterostome context of chordate origins." *Nature* 520: 456–465.

Lundin, Kennet. 1998. "The epidermal ciliary rootlets of *Xenoturbella bocki* (Xenoturbellida) revisited: New support for a possible kinship with the Acoelomorpha (Platyhelminthes)." *Zoologica Scripta* 27: 263–270.

Luo, Huilin, Hu, Shixue, and Chen, Liangzhong. 2001. "New Early Cambrian chordates from Haikou, Kunming." *Acta Geologica Sinica* 75: 345–348.

Lüter, Carsten. 2000. "The origin of the coelom in Brachiopoda and its phylogenetic significance." *Zoomorphology* 120: 15–28.

Luttrell, S., et al. 2012. "Ptychoderid hemichordate neurulation without a notochord." *Integrative and Comparative Biology* 52: 829–834.

Malicki, Jarema, Cianetti, Luciano C., Peschle, Cesare, and McGinnis, William. 1992. "A human HOX4B regulatory element provides head-specific expression in *Drosophila* embryos." *Nature* 358: 345–347.

Mallatt, Jon, and Chen, Jun-Yuan. 2003. "Fossil sister group of craniates: Predicted and found." *Journal of Morphology* 258: 1–31.

Mallatt, J., and Sullivan, J. 1998. "28S and 18S rDNA sequences support the monophyly of lampreys and hagfishes." *Molecular Biology and Evolution* 15: 1706–1718.

Mallet, James. 2007. "Hybrid speciation." *Nature* 446: 279–283.

Manigai, Erica K. O., Kenny, Nathan J., and Shimeld, Sebastian M. 2014. "Right across the tree of life: The evolution of left-right asymmetry in the Bilateria." *Genesis* 52: 458–470.

Manni, Lucia, Agnoletto, Alberto, Zaniolo, Giovanna, and Burighel, Paolo. 2005. "Stomodeal and neurohypophysial placodes in *Ciona intestinalis*: Insights into the origin of the pituitary gland." *Journal of Experimental Zoology* 304B. doi:10.1002/jez.b.21039.

Marlétaz, F., et al. 2006. "Chaetognath phylogenomics: a protostome with deuterostome-like development." *Current Biology* 16: R577–R587.

Marlétaz, F., et al. 2008. "Chaetognath transcriptome reveals ancestral and unique features among bilaterians." *Genome Biology* 9. doi:10.1186/gb-2008-9-6-r94.

Maroto, Miguel, Bone, Robert A., and Dale, Kim. 2012. "Somitogenesis." *Development* 139: 2453–2456.

Martín-Durán, José M. 2012. "Deuterostomic development in the protostome *Priapulus caudatus*." *Current Biology* 22: 2161–2166.

Martín-Durán, José M., Passamaneck, Yale J. , Martindale, Mark Q., and Hejnol, Andreas. 2016. "The developmental basis for the recurrent evolution of deuterostomy and protostomy." *Nature Ecology and Evolution* 1. doi:10.1038/s41559-016-0005.

Martinez, Pedro, et al. 1999. "Organization of an echinoderm Hox gene cluster." *Proceedings of the National Academy of Sciences of the United States of America* 96: 1469–1474.

Matsumoto, H. 1929. "Outline of a classification of the echinodermata." *Science Reports of Tohoku University, Sendai (Geology)* 13: 27–33.

Matsuoka, Toshiyuki, et al. 2005. "Neural crest origins of the neck and shoulder." *Nature* 436: 347–355.

Matthysse, Ann G., et al. 2004. "A functional cellulose synthase from ascidian epidermis." *Proceedings of the National Academy of Sciences of the United States of America* 101: 986–991.

Matus, David Q., et al. 2006. "Broad taxon and gene sampling indicate that chaetognaths are protostomes." *Current Biology* 16: R575–R576.

Matus, David Q., Halanych, Kenneth M., and Martindale, Mark Q. 2007. "The Hox gene complement of a pelagic chaetognath, *Flaccisagitta enflata*." *Integrative and Comparative Biology* 47: 854–864.

Maxmen, Amy. 2011. "A can of worms." *Nature* 470: 161–162.

Mazet, Françoise, Hutt, James A., Millard, John, and Shimeld, Sebastian M. 2003. "*Pax* gene expression in the developing central nervous system of *Ciona intestinalis*." *Gene Expression Patterns* 3: 743–745.

Mazet, Françoise, and Shimeld, Sebastian M. 2003. "Characterisation of an amphioxus Fringe gene and the evolution of the vertebrate segmentation clock." *Development Genes and Evolution* 213: 505–509.

———. 2005. "Molecular evidence from ascidians for the evolutionary origin of vertebrate cranial sensory placodes." *Journal of Experimental Zoology* 304B: 340–346.

McCoy, Victoria E., et al. 2016. "The 'Tully monster' is a vertebrate." *Nature* 532: 496–499.

McGinnis, W., et al. 1984. "A conserved DNA sequence in homeotic genes of the *Drosophila* Antennapedia and bithorax complexes." *Nature* 308: 428–433.

McHugh, Damhnait. 1997. "Molecular evidence that echiurans and pogonophorans are derived annelids." *Proceedings of the National Academy of Sciences of the United States of America* 94: 8006–8009.

Mehta, Tarang K., et al. 2013. "Evidence for at least six Hox clusters in the Japanese lamprey (*Lethenteron japonicum*)." *Proceedings of the National Academy of Sciences of the United States of America* 110: 16044–16049.

Meier, S., and Packard, D. S. 1984. "Morphogenesis of the cranial segments and distribution of neural crest in the embryos of snapping turtle, *Chelydra serpentina*." *Developmental Biology* 102: 309–323.

Méndez, Ana T., et al. 2000. "Idenitification of Hox gene sequences in the sea cucumber *Holothuria glaberrima* Selenka (Holothuroidea: Echinodermata.)" *Marine Biotechnology* 2: 231–240.

Metchnikoff, V. E. 1881. "*Über die systematische Stellung von Balanoglossus*." *Zoologischer Anzeiger* 4: 139–157.

Meyer, Axel, and Van de Peer, Yves. 2005. "From 2R to 3R: Evidence for a fish-specific genome duplication (FSGD)." *Bioessays* 27: 937–945.

Miller, David J., and Ball, Eldon E. 2008. "Cryptic complexity captured: The *Nematostella* genome reveals its secrets." *Trends in Genetics* 24: 1–4.

Miller, Douglas. 1988. *Goethe: The Collected Works*. Vol. 12, *Scientific Studies*. Princeton: Princeton University Press.

Mitchell, Charles E., Melchin, Michael J., Cameron, Chris B., and Maletz, Jörg. 2013. "Phylogenetic analysis reveals that *Rhabdopleura* is an extant graptolite." *Lethaia* 46: 34–56.

Mitchell, Emily G., et al. 2015. "Reconstructing the reproductive mode of an Ediacaran macro-organism." *Nature* 524: 343–346. doi:10.1038/nature14646.

Mito, Taro, and Endo, Kazuyoshi. 2000. "PCR survey of Hox genes in the crinoid and ophiuroid: Evidence for anterior conservation and posterior expansion in the echinoderm Hox cluster." *Molecular Phylogenetics and Evolution* 14: 375–388.

Miyamoto, Norio, Nakajima, Yoko, Wada, Hiroshi, and Saito, Yasunori. 2010. "Development of the nervous system in the acorn worm *Balanoglossus simodensis*: Insights into nervous system evolution." *Evolution and Development* 12: 416–424.

Miyamoto, Norio, and Wada, Hiroshi. 2013. "Hemichordate neurulation and the origin of the neural tube." *Nature Communications* 4: 2713. doi:10.1038/ncomms3713.

Miyawaki, Kyoji, et al. 2003. "Nuclear localization of β-catenin in vegetal pole cells during early embryogenesis of the starfish *Asterina pectinifera*." *Development, Growth and Differentiation* 45: 121–128.

Mizutani, Claudia Mieko, and Bier, Ethan. 2008. "EvoD/Vo: The origins of BMP signalling in the neurectoderm." *Nature Reviews Genetics* 9: 663–677.

Molina, M. Dolores, de Crozé, Noémie, Haillot, Emmanuel, and Lepage, Thierry. 2013. "Nodal: Master and commander of the dorsal-ventral and left-right axes in the sea urchin embryo." *Current Opinion in Genetics and Development* 23: 445–453.

Monahan-Earley, R., Dvorak, A. M., and Aird, W. C. 2013. "Evolutionary origins of the blood vascular system and endothelium." *Journal of Thrombosis and Haemostasis* 11: 46–66.

Moret, Frédéric, et al. 2005. "Regulatory gene expressions in the ascidian ventral sensory vesicle: Evolutionary relationships with the vertebrate hypothalamus." *Developmental Biology* 277: 567–579.

Moroz, Leonid L., et al. 2014. "The ctenophore genome and the evolutionary origins of neural systems." *Nature* 510: 109–114.

Müller, W. 1873. "*Über die Hypobranchialrinne der Tunikaten unde den Vorhandsein bei Amphioxus und den Cyclostomen.*" *Zeitschrift Med. Naturwissenschaften* 7: 327–332.

Müller, Werner E. G. 1995. "Molecular phylogeny of metazoa (animals): Monophyletic origin." *Naturwissenschaften* 82: 321–329.

Murdock, Duncan J. E., et al. 2013. "The origin of conodonts and of vertebrate mineralized skeletons." *Nature* 502: 546–549.

Musser, Melissa A., and Southard-Smith, E. Michelle. 2013. "Balancing on the crest—evidence for disruption of the enteric ganglia via inappropriate lineage segregation and consequences for gastrointestinal function." *Developmental Biology* 382: 356–364.

Nakano, Hiroaki, Nakajima, Yoko, and Amemiya, Shonan. 2009. "Nervous system development of two crinoid species, the sea lily *Metacrinus rotundus* and the feather star *Oxycomanthus japonicus*." *Development Genes and Evolution* 219: 565–576.

Nakano, H., et al. 2013. "*Xenoturbella bocki* exhibits direct development with similarities of Aceolomorpha." *Nature Communications* 4. doi:10.1038/ncomms2556.

Nakashima, Keisuke, et al. 2004. "The evolutionary origin of animal cellulose synthase." *Development Genes and Evolution* 214: 81–88.

Nanglu, Karma, Caron, Jean-Bernard, Conway Morris, Simon, and Cameron,

Christopher B. 2016. "Cambrian suspension-feeding tubicolous hemichordates." *BMC Biology*. doi:10.1186/s12915-016-0271-4.

Narbonne, Guy M. 2005. "The Ediacara biota: Neoproterozoic origin of animals and their ecosystems." *Annual Review of Earth and Planetary Sciences* 33: 421–442.

Newgreen, Donald F., Dufour, Sylvie, Howard, Marthe J., and Landman, Kerry A. 2013. "Simple rules for a 'simple' nervous system? Molecular and biomathematical approaches to enteric nervous system formation and malformation." *Developmental Biology* 382: 305–319.

Nibu, Yutaka, José-Edwards, Diana S., and Di Gregorio, Anna. 2013. "From notochord formation to hereditary chordoma: The many roles of *Brachyury*." *BioMed Research International*. doi:10.1155/2013/826435.

Nichols, D. 1967. *Echinoderms*. 3rd ed. London: Hutchinson University Library.

Niehrs, Christof. 2004. "Regionally specific induction by the Spemann-Mangold Organizer." *Nature Reviews Genetics* 5: 425–434.

———. 2010. "On growth and form: A Cartesian coordinate system of *Wnt* and BMP signaling specifies bilaterian body axes." *Development* 127: 845–857.

Nielsen, Claus. 2010. "After all: *Xenoturbella* is an acoelomorph!" *Evolution and Development* 12: 241–243.

Noden, Drew M., and Trainor, Paul A. 2005. "Relations and interactions between cranial mesoderm and neural crest populations." *Journal of Anatomy* 207: 575–601.

Nomaksteinsky, Marc, et al. 2009. "Centralization of the deuterostome nervous system predates chordates." *Current Biology* 19: 1264–1269.

Nomaksteinsky, Marc, et al. 2013. "Ancient origin of somatic and visceral neurons." *BMC Biology* 11. doi:10.1186/1741-7007-11-53.

Norén, Michael, and Jondelius, Ulf. 1997. "*Xenoturbella*'s molluskan relatives . . ." *Nature* 390: 31–32.

Nosenko, Tetyana, et al. 2013. "Deep metazoan phylogeny: When different genes tell different stories." *Molecular Phylogenetics and Evolution* 67: 223–233.

Nübler-Jung, K., and Arendt, D. 1994. "Is ventral in insects dorsal in vertebrates?" *Roux's Archives in Developmental Biology* 203: 357–366.

———. 1999. "Dorsoventral axis inversion: Enteropneust anatomy links invertebrates to chordates turned upside down." *Journal of Zoological Systematics and Evolutionary Research* 37: 93–100.

Nüsslein-Volhard, Christiane, and Wieschaus, Eric. 1980. "Mutations affecting segment number and polarity in *Drosophila*." *Nature* 287: 795–801.

Oda-Ishii, Izumi, et al. 2005. "Making very similar embryos with divergent genomes: Conservation of regulatory mechanisms of *Otx* between the ascidians *Halocynthia roretzi* and *Ciona intestinalis*." *Development* 132: 1663–1674.

Ogasawara, M., Wada, H., Peters, H., and Satoh, N. 1999. "Developmental expression of *Pax1/9* genes in urochordate and hemichordate gills: Insight into function and evolution of the pharyngeal epithelium." *Development* 126: 2539–2550.

Ohno, Susumu. 1999. "Gene duplication and the uniqueness of vertebrate genomes circa 1970–1999." *Seminars in Cell and Developmental Biology* 10: 517–522.

Oisi, Yasuhiro, et al. 2013. "Craniofacial development of hagfishes and the evolution of vertebrates." *Nature* 493: 175–180.

Olsson, Ragnar, Yulis, Roberto, and Rodrígues, Estéban M. 1994. "The infundibular organ of the lancelet (*Branchiostoma lanceolatum*, Acrania): An immunocytochemical study." *Cell and Tissue Research* 277: 107–114.

Onai, Takayuki, et al. 2009. "Retinoic acid and *Wnt*/β-catenin have complementary roles in anterior/posterior patterning embryos of the basal chordate amphioxus." *Developmental Biology* 332: 223–233.

Onai, Takayuki, et al. 2010. "Opposing Nodal/Vg1 and BMP signals mediate axial patterning in embryos of the basal chordate amphioxus." *Developmental Biology* 344: 377–389.

Onai, Takayuki, et al. 2015a. "Ancestral mesodermal reorganization and evolution of the vertebrate head." *Zoological Letters* 1:29. doi:10.1186/s40851-015-0030-3.

———. 2015b. "On the origin of vertebrate somites." *Zoological Letters* 1:33. doi:10.1186/s40851-015-0033-0.

Onai, Takayuki, Irie, Naoki, and Kuratani, Shigeru. 2014. "The evolutionary origin of the vertebrate body plan: The problem of head segmentation." *Annual Review of Genomics and Human Genetics* 15: 21.1–21.17.

Ono, Hiroki, Kozmik, Zbynek, Yu, Jr-Kai, and Wada, Hiroshi. 2014. "A novel N-terminal motif is responsible for the evolution of neural crest-specific gene-regulatory activity in vertebrate FoxD3." *Developmental Biology* 385: 396–404.

Osborn, Karen J., et al. 2012. "Diversification of acorn worms (Hemichordata, Enteropneusta) revealed in the deep sea." *Proceedings of the Royal Society of London B* 279: 1646–1654.

Ota, Kinya G., Kuraku, Shigehiro, and Kuratani, Shigeru. 2007. "Hagfish embryology with reference to the evolution of the neural crest." *Nature* 446: 672–675.

Ota, Kinya G., Fujimoto, Satoko, Oisi, Yasuhiro, and Kuratani, Shigeru. 2011. "Identification of vertebra-like elements and their possible differentiation from sclerotomes in the hagfish." *Nature Communications* 2. doi:10.1038/ncomms1355.

Ou, Qiang, et al. 2012. "Evidence for gill slits and a pharynx in Cambrian vetulicolians: Implications for the early evolution of deuterostomes." *BMC Biology* 10: 10.1186/1741-7007-10-81.

Owen, Richard. 2007. *On the Nature of Limbs*. Edited by R. Amundson. Chicago: University of Chicago Press.

Palmer, D. A., and Rickards, B. A. 1991. *Graptolites*. Woodbridge, England: Boydell and Brewer.

Pancer, Z., et al. 2004. "Somatic diversification of variable lymphocyte receptors in the agnathan sea lamprey." *Nature* 430: 174–180.

Pancer, Z., et al. 2005. "Variable lymphocyte receptors in hagfish." *Proceedings of the National Academy of Sciences of the United States of America* 102: 9224–9229.

Pani, Ariel M., et al. 2012. "Ancient deuterostome origins of vertebrate brain signalling centres." *Nature* 483: 289–294.

Papaioannou, Virginia E., and Silver, Lee M. 1998. "The T-box gene family." *BioEssays* 20: 9–19.

Papillon, Daniel, et al. 2003. "*Hox* gene survey in the chaetognath *Spadella cephaloptera*: evolutionary implications." *Development, Genes and Evolution* 213: 142–148.

Papillon, D., Perez, Yvon, Caubit, Xavier, and Le Parco, Yannick. 2004. "Identification of chaetognaths as protostomes is supported by the analysis of their mitochondrial genome." *Molecular Biology and Evolution* 21: 2122–2129.

Paps, Jordi, Baguñà, Jaume, and Riutort, Marta. 2009. "Bilaterian phylogeny: A broad sampling of 13 nuclear genes provides a new lophotrochozoa phylogeny and supports a paraphyletic basal acoelomorpha." *Molecular Biology and Evolution* 26: 2397–2406.

Paps, Jordi, et al. 2013. "Molecular phylogeny of Unikonts: New insights into the position of apusomonads and ancyromonads and the internal relationships of opisthokonts." *Protist* 164: 2–12.

Paris, Mathilde, et al. 2008b. "The amphioxus genome enlightens the evolution of the thyroid hormone signaling pathway." *Development Genes and Evolution* 218: 667–680.

Pascual-Anaya, Juan, et al. 2013. "The evolutionary origins of chordate hematopoiesis and vertebrate endothelia." *Developmental Biology* 375: 182–192.

Passamaneck, Yale, Hadjantonakis, Anna-Katerina, and Di Gregorio, Anna. 2007. "Dynamic and polarized muscle cell behaviors accompany tail morphogenesis in the ascidian *Ciona intestinalis*." *PLOS ONE* 2: e714. doi:10.1371/journal.pone.0000714.

Passamaneck, Yale, and Halanych, Kenneth M. 2006. "Lophotrochozoan phylogeny assessed with LSU and SSU data: Evidence of lophophorate polyphyly." *Molecular Phylogenetics and Evolution* 40: 20–28.

Patthey, Cedric, Schlosser, Gerhard, and Shimeld, Sebastian M. 2014. "The evolutionary history of vertebrate cranial placodes, I: Cell type evolution." *Developmental Biology* 389: 82–97.

Pedersen, Knud Jørgen, and Pedersen, Lars Ryde. 1986. "Fine structural observations on the extracellular matrix (ECM) of *Xenoturbella bocki* Westblad, 1949." *Acta Zoologica* 67: 103–113.

———. 1988. "Ultrastructural observations on the epidermis of *Xenoturbella bocki* Westblad, 1949; with a discussion of epidermal cytoplasmic filament systems of invertebrates." *Acta Zoologica* 69: 231–246.

Perseke, Marleen, et al. 2007. "The mitochondrial DNA of *Xenoturbella bocki*: Genomic architecture and phylogenetic analysis." *Theory in Biosciences* 126: 35–42.

Peterson, Kevin J. 1995. "A phylogenetic test of the calcichordate scenario." *Lethaia* 28: 25–38.

Peterson, Kevin J., and Eernisse, Douglas J. 2001. "Animal phylogeny and the ancestry of bilaterians: Inferences from morphology and 18S rDNA gene sequences." *Evolution and Development* 3: 170–205.

Peterson, Kevin J., Harada, Yoshito, Cameron, R. Andrew, and Davidson, Eric H. 1999a. "Expression pattern of *Brachyury* and *Not* in the sea urchin: Comparative implications for the origins of mesoderm in the basal deuterostomes." *Developmental Biology* 207: 419–431.

Peterson, Kevin J., et al. 1999b. "A comparative molecular approach to mesodermal patterning in basal deuterostomes: The expression pattern of *Brachyury* in the enteropneust hemichordate *Ptychodera flava*." *Development* 126: 85–95.

Peterson, Kevin J., et al. 2004. "Estimating metazoan divergence times with a molecular clock." *Proceedings of the National Academy of Science of the United States of America* 101: 6536–6541.

Philippe, Hervé, Lartillot, Nicolas, and Brinkmann, Henner. 2005. "Multigene analysis of bilaterian animals corroborate the monophyly of Ecdysozoa, Lophotrochozoa and Protostomia." *Molecular Biology and Evolution* 22: 1246–1253.

Philippe, H., et al. 2007. "Acoel flatworms are not platyhelminthes: Evidence from phylogenomics." *PLOS ONE* 2: e717. doi.org/10.1371/journal.pone.0000717.

Philippe, Hervé, et al. 2011. "Aceolomorph flatworms are deuterostomes related to *Xenoturbella*." *Nature* 470: 255–258.

Pisani, Davide, et al. 2015. "Genomic data do not support comb jellies as the sister group to all other animals." *Proceedings of the National Academy of Sciences of the United States of America*. doi/10.1073/pnas.1518127112.

Presnell, Jason S., et al. 2016. "The presence of a functionally tripartite through-gut in Ctenophora has implications for metazoan character trait evolution." *Current Biology* 26: 2814–2820. doi:10.1016/j.cub.2016.08.019.

Prud'homme, Benjamin, and Gompel, Nicolas. 2010. "Genomic hourglass." *Nature* 468: 768–769.

Putnam, Nicholas H., et al. 2007. "Sea anemone genome reveals ancestral Eumetazoan gene repertoire and genomic organization." *Science* 317: 86–94.

———. 2008. "The amphioxus genome and the evolution of the chordate karyotype." *Nature* 453: 1064–1071.

Quiring, R., Walldorf, U., Kloter, U., and Gehring, W. J. 1994. "Homology of the *eye-*

less gene of *Drosophila* to the *Small eye* gene in mice and Aniridia in humans." *Science* 265: 785–789.

Rahman, Imran A., Zamora, Samuel, Falkingham, Peter L., and Phillips, Jeremy C. 2015. "Cambrian cinctan echinoderms shed light on feeding of ancestral deuterostome." *Proceedings of the Royal Society of London B*. doi:10.1098/rspb.2015.1964.

Raikova, Olga I., Reuter, Maria, Jondelius, Ulf, and Gustafsson, Margaretha K. S. 2000. "An immunocytochemical and ultrastructural study of the nervous and muscular systems of *Xenoturbella westbladi* (Bilateria *inc. sed.*)." *Zoomorphology* 120: 107–118.

Ramos, Olivia Mendivil, Barker, Daniel, and Ferrier, David E. K. 2012. "Ghost loci imply Hox and ParaHox existence in the last common ancestor of animals." *Current Biology* 22: 1951–1956.

Raya, Ángel, and Izpisúa Belmonte, Juan Carlos. 2006. "Left-right asymmetry in the vertebrate embryo: From early information to higher-level integration." *Nature Reviews Genetics* 7: 283–293.

Reversade, Bruno, and De Robertis, E. M. 2005. "Regulation of ADMP and BMP2/4/7 at opposite embryonic poles generates a self-regulating morphogenetic field." *Cell* 123: 1147–1160.

Richards, Robert J. 2009. "Haeckel's embryos: Fraud not proven." *Biology and Philosophy* 24: 147–154.

Richardson, Michael K., and Keuck, Gerhard. 2002. "Haeckel's ABC of evolution and development." *Biological Reviews of the Cambridge Philosophical Society* 77: 495–528.

Richter, Daniel J., and King, Nicole. 2013. "The genomic and cellular foundations of animal origins." *Annual Review of Genetics* 47: 509–537.

Riddihough, G. 1992. "Homing in on the homeobox." *Nature* 357: 643–644.

Roch, Graeme J., Tello, Javier A., and Sherwood, Nancy M. 2014. "At the transition from invertebrates to vertebrates, a novel GnRH-like peptide emerges in Amphioxus." *Molecular Biology and Evolution* 31: 765–778.

Roessler, E., et al. 1996. "Mutations in the human *Sonic Hedgehog* gene cause holoprosencephaly." *Nature Genetics* 14: 357–360.

Romer, Alfred Sherwood. 1972. "The vertebrate as a dual animal—somatic and visceral." *Evolutionary Biology* 6: 121–156.

Röttinger, Eric, DuBuc, Timothy Q., Amiel, Aldine R., and Martindale, Mark Q. 2015. "Nodal signaling is required for mesodermal but not for dorsal fates in the indirect developing hemichordate, *Ptychodera flava*." *Biology Open*. doi:10.1242/bio.011809.

Röttinger, Eric, and Lowe, Christopher J. 2012. "Evolutionary crossroads in developmental biology: Hemichordates." *Development* 139: 2463–2475.

Röttinger, Eric, and Martindale, Mark Q. 2011. "Ventralization of an indirect developing hemichordate by NiCl2 suggests a conserved mechanism of dorso-ventral (D/V) patterning in Ambulacraria (hemichordates and echinoderms)." *Developmental Biology* 354: 173–190.

Rouse, Greg W., et al. 2016. "New deep-sea species of *Xenoturbella* and the position of Xenacoelomorpha." *Nature* 530: 94–97.

Ruiz-Trillo, Iñaki, and Paps, Jordi. 2015. "Aceolomorpha: Earliest branching bilaterians or deuterostomes?" *Organisms Diversity and Evolution* 16: 391–399. doi:10.1007/s13127-015-0239-1.

Ruiz-Trillo, Iñaki, et al. 1999. "Acoel flatworms: Earliest extant bilaterian metazoans, not members of Platyhelminthes." *Science* 283: 1919–1923.

Ruppert, Edward E. 1990. "Structure, ultrastructure and function of the neural gland complex of *Ascidia interrupta* (Chordata, Ascidiacea): Clarification of hypotheses regarding the evolution of the vertebrate anterior pituitary." *Acta Zoologica* 71: 135–149.

———. 1996. "Morphology of Hatschek's nephridium in larval and juvenile stages of *Branchiostoma virginiae* (Cephalochordata)." *Israel Journal of Zoology* 42: 161–182.

———. 2005. "Key characters uniting hemichordates and chordates: Homologies or homoplasies?" *Canadian Journal of Zoology* 83: 8–23.

Ryan, Joseph F., and Chiodin, Marta. 2015. "Where is my mind? How sponges and placozoans may have lost neural cell types." *Philosophical Transactions of the Royal Society of London B* 370. doi:10.1098/rstb.2015.0059.

Ryan, Joseph F., et al. 2007. "Pre-bilaterian origins of the Hox cluster and the Hox code: Evidence from the sea anemone, *Nematostella vectensis*." *PLOS ONE* 2: e153. doi:10.1371/journal.pone.0000153.

Ryan, Joseph F., et al. 2013. "The genome of the ctenophore *Mnemiopsis leidyi* and its implications for cell type evolution." *Science* 342. doi:10.1126/science.1242592.

Ryan, Kerrianne, Lu, Zhiyuan, and Meinertzhagen, Ian A. 2016. "The CNS connectome of a tadpole larva of *Ciona intestinalis* (L.) highlights sidedness in the brain of a chordate sibling." *eLife* 5: e16962.

Ryan, Tomás J., and Grant, Seth G. N. 2009. "The origin and evolution of synapses." *Nature Reviews Neuroscience* 10: 701–712.

Rychel, Amanda L., Smith, Shannon E., Shimamoto, Heather T., and Swalla, Billie J. 2006. "Evolution and development of the chordates: Collagen and pharyngeal cartilage." *Molecular Biology and Evolution* 23: 541–549.

Rychel, Amanda L., and Swalla, Billie J. 2007. "Development and evolution of chordate cartilage." *Journal of Experimental Zoology* 308B: 325–335.

Sambasivan, Ramkumar, Kuratani, Shigeru, and Tajbakhsh, Shahragim. 2011. "An eye on the head: The development and evolution of craniofacial muscles." *Development* 138: 2401–2415.

Sansom, Robert S. 2014. "Experimental decay of soft tissues." In *Reading and Writing of the Fossil Record: Preservational Pathways to Exceptional Fossilization*, edited by Marc Laflamme, James D. Schiffbauer, and Simon A. F. Darroch. Palaeontological Society Papers 20. https://www.research.manchester.ac.uk/portal/files/32469058/FULL_TEXT.PDF

Sansom, Robert S., Gabbott, Sarah E., and Purnell, Mark A. 2010. "Non-random decay of chordate characters causes bias in fossil interpretation." *Nature* 463: 797–800.

Sasakura, Yasunori, et al. 2005. "Transposon-mediated insertional mutagenesis revealed the functions of animal cellulose synthase in the ascidian *Ciona intestinalis*." *Proceedings of the National Academy of Sciences of the United States of America* 102 (3005): 15134–15139.

Sato, Atsuko, Bishop, John D. D., and Holland, Peter W. H. 2008. "Developmental biology of pterobranch hemichordates: History and perspectives." *Genesis* 46: 587–591.

Satoh, Noriyuki. 1994. *Developmental Biology of Ascidians*. Cambridge: Cambrdge University Press.

Satoh, Noriyuki, et al. 2014. "On a possible evolutionary link of the stomochord of hemichordates to pharyngeal organs of chordates." *Genesis* 52: 925–934.

Schaeffer, B. 1987. "Deuterostome monophyly and phylogeny." *Evolutionary Biology* 21: 179–235.

Schilling, Thomas F., and Knight, Robert D. 2001. "Origins of anteroposterior patterning and Hox gene regulation during chordate evolution." *Philosophical Transactions of the Royal Society of London B* 356. doi:10.1098/rstb.2001.0918.

Schlosser, Gerhard. 2005. "Evolutionary origins of vertebrate placodes: Insights from developmental studies and from comparisons with other deuterostomes." *Journal of Experimental Zoology B* 304B: 347–399.

———. 2008. "Do vertebrate neural crest and cranial placodes have a common evolutionary origin?' *BioEssays* 30: 659–672.

———. 2010. "Making senses: Development of vertebrate cranial placodes." *International Review of Cell and Molecular Biology* 283: 129–234.

Schlosser, Gerhard, Patthey, Cedric, and Shimeld, Sebastian M. 2014. "The evolutionary history of vertebrate cranial placodes, II: Evolution of ectodermal patterning." *Developmental Biology* 389: 98–119.

Scholz, Corinna B., and Technau, Ulrich. 2003. "The ancestral role of *Brachyury*: Expression of NemBra1 in the basal cnidarian *Nematostella vectensis* (Anthozoa)." *Development Genes and Evolution* 212: 563–570.

Schubert, M., Holland, L. Z., Stokes, M. D., and Holland, N. D. 2001. "Three amphioxus Wnt genes (*AmphiWnt3*, *AmphiWnt5*, and *AmphiWnt6*) associated with the tail bud: The evolution of somitogenesis in chordates." *Developmental Biology* 240: 262–273.

Scofield, Virginia L., Schlumpberger, Jay M., and Weissman, Irving L. 1982. "Colony specificity in the colonial tunicate *Botryllus* and the origins of vertebrate immunity." *American Zoologist* 22: 783–794.

Seo, Hee-Chan, et al. 2004. "Hox cluster disintegration with persistent anteroposterior order of expression in *Oikopleura dioica*." *Nature* 431: 67–71.

Šestak, Martin Sebastijan, et al. 2013. "Phylostratigraphic profiles reveal a deep evolutionary history of the vertebrate head sensory systems." *Frontiers in Zoology* 10: 18. doi:10.1186/1742-9994-10-18.

Šestak, Martin Sebastijan, and Domazet-Lošo, Tomislav. 2015. "Phylostratigraphic profiles in zebrafish uncover chordate origins of the vertebrate brain." *Molecular Biology and Evolution* 32: 299–312.

Shawlot, William, and Behringer, Richard R. 1994. "Requirement for *Lim1* in head-organizer function." *Nature* 374: 425–430.

Shen, Xin, et al. 2015. "Phylomitogenomic analyses strongly support the sister relationship of the Chaetognatha and Protostomia." *Zoologica Scripta*. doi:10.1111/zsc.12140.

Shimeld, Sebastian. 1999. "The evolution of the hedgehog gene family in chordates: Insights from amphioxus hedgehog." *Development Genes and Evolution* 209: 40–47.

Shimeld, Sebastian M., and Donoghue, Phillip C. J. 2012. "Evolutionary crossroads in developmental biology: Cyclostomes (lamprey and hagfish)." *Development* 139: 2091–2099.

Shinn, George L. 1994. "Epithelial origin of mesodermal structures in arrow-worms (Phylum Chaetognatha)." *American Zoologist* 34: 523–532.

Shinn, George L., and Roberts, Michael E. 1994. "Ultrastructure of hatchling chaetognaths (*Ferosagitta hispida*): Epithelial arrangement of the mesoderm and its phylogenetic implications." *Journal of Morphology* 219: 143–163.

Shu, D., Zhang, X, and Chen, L. 1996a. "Reinterpretation of *Yunnanozoon* as the earliest known hemichordate." *Nature* 380: 428–430.

Shu, D.-G., Conway Morris, S., and Zhang, X.-L. 1996b. "A *Pikaia*-like chordate from the Lower Cambrian of China." *Nature* 384: 157–158.

Shu, D.-G., Conway Morris, S., Zhang, X-L., Chen, L., Li, Y., and Han, J. 1999a. "A pipiscid-like fossil from the Lower Cambrian of south China." *Nature* 400: 746–749.

Shu, D.-G., Luo, H.-L., Conway Morris, S., Zhang, X.-L., Hu, S.-X., Chen, L., Han, J., Zhu, M., Li, Y., and Chen, L-Z. 1999b. "Lower Cambrian vertebrates from south China." *Nature* 402: 42–46.

Shu, D.-G., Conway Morris, S., Han, J., Chen, L., Zhang, X.-L., Zhang, Z.-F., Liu, H.-Q., Li, Y., and Liu, J.-N. 2001a. "Primitive deuterostomes from the Chengjiang Lagerstätte (Lower Cambrian, China)." *Nature* 414: 419–424.

Shu, D.-G., Chen, L., Han, J., and Zhang, X.-L. 2001b. "An early Cambrian tunicate from China." *Nature* 411: 472–473.

Shu, D.-G., Conway Morris, S., Zhang, Z. F., Liu, J. N., Han, J., Chen, L., Zhang, X. L., Yasui, K., Li, Y. 2003a. "A new species of yunnanozoan with implications for deuterostome evolution." *Science* 299: 1380–1384.

Shu, D.-G., Conway Morris, S., Han, J., Zhang, Z.-F., Yasui, K., Janvier, P., Chen, L., Zhang, X.-L., Liu, J.-N., Li, Y., and Liu, H.-Q. 2003b. "Head and backbone of the Early Cambrian vertebrate *Haikouichthys*." *Nature* 421: 526–529.

Shu, D.-G., et al. 2004. "Ancestral echinoderms from the Chengjiang deposits of China." *Nature* 430: 422–428.

Shu, D.-G., Conway Morris, S., Zhang, Z.-F., and Han, J. 2010. "The earliest history of the deuterostomes: The importance of the Chengjiang Fossil-Lagerstätte." *Proceedings of the Royal Society of London B* 277: 165–174.

Shubin, Neil, Tabin, Cliff, and Carroll, Sean. 2009. "Deep homology and the origins of evolutionary novelty." *Nature* 457: 818–823.

Simakov, Oleg, et al. 2015. "Hemichordate genomes and deuterostome origins." *Nature* 527: 459–465.

Simeone, A., et al. 1993. "A vertebrate gene related to orthodenticle contains a homeodomain of the bicoid class and demarcates anterior neuroectoderm in the gastrulating mouse embryo." *EMBO Journal* 12: 2735–2747.

Simões-Costa, Marcos S., et al. 2005. "The evolutionary origin of cardiac chambers." *Developmental Biology* 277: 1–15.

Singer, J. B., et al. 1996. "*Drosophila brachyenteron* regulates gene activity and morphogenesis in the gut." *Development* 122: 3707–3718.

Slack, J. M. W., Holland, P. W. H., and Graham, C. F. 1993. "The zootype and the phylotypic stage." *Nature* 361: 490–492.

Small, Kerrin S., Brudno, Michael, Hill, Matthew M., and Sidow, Arend. 2007. "A haplome alignment and reference sequence of the highly polymorphic *Ciona savignyi* genome." *Genome Biology* 8. doi:10.1186/gb-2007-8-3-r41.

Smith, Andrew B. 2004. "Echinoderm roots." *Nature* 430: 411–412.

———. 2005. "The pre-radial history of echinoderms." *Geological Journal* 40: 255–280.

———. 2012. "Cambrian problematica and the diversification of deuterostomes." *BMC Biology* 10. 10.1186/1741-7007-10-79.

Smith, Jeramiah J., et al. 2013. "Sequencing of the sea lamprey (*Petromyzon marinus*) genome provides insights into vertebrate evolution." *Nature Genetics* 45: 415–421.

Smith, Jeramiah J., and Keinath, Melissa C. 2015. "The sea lamprey meiotic map improves resolution of ancient vertebrate genome duplications." *Genome Research* 25: 1081–1090.

Smith, Martin R., and Caron, Jean-Bernard. 2015. "*Hallucigenia*'s head and the pharyngeal armature of early ecdysozoans." *Nature* 523: 75–78.

Sobral, Daniel, Tassy, Oliver, and Lemaire, Patrick. 2009. "Highly divergent gene expression programs can lead to similar chordate larval body plans." *Current Biology* 19: 2014–2019.

Sodergren, Erica, et al. 2006. "The genome of the sea urchin *Strongylocentrotus purpuratus*." *Science* 314: 941–952.

Soukup, Vladimir, et al. 2015. "The Nodal signalling pathway controls left-right asymmetric development in amphioxus." *EvoDevo* 6. doi 10.1186/2041-9139-6-5.

Spagnuolo, Antonietta, et al. 2003. "Unusual number and genomic organization of Hox genes in the tunicate *Ciona intestinalis*." *Gene* 309: 71–79.

Spemann, H., and Mangold, Hilde. 1924. "*über Induktion von Embryonalanlagen durch Implantation artfremder Organisatoren*." *Archiv für mikroskopische Anatomie und Entwicklungsmechanik* 100: 599–638.

Sprinkle, James, and Wilbur, Bryan C. 2005. "Deconstructing helicoplacoids: Reinterpreting the most enigmatic Cambrian echinoderms." *Geological Journal* 40: 281–293.

Srivastava, Mansi, et al. 2008. "The *Trichoplax* genome and the nature of placozoans." *Nature* 454: 955–960.

Srivastava, Mansi, et al. 2010. "The *Amphimedon queenslandica* genome and the evolution of animal complexity." *Nature* 466: 720–726.

Stach, Thomas. 2013. "Larval anatomy of the pterobranch *Cephalodiscus gracilis* supports secondarily derived sessility concordant with molecular phylogenies." *Naturwissenschaften* 100: 1187–1191.

Stach, Thomas, et al. 2005. "Nerve cells of *Xenoturbella bocki* (phylum uncertain) and *Harrimania kupfferi* (Enteropneusta) are positively immunoreactive to antibodies raised against echinoderm neuropeptides." *Journal of the Marine Biological Association of the United Kingdom* 85: 1519–1524.

Stach, Thomas, et al. 2008. "Embryology of a planktonic tunicate reveals traces of sessility." *Proceedings of the National Academy of Sciences* 105: 7229–7234.

Stach, Thomas, Gruhl, Alexander, and Kaul-Strehlow, Sabrina. 2012. "The central and peripheral nervous system of *Cephalodiscus gracilis* (Pterobranchia, Deuterostomia)." *Zoomorphology* 131: 11–24.

Stemple, Derek L. 2005. "Structure and function of the notochord: An essential organ for chordate development." *Development* 132: 2503–2512.

Stemple, Derek L., et al. 1996. "Mutations affecting development of the notochord in zebrafish." *Development* 123: 117–128.

Stewart, R. M., and Gerhart, J. C. 1990. "The anterior extent of dorsal development of the *Xenopus* embryonic axis depends on the quantity of organizer in the late blastula." *Development* 109: 363–372.

Stolfi, Alberto, et al. 2010. "Early chordate origins of the vertebrate second heart field." *Science* 329: 565–568.

Stolfi, Alberto, et al. 2014. "Divergent mechanisms regulate conserved cardiopharyngeal development and gene expression in distantly related ascidians." *eLife*. doi:10.7554/eLife.03728.

Stolfi, Alberto, and Brown, Federico D. 2015. "Tunicata." In *Evolutionary Developmental Biology of Invertebrates*, vol. 6: *Deuterostomia*, edited by A. Wanninger, 135–204. Vienna: Springer.

Stolfi, Alberto, Ryan, Kerrianne, Meinertzhagen, Ian A., and Christiane, Lionel. 2015. "Migratory neuronal progenitors arise from the neural plate borders in tunicates." *Nature* 527: 371–374.

Su, Yi-Hsien. 2014. "Telling left from right: Left-right asymmetric controls in sea urchins." *Genesis* 52: 269–278.

Suga, Hiroshi, et al. 2013. "The *Capsaspora* genome reveals a complex unicellular prehistory of animals." *Nature Communications* 4. doi:10.1038/ncomms3325.

Sumoy, Lauro, Keasey, J. Bennett, Dittman, Tyler D., and Kimmelman, David. 1997. "A role for notochord in axial vascular development revealed by analysis of phenotype and the expression of VEGR-2 in zebrafish *flh* and *ntl* mutant embryos." *Mechanisms of Development* 63: 15–27.

Suzuki, Daichi G., Murakami, Yasunori, Escrivà, Hector, and Wada, Hiroshi. 2015. "A comparative examination of neural circuit and brain patterning between the lamprey and amphioxus reveals the evolutionary origin of the vertebrate visual center." *Journal of Comparative Neurology* 523: 251–261.

Swalla, Billie J., Cameron, Chris B., Corley, Laura S., and Garey, James R. 2000. "Urochordates are monophyletic within the deuterostomes." *Systematic Biology* 49: 52–64.

Swalla, Billie J., and Jeffery, William R. 1990. "Interspecific hybridization between an anural and urodele ascidian: Differential expression of urodele features suggests multiple mechanisms control anural development." *Developmental Biology* 142: 319–334.

———. 1996. "Requirement of the *Manx* gene for expression of chordate features in a tailless ascidian larva." *Science* 274: 1205–1208.

Swalla, Billie J., and Smith, Andrew B. 2008. "Deciphering deuterostome phylogeny: Molecular, morphological and palaeontological perspectives." *Philosophical Transactions of the Royal Society B* 363: 1557–1568.

Szaniawski, Hubert. 1982. "Chaetognath grasping spines recognized among Cambrian protoconodonts." *Journal of Paleontology* 56: 806–810.

Tabin, Cliff. 2006. "The key to left-right asymmetry." *Cell* 127: 27–32.

Tagawa, Kunifumi, Humphreys, Tom, and Satoh, Noriyuki. 1998. "Novel pattern of *Brachyury* gene expression in hemichordate embryos." *Mechanisms of Development* 75: 139–143.

Tarazona, Oscar A., et al. 2016. "The genetic program for cartilage development has deep homology within Bilateria." *Nature*. doi:10.1038/nature17398.

Technau, Ulrich. 2001. "*Brachyury*, the blastopore and the evolution of the mesoderm." *BioEssays* 23: 788–794.

Telford, Maximilian J. 2008. "Xenoturbellida: The fourth deuterostome phylum and the diet of worms." *Genesis* 46: 580–586.

Telford, Maximilian, and Copley, Richard R. 2016. "War of the Worms." *Current Biology* 26: R335–R337.

Telford, M. J., and Holland, P. W. 1993. "The phylogenetic affinities of the chaetognaths: A molecular analysis." *Molecular Biology and Evolution* 10: 660–676.

Telford, M. J., Lockyer, A. E., Cartwright-Finch, C., and Littlewood, D. T. J. 2003. "Combined large and small subunit RNA phylogenies support a basal position of the acoelomorph flatworms." *Proceedings of the Royal Society of London B* 270: 1077–1083.

Theobold, Douglas L., 2010. "A formal test of the theory of universal common ancestry." *Nature* 465: 219–222.

Tolkin, Theadora, and Christiaen, Lionel. 2012. "Development and evolution of the ascidian cardiogenic mesoderm." In *Heart Development*, edited by Benoit G. Bruneau, 107–142. San Diego: Academic.

Tonegawa, Susumu. 1983. "Somatic generation of antibody diversity." *Nature* 302: 575–581.

Toresson, Håkan, et al. 1998. "Conservation of BF-1 expression in amphioxus and zebrafish suggests evolutionary ancestry of anterior cell types that contribute to the vertebrate telencephalon." *Development Genes and Evolution* 208: 431–439.

Tsagkogeorga, Georgia, et al. 2009. "An updated 18S rRNA phylogeny of tunicates based on mixture and secondary structure models." *BMC Evolutionary Biology* 9: 187. doi:10.1186/1471-2148-9-187.

Tung, T. C., Wu, S. C., and Tung, Y. Y. F. 1962. "Experimental studies on the neural induction in amphioxus." *Scientia Sinica* 9: 805–820.

———. 1965. "Differentiation of the prospective ectodermal and entodermal cells after transplantation to new surroundings in amphioxus." *Scientia Sinica* 14: 1785–1794.

Turbeville, J. M., Schulz, J. R., and Raff, R. A. 1994. "Deuterostome phylogeny and the sister group of the chordates: Evidence from molecules and morphology." *Molecular Biology and Evolution* 11: 648–655.

Turner, Susan, et al. 2010. "False teeth: Conodont-vertebrate phylogenetic relationships revisited." *Geodiversitas* 32: 545–594.

Tzahor, Eldad. 2009. "Heart and craniofacial muscle development: A new developmental theme of distinct myogenic fields." *Developmental Biology* 327: 273–279.

Tzahor, Eldad, and Evans, Sylvia M. 2011. "Pharyngeal mesoderm development during embryogenesis: Implications for both heart and head myogenesis." *Cardiovascular Research* 91: 196–202.

Urata, Makoto, Iwasaki, Sadaharu, Ohtsuka, Susumu, and Yamaguchi, Masaaki. 2014. "Development of the swimming acorn worm *Glandiceps hacksi*: Similarity to holothuroids." *Evolution and Development* 16: 149–154.

Vannier, J., et al. 2007. "Early Cambrian origin of modern food webs: Evidence from predator arrow worms." *Proceedings of the Royal Society of London B* 274: 627–633.

Veeman, Michael T., Newman-Smith, Erin, El-Nachef, Danny, and Smith, William C. 2010. "The ascidian mouth opening is derived from the anterior neuropore: Reassessing the mouth/neural tube relationship in chordate evolution." *Developmental Biology* 344: 138–149.

Veis, Arthur, et al. 2002. "Mineral-related proteins of sea urchin teeth: *Lytechinus variegatus*." *Microscopy Research and Technique* 59: 342–351.

Venkatesh, Byrappa, et al. 2007. "Survey sequencing and comparative analysis of the elephant shark (*Callorhinchus milii*) genome." *PLOS Biology*. http://dx.doi.org/10.1371/journal.pbio.0050101

Venkatesh, Byrappa, et al. 2014. "Elephant shark genome provides unique insights into gnathostome evolution." *Nature* 505: 174–179.

Vinson, Jade P., et al. 2005. "Assembly of polymorphic genomes: Algorithms and application to *Ciona savignyi*." *Genome Research* 15: 1127–135.

Vinther, Jakob, Smith, M. Paul, and Harper, David A. T. 2011. "Vetulicolians from the Lower Cambrian Sirius Passet Lagerstätte, North Greenland, and the polarity of morphological characters in basal deuterostomes." *Palaeontology* 54: 711–719.

Vopalensky, Pavel, et al. 2012. "Molecular analysis of the amphioxus frontal eye unravels the evolutionary origin of the retina and pigment cells of the vertebrate eye." *Proceedings of the National Academy of Sciences of the United States of America* 109: 15383–15388.

Voskoboynik, Ayelet, et al. 2013. "The genome sequence of the colonial chordate, *Botryllus schlosseri*." *eLife*. doi:http://dx.doi.org/10.7554/eLife.00569.

Wada, H. 1998. "Evolutionary history of free-swimming and sessile lifestyles in urochordates as deduced from 18S rDNA molecular phylogeny." *Molecular Biology and Evolution* 15: 1189–1194.

———. 2001. "Origin and evolution of the neural crest: A hypothetical reconstruction of its evolutionary history." *Development, Growth and Differentiation* 43: 509–520.

Wada, H., Saiga, H., Satoh, N., and Holland, P. W. 1998. "Tripartite organization of the ancestral chordate brain and the antiquity of placodes: Insights from ascidian *Pax-2/5/8*, *Hox* and *Otx* genes." *Development* 125: 1113–1122.

Wada, H., and Satoh, N. 1994. "Details of the evolutionary history from invertebrates to vertebrates, as deduced from sequences of 18S rDNA." *Proceedings of the National Academy of Science of the United States of America* 91: 1801–1804.

Wada, Hiroshi, Garcia-Fernàndez, J., and Holland, Peter W. H. 1999. "Colinear and segmental expression of amphioxus Hox genes." *Developmental Biology* 213: 131–141.

Wagner, Gunter P., Amemiya, Chris, and Ruddle, Frank. 2003. "Hox cluster duplications and the opportunity for evolutionary novelties." *Proceedings of the National Academy of Sciences of the United States of America* 100: 14603–14606.

Wainright, P. O., Hinkle, G., Sogin, M. L., and Stickel, S. K. 1993. "Monophyletic origins of the metazoa: An evolutionary link with fungi." *Science* 260: 340–342.

Warner, Jacob F., and McClay, David R. 2014. "Left-right asymmetry in the sea urchin." *Genesis* 52: 481–487.

Watanabe, Hiroshi, et al. 2014. "Nodal signalling determines biradial asymmetry in Hydra." *Nature* 515: 112–115.

Welker, Frido, et al. 2015. "Ancient proteins resolve the evolutionary history of Darwin's South American ungulates." *Nature* 522: 81–84.

Wellik, Deneen M. 2007. "Hox patterning of the vertebral axial skeleton." *Developmental Dynamics* 236: 2454–2463.

Wicht, Helmut, and Lacalli, Thurston C. 2005. "The nervous system of amphioxus: Structure, development, and evolutionary significance." *Canadian Journal of Zoology* 83: 122–150.

Wilkinson, D. G., Bhatt, Sangita, and Herrmann, Bernhard G. 1990. "Expression pattern of the mouse T gene and its role in mesoderm formation." *Nature* 343: 657–659.

Willey, A. 1893. "Studies on the Protochordata, III: On the position of the mouth in the larvae of the ascidians and amphioxus, and its relation to the neuroporus." *Quarterly Journal of Microscopical Science* 35: 316–333.

Williams, Frederike, Tew, Hannah A., Paul, Catherine E., and Adams, Josephine C. 2014. "The predicted secretomes of *Monosiga brevicollis* and *Capsaspora owczarzaki*, close unicellular relatives of metazoans, reveal new insights into the evolution of the metazoan extracellular matrix." *Matrix Biology* 37: 60–68.

Williams, N. A., and Holland, P. W. 1998. "Gene and domain duplication in the chordate Otx gene family: Insights from amphioxus *Otx*." *Molecular Biology and Evolution* 15: 600–607.

Willmer, Pat. 1990. *Invertebrate Relationships: Patterns in Animal Evolution*. Cambridge: Cambridge University Press.

Wills, Matthew A., Briggs, Derek E. G., and Fortey, Richard A. 1994. "Disparity as an evolutionary index: A comparison of Cambrian and Recent arthropods." *Paleobiology* 20: 93–130.

Wilson, V., Manson, L., Skarnes, W. C., and Beddington, R. S. 1995. "The T gene is necessary for normal mesodermal morphogenetic cell movements during gastrulation." *Development* 121: 877–886.

Wilt, Fred H., Killian, Christopher E., and Livingston, Brian T. 2003. "Development of calcareous skeletal elements in invertebrates." *Differentiation* 71: 237–250.

Winchell, Christopher J., et al. 2002. "Evaluating hypotheses of deuterostome phylogeny and chordate evolution with new LSU and SSU ribosomal DNA data." *Molecular Biology and Evolution* 19: 762–776.

Worsaae, Katrine, et al. 2012. "An anatomical description of a miniaturized acorn worm (Hemichordata, Enteropneusta) with asexual reproduction by paratomy." *PLOS ONE* 7: e48529. doi:10.1371/journal.pone.0048529.

Wray, Gregory A. 2015. "Molecular clocks and the early evolution of metazoan nervous systems." *Philosophical Transactions of the Royal Society of London B* 370. doi:10.1098/rstb.2015.0046.

Xiao, Shuhai, and Laflamme, Marc. 2009. "On the eve of animal radiation: Phylogeny, ecology and evolution of the Ediacara biota." *Trends in Ecology and Evolution* 24: 31–40.

Yalden, Derek W. 1985. "Feeding mechanisms as evidence for cyclostome monophyly." *Zoological Journal of the Linnean Society* 84: 291–300.

Yamada, Atsuko, Martindale, Mark Q., Fukui, Akimasa, and Tochinai, Shin. 2010. "Highly conserved functions of the *Brachyury* gene on morphogenetic movements: Insight from the early-diverging phylum Ctenophora." *Developmental Biology* 339: 212–222.

Yao, Y., et al. 2016. "Cis-regulatory architecture of a brain signaling center predates the origin of chordates." *Nature Genetics* 48: 575–580.

Yasui, K., Zhang, S., Uemura, M., and Saiga, H. 2000. "Left-right asymmetric expression of BbPtx, a Ptx-related gene, in a lancelet species and the developmental left-sidedness in deuterostomes." *Development* 127: 187–195.

Yasui, Kinya, Kaji, Takao, Morov, Arseniy R., and Yonemura, Shigenobu. 2014. "Development of oral and branchial muscles in lancelet larvae of *Branchiostoma japonicum*." *Journal of Morphology* 275: 465–477.

Yasuo, Hitoyoshi, and Satoh, Noriyuki. 1993. "Function of vertebrate T gene." *Nature* 364: 582–583.

———. 1994. "An ascidian homolog of the mouse *Brachyury* (T) gene is expressed exclusively in notochord cells at the fate restricted stage." *Development, Growth and Differentiation* 36: 9–18.

———. 1998. "Conservation of the developmental role of *Brachyury* in notochord formation in a urochordate, the ascidian *Halocynthia roretzi*." *Developmental Biology* 200: 158–170.

Yoshiba, Satoko, and Hamada, Hiroshi. 2014. "Roles of cilia, fluid flow and Ca^{2+} signalling in breaking of left-right symmetry." *Trends in Genetics* 30: 10–17.

Young, Gavin C., Karatajute-Talimaa, Valya N., and Smith, Moya M. 1996. "A possible Late Cambrian vertebrate from Australia." *Nature* 383: 810–812.

Yu, Jr-Kai, Holland, Linda Z., and Holland, Nicholas D. 2002a. "An amphioxus nodal gene (AmphiNodal) with early symmetrical expression in the organiser and mesoderm and later asymmetrical expression associated with left-right axis formation." *Evolution and Development* 4: 418–425.

Yu, Jr-Kai, Holland, Nichols D., and Holland, Linda Z. 2002b. "An amphioxus winged helix/forkhead gene, AmphiFoxD: Insights into vertebrate neural crest evolution." *Developmental Dynamics* 225: 289–297.

Yu, Jr-Kai, et al. 2007. "Axial patterning in cephalochordates and the evolution of the organizer." *Nature* 445: 613–617.

Yu, Jr-Kai, Meulemans, Daniel, McKeown, Sonja J., and Bronner-Fraser, Marianne. 2008. "Insights from the amphioxus genome on the origin of vertebrate neural crest." *Genome Research* 18: 1127–1132.

Yuan, Shaochun, Ruan, Jie, Huang, Shengfeng, Chen, Shangwu, and Xu, Anlong. 2015. "Amphioxus as a model for investigating evolution of the vertebrate immune system." *Developmental and Comparative Immunology* 48: 297–305.

Yue, Jia-Xing, Yu, Jr-Kai, Putnam, Nicholas H., and Holland, Linda Z. 2014. "The transcriptome of an amphioxus, *Asymmetron lucayanum*, from the Bahamas: A window into chordate evolution." *Genome Biology and Evolution* 6: 2681–2696.

Zaffran, Stéphane, Das, Gishnu, and Frasch, Manfred. 2000. "The NK-2 homeobox gene *scarecrow* (*scro*) is expressed in pharynx, ventral nerve cord and brain of *Drosophila* embryos." *Mechanisms of Development* 94: 237–241.

Zamora, Samuel, and Rahman, Imran. 2014. "Deciphering the early evolution of Echinoderms with Cambrian fossils." *Palaeontology* 57: 1105–1119.

Zeng, Liyun, Jacobs, Molly W., and Swalla, Billie J. 2006. "Coloniality has evolved once in stolidobranch ascidians." *Integrative and Comparative Biology* 46: 255–268.

Zeng, Liyun, and Swalla, Billie J. 2005. "Molecular phylogeny of the protochordates: Chordate evolution." *Canadian Journal of Zoology* 83: 24–33.

Zhang, Jian, et al. 1998. "The role of maternal VegT in establishing the primary germ layers in *Xenopus* embryos." *Cell* 94: 515–524.

Zhang, Yanni, et al. 2014. "An amphioxus RAG1-like DNA fragment encodes a functional central domain of vertebrate core RAG1." *Proceedings of the National Academy of Sciences of the United States of America* 111: 397–402.

Zhu, Min, et al. 2013. "A Silurian placoderm with osteichthyan-like marginal jaw bones." *Nature* 502: 188–193.

Zintzen, Vincent, et al. 2011. "Hagfish predatory behaviour and slime defence mechanism." *Scientific Reports* 1. doi:10.1038/srep00131.

Index

The letter *f* following a page number denotes a figure.

abducens nerves, 106, 107
Acanthaster planci, 72
acanthodians, 114
acoels (acoelomorphs), 29, 30, 35–37, 118–20, 239n1
acorn worms. *See* enteropneusts
acraniates. *See* amphioxus
adenohypophysis, 81, 106, 177
agnathans, 103, 110, 112, 150; fossil record of, 103
ambulacra, 57, 61
ambulacraria, 6–7, 31, 204, 220; failure of larval nervous system to persist to adulthood, 204; Hox genes specific to, 72, 204; interrelationships of, 24, 53; and mesocoels, 204; in solutes, 60; stalked habit, 219
ameloblasts, 159
Amiskwia sagittiformis, 121
ammocoete, 93, 102–3, 115, 173
amniotes, 173
AmphiHox3, 78
Amphioxides, 235n35
amphioxus, 5, 6, 7, 11, 52, 75–83, 86, 102–6, 110, 123, 138–41, 148, 150, 154, 158, 160, 171, 172–73, 193, 203, 210–11, 215, 222, 242n58; cartilage in, 154, 160; compared with *Pikaia*, 193; distinctive features of, 210, 222; early development of, 79, 81; excretory system in, 173; fossil evidence, 83; genome of, 77, 210, 222; "head" in, 138, 142; interrelationships of, 24; as a model for segmentationist view, 138–39, 140–41; notochord in, 130–31; organizer in, 128; replication of Mangold's experiments in, 127–28; rostral somites of, 135, 136, 138, 140, 142, 203
AmphiPax2/5/8, 234n20
ancestry and descent, 18, 19f, 23, 24, 25
anenkephaly, 148

annelids, 30, 31–32, 37, 233n3
Anomalocaris, 188
antennapedia mutation, 42
anterior dorsal unit (in *Pikaia*), 192
anterior neural ridge (ANR), 70, 78–79, 149, 150, 158, 204, 208
anterior-posterior (AP) axis, 44, 203; in amphioxus, 80; in chordates, 130; in deuterostomes, 128, 203; in enteropneusts, 69, 128; in tunicate and vertebrate nervous systems, 92, 149
Anti-Dorsalizing Morphogenetic Protein (*Admp*), 127–29
antisegmentationist view, 125, 138, 140
aorta: dorsal, 109; ventral, 109
aorta-gonads-mesonephros (AGM) domain, 171, 211
aplousobranch ascidians, 96
appendicularians. *See* larvaceans
archenteron, 28, 29, 51, 110, 115, 125, 133
arcualia (in lampreys), 104, 115
Aristotle, 85
arrow worms. *See* chaetognaths
arthropods, 2, 31–32, 37, 38, 44
ascidians, 86f, 87, 88, 89, 90, 91, 93, 96, 131, 154
asteroids (sea stars, starfish), 55
Asymmetron, 75, 233n1
atrial (excurrent) siphon in tunicates, 87, 88, 98; siphon primordia, 159, 171
atrium, 76, 87, 208; in amphioxus, 76; in ancestral chordates, 208; homology between primordia and posterior placodes, 211; in tunicates, 87
Atubaria, 73
auditory nerves, 105, 106, 107, 116
axial complex, 68, 203
axochord, 132, 209, 233n3
axocoel, 57
axons, 145

Balanoglossus simodensis, 152
Balfour, Francis Maitland, 139, 141–42, 143, 242n47
Banffia, 187, 188
bare-handed hagfish-catching as an Olympic sport, 110
Bateson, William, 41–42, 48; *Materials for the Study of Variation*, 42
bauchstück, 126
Bayesian statistics, 23
Before the Backbone, ix–xi, 218
β-catenin, 54, 71, 80, 89, 127, 203, 214; role in the Organizer, 127, 203
bilateria, 30, 34, 36, 43, 118, 119, 146, 154; last common ancestor of, 132
biomineralization, 57, 58, 159, 202; as an ancestral deuterostome trait, 202; in echinoderms, 57, 64; in hemichordates, 58, 62
bithorax mutation, 42
blastocoel, 28, 29, 30, 125
blastopore, 29, 31, 37, 51, 125, 126, 132, 145; association with *brachyury*, 132; dorsal lip of, 80, 110, 115, 116, 125, 126–29, 131
blastula, 28, 29, 115
bone, 5, 57, 103, 114, 130, 159, 160, 214; dermal bone, 156; formation from cartilage, 130, 160
Bone Morphogenetic Proteins (BMPs), 70, 80, 89, 126, 130, 145–47, 152, 205, 212, 214; antagonists of, 130, 152, 205; in neurogenesis, 145–47, 152; role in the Organizer, 127–29

bony fishes. *See* osteichthyans (bony fishes)
Botryllus schlosseri, 88, 96; genome of, 96–97, 236n23
Bowman's capsules, 173
brachiopods, 31, 32, 36, 37, 52, 117, 225n5
brachyury, 76, 131–32, 212, 222, 233n3; in amphioxus, 76, 131, 222; association with the blastopore, 132; duplication in chordoma, 131; in echinoderms, 132; in enteropneusts, 132; function in mesoderm specification, 131; in *Oikopleura*, 94, 212; in tunicates, 89, 92, 94, 212
brain, 7; in amphioxus, 78, 140, 148, 149; in chordates, 147, 150, 207; compared with invertebrates, 7–8; comparisons between amphioxus and vertebrates, 78, 140, 143, 148; comparisons between enteropneusts and other deuterostomes, 70; comparisons between tunicates and vertebrates, 98; in gnathostomes, 115; in hagfishes, 111; in lampreys, 104, 105; organizing centers of, 78–79, 149, 150; in protostomes, 147; in vertebrates, 7, 78, 105–7, 213
branchial skeleton. *See under* skeleton
branchial slits. *See* pharyngeal slits
branchiomeric musculature (in head and heart development), 169, 171
Branchiostoma, 75, 233n1; *B. belcheri*, 75; *B. floridae*, 75; *B. japonicum*, 75
brittle stars, 55
bryozoans (moss animals), 31, 36, 37, 117
Burgess Shales, 38, 72, 83, 121, 184, 185, 192, 195, 198
Caenorhabditis elegans, 32, 36
calcichordate theory, 63–64, 231n20
Callorhinchus milii, 77, 116
cambroernids, 182, 184–85, 204
cardiac mesoderm, 174, 209
cardio-pharyngeal mesoderm/musculature, 88, 169, 208, 209
cartilage, 5, 114, 115, 130, 153, 154, 159–60, 213; in amphioxus, 160; in bilateria, 154, 159–60; in notochord, 130; in skeleton, 153, 213
cartilaginous fishes, 114
Cathaymyrus, 83, 194, 199; *C. diadexus*, 194; *C. haikouensis*, 194
cellulose (in tunicates), 88, 212
cellulose synthase, 88
cephalochordates. *See* amphioxus
Cephalodiscus, 73, 206
cerberus, 127
cerebellum, 106, 111
cerebral vesicle. *See* brain: in amphioxus
chaetognaths, 31, 36, 37, 52, 117, 120–22, 196; fossil record, 121; similarity with Jack Russell terriers, 121
Chengjiang fauna, 185, 186, 188, 190, 194, 198
Cheungkongella, 99
chicken, topology of, 165, 166
chitin, 246n30
choanoflagellates, 34
chondrichthyans, 114
chondrocytes, 159
chordates, 5, 6, 31, 52, 150, 206–10, 217; definition of, 217 (points 2, 3); earliest, 150, 221; segmentation in, 32, 207
chordin, 44–45, 70, 127–29, 146
chordoma, 131
chromaffin cells, 153

ciliated pit (in amphioxus), 81, 143
cinctans, 59–61, 185
Ciona, 81, 94, 95, 96, 99, 155, 157, 161; *C. intestinalis*, 87, 88, 92, 149; *C. intestinalis*, genome of, 96, 98; *C. savignyi*, genome of, 96, 98
circulatory system, 87, 165; in hagfishes, 111; in lampreys, 109; in tunicates, 87; in vertebrates, 165
cladistics, 21, 23, 24, 33, 167
cladograms, 21, 22, 23
Clavelina, 96
cleavage, 32, 34, 37, 51
"clock-and-wavefront" model of somitogenesis, 133–34, 135, 145, 207, 214
club-shaped gland (in amphioxus), 82
cnidaria, 36
coelacanths, notochord in, 130
coelom, 29, 31, 108, 203, 211; common ancestry of tunicates and vertebrates, 211; in deuterostomes, 203; by enterocoely, 30, 31, 32, 51; in enteropneusts, 66, 67; formation by amphistomy, 32; in gnathostomes, 115; in hagfishes, 111; in lampreys, 104, 105; by schizocoely, 30, 32, 51; visceral compartment, 165
coelomates, 30, 34
collagen, 130, 160, 161; association with Hox clusters, 161; type II, 130; type X, 130
collar cord, 69, 152, 205, 220
conodonts, 150, 182, 195–96, 199
conus arteriosus, 170
convergence (evolutionary), 20, 21, 23, 119
convergent extension, 48, 126, 129, 145, 147
Conway Morris, Simon, 38
cornutes, 61
Cothurnocystis, 63, 182
cranial nerves, 106–7, 139, 213
cranial placodes. *See* placodes
Craniata, 112
craniofacial mesoderm/musculature, 88, 170; homologs in tunicates, 88, 170
cranium. *See* skull
creationists, 21
crinoids, 55, 232n23
crown group, 179–81
ctenocystoids, 59, 61
ctenophores, 34, 35, 36, 228n5, 243n61
Cuvier, Georges, 15, 16, 226n1; *embranchements*, 15, 16, 33, 34
cycliophorans, 37
cycloneuralians, 37, 38, 188
cyclopia, 150
cyclostomes, 103, 111–12, 115, 154, 156, 157, 159, 176, 197, 198, 214

Danio rerio, 129
Darwin, Charles, 17, 23, 24, 41, 46, 47; descent with modification, 17; theory of evolution by natural selection, 17, 27
decapentaplegic (*dpp*), 45, 145
deep homology, 47
dendrites, 145
dentine, 5, 57, 114, 115, 160, 214
dermal skeleton. *See under* skeleton
dermatome, 135
descent with modification. *See under* Darwin, Charles
deuterostomes, 7, 34, 39, 44, 51, 52, 58, 71, 72, 101, 166, 201–4, 218; asymmetry of internal organs, 58; blastopore formation in, 31; coelomic spaces in, 51–52, 203; definition of,

51, 201–4, 217 (point 1); fossils of, 179–200; interrelationships of, 24, 86, 87, 199; latest common ancestor of, 37, 166; left-right asymmetry in, 53–54, 70, 76; monophyly of, 52; phylogeny of, 218
development, physical forces in, 48
dickkopf, 127
diencephalon, 78, 105, 106, 140, 143, 148, 150, 207, 208, 213, 215
dipleurula (larva of echinoderms), 71
diploblasts, 28, 29, 34, 36
disparity, 247n1
Dobzhansky, Theodosius, 17
doliolids, 95
dorsal lamina (in tunicates), 87
dorsal lip. *See under* blastopore
dorsal nerve cord. *See* nervous system
dorsal organ (in *Pikaia*), 192
dorsal sac (in echinoderms), 57
dorso-ventral axis, in enteropneusts, 70; comparison with neurogenic induction, 146; inversion of (*see under* Geoffroy Inversion)
Drosophila, 36, 42, 47, 128, 137, 154, 235n42

ecdysozoa, 37, 38, 188, 219
echinoderms, 6, 55–64, 123, 185, 198, 204–5, 219; asymmetries of, 57, 58, 61, 205; biomineralization in, 57–58, 123, 185, 204–5; classes of, 55; definition of, 55; endoskeleton of, 56; fossils of, 58–64; interrelationships of, 24, 52, 53, 62; nervous system, 56, 57, 69, 151; Organizer in, 128–29; pentameral symmetry, 58; pharyngeal slits in, 60–61, 220; water-vascular system, 57, 58, 61
echinoids, 55
ectoderm, 28, 29, 67, 71, 110, 158, 203, 227n3
Ediacara fauna, 226n18
eldoniids, 182, 184, 185
elephants, 62, 247; distinction from table-tennis balls, 247n15; quadrupedal status of, 62; in the room, 62
embranchements. *See under* Cuvier, Georges
embryology, 27, 28, 29f
Emu Bay Shales, 186
enamel (enameloid), 5, 103, 114, 115, 156, 159, 160, 214
endoderm, 28, 29, 51, 71, 89, 90, 174, 227n3
endostyle, 6, 68, 189, 205, 208; in amphioxus, 76, 81–82; similarity with stomochord, 68, 205; transformation into thyroid gland, 6; in tunicates, 87, 90, 92
endothelium, 87, 166, 174, 213
Entelognathus, 216
enteric nervous system, 153, 166, 167–68, 173, 245n8
enterocoely. *See under* coelom
enteropneusts, 6, 58, 65, 66, 108, 189, 203, 206, 220; anatomy of, 67; biomineralization in, 58; *brachyury*, 132; families of, 65, 66; feeding, 67; fossil record of, 72, 73; genomes of, 71, 72; nervous system, 69–70, 152
Epigonichthys, 233n1
Eptatretus, 111
esophagocutaneous duct, 111
euconodonts, 196
eukaryotes, 34, 35
evolution, 17, 23; of horses, 18, 20, 21; misunderstanding of, 25; by natural selection, 17, 25; pattern generated

evolution (*continued*)
by, 21; process of, 25; in terms of interacting genetic networks, 47
evolutionary developmental biology (EvoDevo), 47
excretory system. *See* urogenital system
Eya, 158
eyeless, 47
eye muscles, 106, 107
eyes, in earliest chordates and vertebrates, 150, 191, 195, 196, 213, 215, 216, 223
eyespot (in amphioxus), 79, 140, 148, 150, 154; in ancestral chordate, 208, 213, 215; homology with vertebrate eyes, 140, 148, 150, 216

face, evolution and formation of, 3–4, 7, 150, 197, 198, 215–16; neural crest involvement in, 155; skeleton of, 153
facial nerves, 106, 107, 116
feather stars, 55, 232n23
Fibroblast Growth Factor (FGF), 71, 89, 133, 135, 203; restriction to rostral somites in *amphioxus*, 135, 203
fin-ray boxes (in amphioxus), 193
first heart field (FHF), 170, 171
forebrain, 78, 105, 149, 150, 207; separation into hemispheres, 150, 213, 215
FoxD, 80, 81, 154; *FoxD3*, 81
FoxE, 68

Garstang, Walter, ix, 151, 152, 166, 225n1
gastrula, 28, 79, 125
gastrulation, 28, 29, 30, 126, 129, 133, 145; in amphioxus, 80; in enteropneusts, 71; in gnathostomes, 115–16; in lampreys, 110; in tunicates, 89, 90; in vertebrates, 126, 133
genital ridges, 171
genome, 8; in amphioxus, 77; duplication in vertebrates, 8, 49, 77, 82, 113, 149, 155–58, 161, 213, 222; in enteropneusts, 68, 120; in gnathostomes, 109, 116; in lampreys, 113; in teleost fishes, 8; in tunicates, 96–98, 212; in vertebrates, 77
Geoffroy (Etienne Geoffroy Saint-Hilaire), 16, 31, 41, 44, 46, 47, 48, 221, 226n1
Geoffroy Inversion (inversion of dorsoventral axis in chordates relative to other animals), 44–46, 49, 70, 147, 152, 205, 206
germ layers, 28
gill pores. *See* pharyngeal slits
glomerular complex. *See* heart-kidney complex
glossopharyngeal nerves, 106, 107, 116
gnathostomes, 103, 104, 105, 107, 109, 114–16, 130, 141, 143, 150, 154, 156, 157, 160, 173, 197, 214–15, 218; definition of, 218 (points 5, 6); early evolution of, 197–98; interrelationships between gnathostomes and cyclostomes, 111–12, 198; paired nostrils of, 150, 197
Goethe, Johann Wilhelm von, 27, 31, 41, 44, 48, 138, 142
Goldblum, Jeff, 44
gonadotropin-releasing hormone (GnRH), 177
gonads, 172, 173
Goodrich, Edwin Stephen, 139, 142, 163
goosecoid, 127, 129

Gould, Stephen Jay, 83, 192; *Wonderful Life*, 83, 192
graptolites, 74
gravity sensor. *See* statolith
gut, patterning of, 173

Haeckel, Ernst, 27, 45, 46, 49, 85
hagfishes, 8, 103, 107, 109, 110–12, 130, 173; notochord in, 130; pronephros, 173
Haikouella, 187, 191
Haikouichthys, 194, 195, 199; paired eyes of, 195
hairy-enhancer-of-split-related (*HER*) genes, 133
Hallucigenia, 38, 188
Halocynthia, 96; *H. roretzi*, 98
Hardy-Weinberg Equilibrium, 22
Harrimaniidae, 66, 71, 74
Hatschek's diverticulum, 143
Hatschek's nephridium, 81
Hatschek's pit, 158, 177, 208, 213
head: of animals in general, 5; dermal skeleton of, 153; as a distinct region from the trunk, 71, 72, 116, 139, 141, 142, 156, 168, 170, 190, 199, 207–9; Hox genes, 244n101; mesoderm, 136–44, 214; neural crest, 138, 153, 156, 157, 208; posterior boundary of, 208, 209; of vertebrates, 3, 4, 78, 110, 125, 136–44, 153, 168–71, 206, 208, 214
head cavities, 140–41, 143, 215
heart, 82, 87, 107, 109, 168–71, 173, 211; in amphioxus, 82; comparisons between tunicate and vertebrate hearts, 98–99, 169–71, 211; in gnathostomes, 109; in hagfishes, 111; in lampreys, 109; role of notochord in development, 130; in tunicates, 87, 90, 98, 169–71, 211; in vertebrates, 109
heart-kidney complex, 67, 68, 108, 173, 203
hedgehog (family of transcription factor genes), 76, 152
helicoplacoids, 61, 205
hematopoietic domain, 171, 208
hemichordates, 6, 31, 45, 53, 58, 62, 65–74, 123, 128, 151, 205–6, 220; biomineralization in, 58, 62; cartilage in, 154; interrelationships of, 24, 52, 53, 74; lack of Organizer in, 128; nervous system, 69; similarities to *Xenoturbella*, 118; "skin brain" of, 45
Hennig, Willi, 18, 20, 21
Herpetogaster, 184, 185, 204
heterochrony, 99
heterostraci, 161, 197
hindbrain, 78, 79, 92, 98, 106, 116, 137, 141, 149, 150, 156, 207, 209, 213
Hirschsprung's Disease, 168
Holland, Linda, 140, 142, 144
holoprosencephaly, 150, 151
holothurians, 55, 56*f*, 185, 186
homeobox, 42
homeosis, 41, 42
homeotic mutation, 42
homology, 20, 21, 46, 47, 49, 143; "deep" homology, 47; distinction between homology and analogy, 46–47
horizontal gene transfer, 72, 88
horses, evolution of. *See under* evolution
Hox clusters, 43, 47, 48, 64, 113, 155, 156, 161, 212; in amphioxus, 77; in chaetognaths, 121; in echinoderms,

Hox clusters (*continued*)
64; in enteropneusts, 72; in lampreys, 113
Hox genes, x, 42, 43, 44, 48, 78, 150, 155, 204, 207; anterior expression boundaries, 78, 155, 156, 207; chordates, 207; collinearity, 155; in echinoderms, 64; in enteropneusts, 72; expression in the nervous system, 208; *Oikopleura*, 97, 155; role in regional variation of vertebrae, 135; in specification of hindbrain regions, 44, 150, 154; in trunk, 156; in tunicates, 95, 155; in vertebrates, 155
Huxley, Thomas Henry, 85
Hydra, 29, 54, 146
hydrocoels, 52, 57, 59
hydropore, 68
hydroxyapatite (form of calcium phosphate), 5, 114, 160
hyoid bone, 155
hyoidean arch, 115
hypoglossal nerves, 106, 107, 116
hypophysis, 104, 143, 215, 223
hypothalamus, 78, 98, 177

immunity, 175–77, 214; acquired (adaptive), 8, 77, 109, 166, 175, 215; in amphioxus, 77, 176; in echinoderms, 176; in gnathostomes, 215; in hagfishes and lampreys, 109, 111, 175–76, 214; innate, 8, 77, 92, 175; in tunicates, 92, 109
immunoglobulins, 109, 175
incomplete lineage sorting, 226n5
infundibulum, 78, 106, 143, 208
insects, 1–2
intermediate mesoderm, 172
intervertebral discs, 130
invertebrates, everyday examples of, 1–3
iodine-containing hormones as a deuterostome trait, 203

Jaekelocarpus, 182

kidneys, 8, 81, 108–9, 125, 171–74
kinorhynchs, 37, 188
Kuratani, Shigeru, 112, 140–42, 144

Lamarck, Jean-Baptiste, 16, 85; *Philosophie Zoologique*, 16
lamellar body, 78, 148, 208
lampreys, 8, 103, 104, 109, 110, 140, 141, 169, 173, 216, 245; development, 110; enteric nervous system in, 245n8; fossil record, 112; genome, 113; head cavities, 140, 141; life cycle of, 104; mesonephros, 173; pronephros, 173
lancelets. *See* amphioxus
larvaceans, 43, 85, 94, 131, 212
lateral gene transfer. *See* horizontal gene transfer
lateral line system, 105, 106, 111, 114, 213
lateral plate mesoderm, 116, 172–73
Le Douarin, Nicole, 244n105, 246n21
left-right asymmetry, 53, 70, 203; in amphioxus, 76, 81; in deuterostomes, 203; role of retinoic acid, 134
Lefty, 80, 129
Lethenteron japonicum, 113
Lewis, Edward B., 42, 48
Lim1, 127, 129
Linnaeus, 227n13
litopterns, 21, 227n6

liver, role of notochord in development of, 130, 174
lobopodians, 38, 188
long branch problem, 10, 13f, 24f, 25, 120, 122
lophenteropneusts, 65
lophophorate phyla, 36, 218, 225n8
lophophore, 31, 37, 204; feeding tentacles in pterobranchs, 6, 31, 59, 73; homology with water vascular system, 60, 204; in lophophorate phyla, 36, 60
lophotrochozoa, 37, 218, 219
loriciferans, 37
lymphoid tissue, 176, 177, 215
Lytechinus variegatus, 132

Malpighian tubules, 108
mandibular arch, 115
Mangold, Hilde, 125–27, 240n2, 244n105
manifest destiny, 166
Manx, 99
Mars, surface of, 65
maximum likelihood, 23
Meckel's cartilage, 155
medulla oblongata, 106
meiofauna, 183–84, 247n7
mesencephalon. *See* midbrain
mesenteric artery, 109
mesocoels, 52, 57, 59, 67, 204
mesoderm, 28, 29, 51, 71, 77, 79, 90, 110, 131, 228n4; role of *brachyury*, 131; in the vertebrate head, 136–44, 214
mesonephros, 173
mesosome, 67, 73
metacoels, 52, 57
metameric segmentation. *See under* segmentation
metamorphosis, 82, 92, 103, 207, 210; in amphioxus, 82; in enteropneusts, 207; in lampreys, 103; in tunicates, 92, 93
metanephros, 173–74
metasome, 67
Metaspriggina, 150, 187, 195, 199, 215, 222
metazoa, 34, 35, 36; diphyletic origin of, 34; monophyly of, 34; zootype as defining mark of, 43
metencephalon, 106
Mexican stand-off, 127
midbrain, 78, 106, 148, 150, 207
midbrain-hindbrain boundary (MHB), 70, 79, 140, 149, 150, 158, 204, 208, 234n20
middle ear, 155
"missing link," 10, 103, 226n17
mitrates, 61; pharyngeal filter-feeding system in, 64
Molgula oculata, 99; *M. occulta*, 99
molgulids, 99
mollusks, 31, 32, 37, 118
myelin, 111, 114, 215
Myllokunmingia, 187, 194, 199
myopterygia, 111
myotome, 135
Myxine, 111

Nash, Ogden, 103, 115; *An Introduction to Dogs*, 103
nasohypophysial duct, 105, 111, 178, 215; in hagfishes, 111; in lampreys, 105; precursor of pituitary, 178
natural selection, 17
nature philosophy (*Naturphilosophie*), 27, 28, 32, 39, 139, 229n43
nematodes, 37, 188
Nematostella vectensis, 36, 131

nephrostome, 172, 173, 201
nerve nets, 146–47, 152, 153
nervous system, 7, 70, 144–52, 207; in amphioxus, 76–77, 79, 140, 143, 148, 150; in chordates, 150, 207, 208; in echinoderms, 56, 57, 161, 232n21; enteric, 153; in enteropneusts, 69, 151, 152; in hemichordates, 150; in pterobranchs, 73; role of notochord in development of, 7; similarity between tunicate and vertebrate nervous systems, 92; similarity of enteropneust collar cord and vertebrate dorsal nerve cord, 69; sympathetic, 153; in tunicates, 85, 88, 92, 149; in vertebrates, 7, 140
neural crest, 7, 9, 77, 80–81, 115, 139, 153–59, 168, 170, 208, 211, 213; absence in amphioxus, 77, 154; association with placodes, 157; association with rhombomeres, 138; association with skeleton, 161; in connection with body size, 153, 165; difference in neural crest behavior between head and trunk, 116, 139, 156, 168, 170, 208; evolution of, 156, 157; gene regulatory network, 154; in gnathostomes, 115, 116; in hagfishes, 112; in head formation, 138, 139, 155, 157, 161; in tunicates, 7, 93, 211; vagal neural crest, 168
neural ectoderm, division of, 146
neural gland (in tunicates), 88, 178, 208, 213
neural plate, 79, 130, 147, 148, 153, 158, 208, 212; in amphioxus, 79, 148; lateral border of, 153, 157, 212; role of notochord in patterning, 130
neural tube closure, 148
neural tube defects, 148
neurenteric canal, 110, 116
neurocranium, 161
neurogenic field (neural ectoderm), 145–46, 152, 158
neurogenic induction, 145, 146; comparison with dorsoventral axis patterning, 146
neurogenic placodes. *See* placodes
neurohypophysis, 106
neurons, 145; distinction between somatic and visceral, 245n2
neuropore, 88, 110, 116
neurulation, 80, 147, 148, 150, 152; in amphioxus, 80, 148, 210; in chordates, 150; in enteropneusts, 152; in gnathostomes, 116; in lampreys, 110; secondary, 148; in tunicates, 90, 148
NK (family of transcription factors), 43
Nodal, 54, 58, 59, 70, 80, 127, 133, 203, 205, 206, 212, 214; in amphioxus, 80; in deuterostomes, 203; in echinoderms, 58, 59, 205; role in specifying asymmetry in bilateria, 54, 133; role in the Organizer, 127–29, 206; in tunicates, 89, 212
noggin, 127
non-neurogenic field (non-neural ectoderm), 145, 146, 152, 158, 208
nostrils, 150, 178, 197, 213, 215
Notch signaling, 133
notochord, 5, 6, 7, 9, 76, 103, 114, 116, 125, 126, 128, 129–33, 143, 144, 174, 189, 193, 205, 206, 209–10, 213; in ammocoetes, 103; in amphioxus, 76, 77, 79, 130, 143, 189, 209, 210, 242n58; cartilage in, 130; in gnathostomes, 114, 116; in hagfishes, 110; in lampreys, 104, 110; as a

link between anterior-posterior, dorso-ventral and left-right axes in chordates, 130, 206; in *Pikaia*, 193; as a source of *Sonic hedgehog*, 130, 147–48; in tunicates, 85, 89, 92, 95, 128, 130; in vertebrates, 130, 143; in vetulicolians, 189, 209
nuclei pulposi, 130
Nüsslein-Volhard, Christiane, 42, 48

Occam's Razor, 22
ocellus (in tunicates), 92, 93, 149, 154, 212
oculomotor nerves, 106, 107
odontodes, 160
Oesia, 73, 121
Oikopleura dioica, 94–98, 155, 212; genome of, 96
oikosins, 94, 212
olfactory bulbs, 78, 148, 150
olfactory capsule, 104, 114, 213, 215, 222; in gnathostomes, 114; in lampreys, 104, 105
olfactory nerves, 105, 107
"ontogeny recapitulates phylogeny," 28, 33
onychophores (velvet worms), 31, 37, 38
ophiuroids, 55
opisthokonts, 34, 131
opisthonephros, 173
optic nerves, 105, 106, 107
optic tectum, 106
oral (incurrent) siphon (in tunicates), 87, 88, 98; homology between primordia and anterior placodes, 211; siphon primordia, 159
Organizer, 80, 98, 125–29, 148, 203, 212, 234n29; in amphioxus, 128; association with dorsoventral axis, 129; in deuterostomes, 129, 203; in echinoderms, 128–29
orobranchial musculature (in amphioxus), 82
Orsten, 247n7
orthodenticle. See *Otx*
osteichthyans (bony fishes), 109, 114
osteoblasts, 159
osteostraci, 197
"ostracoderms," 160, 197, 216
Ota, Kinya, 112
otic capsules/vesicles, 105, 114, 133, 141, 142
otolith. *See* statolith
Otx, 89, 92, 244n101
Owen, Richard, 47, 49; *On the Nature of Limbs*, 47

paedomorphosis, 93, 94, 167, 220
pancreas: homology with tunicate pyloric gland, 211; role of notochord in development, 130, 174
paraconodonts, 196
ParaHox (family of transcription factors), 43, 97
paraxial mesoderm, 77, 110, 116, 133–34, 136, 137, 142, 172, 206, 214; in amphioxus, 77; in gnathostomes, 115; in lampreys, 110; in vertebrates, 133
parsimony, principle of, 22, 23, 25
Pax1/9, 72
Pax6, 47, 81
Pax2/5/8, 158
pentameral symmetry (in echinoderms), 58, 61, 186, 205
pentastomids, 225n2
pericardium (pericardial coelom), 67, 171, 173, 203, 209, 211; in enteropneusts, 67; in gnathostomes, 115;

pericardium (*continued*)
in hagfishes, 111; in lampreys, 104; in tunicates, 88, 90, 211
Petromyzon marinus, genome of, 113
pharyngeal arches, 155, 208; first pharyngeal arch derivatives, 155, 167; neural crest population of, 155, 167; second pharyngeal arch derivatives, 155, 169; skeleton, 161, 167, 214; third pharyngeal arch derivatives, 155
pharyngeal slits, 6, 8, 37, 52, 62, 66, 102, 115, 126, 154, 182, 185, 188, 202, 205, 213, 218, 220; in ammocoetes, 103; in amphioxus, 68, 76; cartilage in, 154, 213; in chordates, 137; in enteropneusts, 66, 67, 68, 137; in fossil echinoderms, 60–61, 64, 182, 185, 205, 220; gene complex specifying their formation in deuterostomes, 68, 72, 202; in gnathostomes, 115; influence of neural crest on development, 138; in lampreys, 104; in pterobranchs, 73; segmentation, 137; skeleton, 138, 161; in tunicates, 87
pharyngophores, 229n33
Pharyngotremata, 220
pharynx. *See* pharyngeal slits
Philosophie Zoologique. *See under* Lamarck, Jean Baptiste
phlebobranch ascidians, 96
Phlogites, 99, 184, 185
phoronids, 31, 36, 37, 117
phylogenetic systematics, 18
phylogeny, 25, 34
phylostratigraphy, 46, 223
phylotypic stage, 45, 46, 49
pigment cells (related to neural crest), 153, 154
Pikaia gracilens, 83, 187, 192–93, 199, 208
pineal gland (or "eye"), 78, 148, 208; in lampreys, 104
Pipiscius, 113
pituitary gland, 78, 81, 88, 106, 158, 177–78, 208, 213, 215
Pitx, 54, 58, 59, 81, 89
placoderms, 114, 160, 197, 198, 214; internal fertilization in, 114; skeleton of, 160
placodes, 81, 93, 98, 153–59, 168, 171, 211, 213, 215; acoustic (otic), 105, 157, 159, 171, 213; adenohypophysial, 157, 158; in amphioxus, 157; anterior, 157, 215; association with neural crest, 157; atrial siphon, 171; in common ancestry of tunicates and vertebrates, 159, 211–12; domain anterior to neural plate, 158, 211; epibranchial, 157, 159, 171; hypobranchial, 157; lateral line, 157, 159, 171; lens, 157, 158, 213, 215; olfactory, 157, 158; origins, 157, 208; paratympanic, 157; posterior, 157, 158–59; trigeminal, 157; in tunicates, 157–58, 171, 211
planarians, 36; smiley faces of, 4
platyhelminthes, 36, 37, 225n3
Platynereis, 132
podia, 57
podocytes, 67, 172, 173
pogonophores (beard worms), 36, 117
prechordal plate, 110, 143
preplacodal ectoderm, 158
priapulids, 52
pronephros, 173, 203, 209
prosome, 67
protocoels, 51–52, 57, 67, 68
protoconodonts, 121, 196

protostomes, 31, 34, 37
protozoa, 34, 35
pseudocoelomates, 37, 228n4
pterobranchs, 6, 31, 65, 73, 74, 205
Ptychodera flava: biomineralization in, 58; expression of *brachyury*, 132; genome of, 71, 72
Ptychoderidae, 66, 71
pyloric gland, 174; homology with pancreas, 211
pyrosomes, 95, 96
Pyura chilensis, 88

RAG genes. *See* recombination-activating genes (*RAG1*, *RAG2*)
Rathke's Pouch, 81, 143, 177
recombination-activating genes (*RAG1*, *RAG2*), 175, 176, 208, 215
retinoic acid signaling, 80, 133–34, 228n10; in left-right asymmetry, 135–36, 214
Rhabdopleura, 73; *R. compacta*, 73; *R. normani*, 73, 74
rhombencephalon, 106
rhombomeres, 137, 138, 141, 142, 155, 156, 158, 213; Hox expression and neural crest, 155, 156, 213; signaling to placodes, 158
Romer, Alfred Sherwood, 139, 166, 167, 168, 190, 206, 207; "somatico-visceral animal," 139, 166, 190, 206, 245n8
Roux, Wilhelm, 28

Saccoglossus, 128
Saccoglossus bromophenolosus, biomineralization in, 58
Saccoglossus kowalevskii: early embryology of 71; genome of, 71, 72; neural development of, 152
Saccorhytus coronarius, 184, 186, 202, 207
Saint-Hilaire, Etienne Geoffroy. *See* Geoffroy (Etienne Geoffroy Saint-Hilaire)
salps, 95
scarecrow, 235n42
schizocoely. *See under* coelom
sciatic nerve, 145
sclerotome, 135
sea cucumbers. *See under* holothurians
sea lilies. *See* crinoids
sea squirts. *See* tunicates
sea stars, 55
sea urchins, 55
second heart field, 170, 171; exposure to neural crest, 170
Secreted Protein Acidic and Rich in Cysteine Gene Family (*SPARC*, *SPARCL1*), 160–61, 211, 214
Secretory Calcium-Binding Phosphoproteins (*SCPP*) gene family, 160
segmentation, 7, 136–44; in insects, 137; in invertebrates, 31; loss in tunicates, 7; metameric, 136–37; in pharyngeal arches, 137
segmentationist view, 125, 138–40
semicircular canals, 105, 111, 114, 213; in gnathostomes, 114; in hagfishes, 111; in lampreys, 105
sensory placodes. *See* placodes
Shankouclava, 99
short gastrulation (*sog*), 44–45, 145
sialic acid, 72, 202
sinus venosus, 109, 170
siphon primordia, 159
sipunculids, 117
Sirius Passet fauna, 186
situs inversus viscerum, 53
Six1, 158

Skeemella, 188
skeleton, 5, 159–61; branchial, 104, 115, 116; dermal, 153, 161, 213; in echinoderms, 56; in gnathostomes, 115; hard tissues of, 5, 159–61
skull, 104, 110, 138, 154, 155; in gnathostomes, 115, 116; in hagfishes, 110; in lampreys, 104; in vertebrates, 154
solutes (fossil echinoderms), 60
somatic mesoderm, 172
"somatico-visceral animal." *See under* Romer, Alfred Sherwood
somatocoels, 57
somatopleure, 30f
somites, 7, 77, 86, 108, 125, 133, 135, 203, 207; absence in tunicates, 86, 136; in amphioxus, 77, 79, 80, 135, 203, 210; evolution of, 52; in gnathostomes, 115, 116; in lampreys, 105, 110; somite formation (*see* somitogenesis)
somitogenesis, 133–36, 203, 214; in amphioxus, 135, 136; pacemaker, 133; role of *FGF*, 134; role of *hairy-enhancer-of-split-related* (*HER*) genes, 134; role of *Notch*, 134; role of retinoic acid, 133; role of *Wnt*, 134; specification of anterior-posterior division, 135, 137; specification of dorso-ventral domains, 135
somitomeres, 242n53
Sonic hedgehog (*Shh*), 47, 76, 130, 147, 150, 152, 206
SoxE, 154
SPARC. *See* Secreted Protein Acidic and Rich in Cysteine Gene Family (*SPARC*, *SPARCL1*)
Spartobranchus, 73, 206
Spemann, Hans, 126, 240n2
Spemann Organizer. *See* Organizer
Spengelidae, 66
spina bifida, 148
spinal accessory nerves, 106, 107
spinal cord, 106
spiracle, 115
splanchnic mesoderm, 172, 174
splanchnopleure, 30
sponges, 34, 35, 36, 43, 145, 227n3
Squalus acanthias, 115
stand-off, Mexican, 127
stapes, 155
starfish, 55
statolith (in tunicates), 92, 93, 149, 154, 211–12, 215
stem group, 179–81
stem lineage, 179, 180
stereom, 56, 57, 185, 205
stolidobranch ascidians, 96
stomochord, 68, 73, 152, 205, 220, 232n13, 233n3
stone canal, 57
Strongylocentrotus purpuratus: biomineralization genes in, 58; Hox genes in, 64; innate immunity in, 176
sturgeons, notochord in, 130
Styela, 96
stylophora, 60, 61, 182, 185, 202, 232n21
synapses, 145

tadpole larva. *See* tunicates: larvae
tail bud, 80, 133, 148, 207; in amphioxus, 80; evolutionary origin of, 207; as successor to blastopore lip/Organizer, 133, 148
tardigrades, 37
T-box (*Tbx*) genes, 131, 141, 210–11
teeth, 114, 115, 160, 196; "outside-in" view of tooth evolution, 196

telencephalon, 78, 105, 140, 148, 149, 150, 213, 215, 222, 223
teleost fishes, 8
tetrapods, 6, 114
thaliaceans, 95–96, 99
"third eye." *See* pineal gland (or "eye")
Thoatherium, 20, 21
thyroid gland, 82, 93
thyroid hormones, 92
thyroxine (T4), 93
tinman, 235n42
tornaria (larva of enteropneusts), 66, 71, 204, 207, 210
Torquaratoridae, 65, 66, 67, 232n13
total group, 180
transcription factors, 42, 43
Transforming Growth Factor Beta (TGFβ), 70
Trichoplax, 28, 227n3
trigeminal nerves, 106, 107, 116
triiodothyroacetic acid (TRIAC), 93
triiodothyronine (T3), 93
triploblasts, 29, 34, 36
Triton, 125; *T. cristatis*, 125; *T. taeniatius*, 125
trochlear nerves, 106, 107
trochophore, 37
trunk (body region), as an evolutionary novelty in chordates, 203, 207, 208, 210, 217 (point 2a), 221
tube feet (of echinoderms), 57
Tullimonstrum, 113, 214
tunicates, 5, 6, 7, 11, 83, 85–99, 102, 103, 123, 128, 130, 131, 154, 168, 199, 211–12, 221–24; common ancestry with vertebrates, 211–12, 221; determinate development in, 32; exotic biochemistry of, 88, 92, 202, 212; fossil record of, 99, 199; genome, 86, 96–98, 212; heart development, 169–71; interrelationships of, 53, 96; lack of Organizer in, 128; larva as model for antisegmentationist view, 139; larvae, 85, 89–92, 98, 149, 189, 236n23; loss of segmentation in, 7; mantle, 85; nerve cord, 85, 149; neural crest, 154; notochord, 85, 130–31; rapid evolution of, 12, 86, 123; rapid life history of, 92, 94; siphon primordia, 159, 211; tunic, 85, 88, 94
twist, 154

unicorns, 22
urochordates. *See* tunicates
urogenital system, 171–74; association with circulatory system, 203

vagus nerves, 106, 107, 109, 116, 168, 209
Variable Lymphocyte Receptors (VLRs), 176
Variable-region Chitin-Binding Proteins (VCBPs), 176
V-D-J recombination, 175, 215
VegT, 89, 214
velvet worms, 31, 37, 38
ventral aorta, 109
vertebrae, 104, 111, 114, 115, 135; formation during somitogenesis, 135
vertebrates, 101–16, 123–78, 213–17; biomineralization in, 57–58; body size, 153, 213; brain, 7; circulatory system, 109; conservatism of genome, 36; definition of, 217 (points 4a, 4b); differences from other animals, 3–4; earliest fossils, 194–95; everyday examples of, 1–2; face, 3, 4; genetic specification

vertebrates (*continued*)
of eyes, 47; genome duplication (*see under* genome); head, 3, 4, 5; heart, 109, 169–71; interrelationships of, 24, 111–12; kidneys, 8, 108–9; life cycle, 102; nervous system, 7; notochord, 130; regional variation, 135; reproductive system, 109; skeleton, 5; stepwise evolution of, 9, 124; tunicates, close relationship with, 11, 167, 168, 199, 211–12, 221, 237n59
Vetulicola, 187, 188
vetulicolians, xi, 71, 182, 184, 185, 186–90, 191, 192, 199, 200, 202, 208–10, 221
Vetulicystis (and vetulicystids), 62, 182, 185–87, 205
viscera, 165, 203, 211
visceral skeleton. *See* pharyngeal arches; skeleton
Volvox, 29
Von Baer, Karl Ernst, 45, 49

water-vascular system, 57, 58, 60, 61, 186, 204, 205; homology with lophophore, 204
Wieschaus, Eric, 42, 48
Wnt signaling pathway, 71, 80, 127–29, 133, 203
Wolffian ducts, 108–9, 172, 173

Xenopus, 44, 126
Xenoturbella, 36, 37, 117–20, 146

yunnanozoans, xi, 71, 185, 190–92, 193, 199, 202, 208–10, 221

Zhongxiniscus intermedius, 194
Zona Limitans Intrathalamica (ZLI), 70, 79, 149, 150, 204, 208
zootype, 43